GREEN ELECTRONICS MANUFACTURING

Creating Environmental Sensible Products

John X. Wang

CRC Press
Taylor & Francis Group
Boca Raton   London   New York

CRC Press is an imprint of the
Taylor & Francis Group, an **informa** business

CRC Press
Taylor & Francis Group
6000 Broken Sound Parkway NW, Suite 300
Boca Raton, FL 33487-2742

© 2013 by Taylor & Francis Group, LLC
CRC Press is an imprint of Taylor & Francis Group, an Informa business

No claim to original U.S. Government works

Printed and bound in Great Britain by TJ International Ltd, Padstow, Cornwall
Version Date: 20120518

International Standard Book Number: 978-1-4398-2664-5 (Hardback)

1006740609

Library of Congress Cataloging-in-Publication Data

Wang, John X., 1962-
 Green electronics manufacturing : creating environmental sensible products / John X. Wang.
  p. cm.
 Includes bibliographical references and index.
 ISBN 978-1-4398-2664-5 (hardback)
 1. Green electronics. I. Title.

TK7836.W36 2012
621.381028'6--dc23   2012015477

Visit the Taylor & Francis Web site at
http://www.taylorandfrancis.com

and the CRC Press Web site at
http://www.crcpress.com

To

my home

by the green woods

of Michigan

Contents

Preface

Green electronics manufacturing refers to multidisciplinary approaches aimed at reducing the energy and material intensiveness of electronics manufacturing processes. Examples of green electronics manufacturing disciplines include scheduling and process optimization, advanced or improved fabrication techniques, minimization of waste stream volume or toxicity, and improved energy efficiency. Going "green" is noticeably becoming a major component of the missions for electronics manufacturers in the United States and worldwide. The broad spectrum of green electronics manufacturing includes

- The implementation of Lean electronics manufacturing techniques to maximize efficiency and minimize waste (see my previous book titled *Lean Manufacturing: Business Bottom-Line Based*, CRC Press, 2010)
- The development of energy-efficient and environmentally friendly electronic products
- The scale-up of next-generation energy production technologies, and the development of technology to answer global energy and environmental challenges

While the goal of green electronics manufacturing seems simplistic, the dilemmas posed by this breadth of the spectrum are daunting, especially as success on one end of the spectrum IS sometimes considered a barrier to success on the other end. Given these dilemmas and the varied driving forces behind this "green" mission, it is becoming clear that if domestic manufacturers are to compete effectively in a global economy, they must take the initiative to drive this critical movement. This initiative should include identifying and developing solutions for existing and future critical needs areas in green manufacturing.

Green electronics manufacturing is a method for electronics manufacturing that minimizes waste and pollution. It slows the depletion of natural resources as well as lowers the extensive amounts of electronics waste that enter landfills. Its emphasis is on reducing parts, rationing materials, and reusing components to help make products more efficient to build. The reason it is such an important tool is because it intertwines with today's electronics manufacturing strategies of global sourcing, concurrent engineering, and total quality. Green electronics manufacturing is implemented through product and process design. Its goal is to achieve sustainability to support future generations and at the same time preserve our natural resources, as you are going to read in this book titled *Green Electronics Manufacturing: Creating Environmental Sensible Products.*

About the Author

Dr. John X. Wang is the founder and chief technology officer of the Green Electronics Manufacturing Institute, directing the creation of environmentally sensible products. As a leading expert in reliability engineering, Lean Six Sigma, and green electronics manufacturing, he has taught courses at Panduit Corp., Maytag Corp., Visteon Corp., and General Electric–Gannon University Co-Op Graduate Programs. Dr. Wang has served as principal systems engineer at Raytheon Company (where he led projects related to lead-free electronics reliability), as principal systems engineer at Rockwell Collins
(where he led projects related to avionics safety and reliability), as reliability engineering and design for Six Sigma manager at Maytag (where he led reliability engineering best practices and Design for Lean Six Sigma training), as a Six Sigma Master Black Belt certified by Visteon (where he led Design for Six Sigma training programs), and as a Six Sigma Black Belt certified by General Electric (where he led Design for Six Sigma projects and finite element modeling for reliability engineering).

Dr. Wang has authored and co-authored five engineering books and numerous professional publications on Lean manufacturing, business communication, decision making under uncertainty, risk engineering and management, Six Sigma, and systems engineering. Having taught engineering and professional courses at Gannon University and National Technological University, he has been speaking and presenting at various international and national engineering conferences, symposia, professional meetings, seminars, and workshops. Dr. Wang is an American Society for Quality Certified Reliability Engineer and an International Quality Federation Certified Master Black Belt. He received a B.A. (1985) and M.S. (1987) from Tsinghua University, Beijing, China, and a Ph.D. (1995) from the University of Maryland at College Park.

1

Green Electronic Assembly: Strategic Industry Interconnection Direction

Metallurgical interactions occur between lead-free solders and electrical pad metals in fabrication process, and these intermetallic compounds continue to grow during service periods. Due to its brittle nature and lattice mismatch, the solder cracks tend to be generated near the compounds, and these cracks affect the mechanical integrity of lead-free solder joints. Therefore, it is important to study the effect of intermetallic compounds development on the mechanical properties of the lead-free solder joints.

Electronic devices make up a large percentage of the hazardous lead waste that is landfilled. As a result, the electronic assembly community is beginning to investigate the option of lead-free soldering for its industry. As lead (Pb) is beginning to be eliminated from electronic assembly, manufacturers must be aware of the solder alloy choices available to them and the fact that not all alloys share the same characteristics. Possible driving forces behind this Pb-free solder movement are legislation, marketing pressures, and/or trade barriers.

Legislation has been passed in Europe to remove Pb from electronics assemblies. But, in addition to component surface finishes, solder containing Pb is also prohibited. Lead-free tin-based solders are still relatively new and are untested in widespread applications. Many of their characteristics have yet to be fully defined. For example, the change from a Pb-bearing solder alloy to Pb-free has a significant influence on the durability of soldering tips in hand soldering applications.

1.1 Starting from Your Personal Electronic Lab: Review the Soldering Process

Soldering is the process of using solder to make a solid mechanical and electrical joint between the electrical leads or parts of an electronic component and a printed circuit board (PCB) or some other electrical component (e.g., a connector). Solder is a soft metal alloy composed of lead (Pb) and tin (Sn).

When used in wire form, such as from a coil or roll, it has a rosin flux core. The ratio of Sn to Pb is usually 60:40; that is, 60 parts Sn to 40 parts Pb. Another popular ratio is 63:37. The latter ratio tends to melt and flow better than the former variety. However, it may be harder to remove should desoldering be necessary. Some solders may have a small amount of silver (Ag). The Ag is added to increase flow and produce a more consistent joining of metals. Again, it is more difficult to remove than the more common 60:40, and it is more expensive. Only rosin core solder should be used on electronic equipment. Water-soluble and acid fluxes should not be used. Water-soluble flux is conductive and is used for mass-produced printed circuit boards that are washed one or more times after soldering to remove the flux. Acid fluxes are suitable only for plumbing as it is corrosive to electronic components. Mixing leaded and no-lead solders can lead to eventual corrosion.

1.1.1 Flux in Soldering

The function of flux in soldering is to clean the surfaces to be soldered. It is usually included as a core inside the solder wire and flows onto the surfaces to be soldered as the solder is melted. A side benefit is that it improves the appearance of the soldered joint. Rosin flux is not conductive and, when used in moderation, there is no need to remove it from the soldered components or printed circuit board. More damage can result from cleaning attempts than from simply leaving the flux on the components. If you feel you must clean the flux, use completely 100% denatured alcohol only. Isopropyl alcohol and rubbing alcohol contain water and will lead to corrosion.

1.1.2 Solder Melting Temperature

60/40 solder melts at 368°F; 63/37 solder melts at 361°F. Solder with silver, 62Sn-36Pb-2Ag, melts at 354°F. As can be seen, there is little actual difference in their melting points. The main difference lies in their workability and appearance after melting. Solder is measured by its diameter. For small delicate work, 0.020-inch solder is good; for larger work, 0.030 inch is acceptable. For RF connectors, 0.040 inches and larger is acceptable.

1.1.3 Tip Temperature

Soldering irons and soldering guns are available in electrical wattages ranging from 15 watts to several hundred watts. Their tip temperatures range from about 600°F up to over 1,000°F. The tip temperature that you select should be influenced by the physical mass of the item you will be soldering and the mass of the soldering iron or gun that you select. For small and delicate soldering such as surface-mount technology electronics, 15 watts with a tip as small as 1/32 inch should be used. These will produce a temperature of 500°F or more, depending on whether they are temperature regulated. PCBs with through-hole components should be soldered with a 30- to 40-watt iron with a 1/16 tip at about 700°F. Coax connectors should be soldered with a 100- to –300-watt iron with a 3/16- to 1/4-inch flat tip at 900°F+. Larger wattages and larger tips will contain more heat in their larger mass and allow it to flow more quickly into the items being soldered. This will assure a quicker soldering process, thereby insuring that delicate and heat-sensitive components have less possibility of damage from overheating.

1.1.4 Soldering Iron

Ensure that your soldering iron is completely heated before you attempt to start. Turn it on 10 to 15 minutes prior to needing the iron. If adjustment is available, set the temperature when you turn it on.

1.1.5 Soldering Guns

Soldering guns were popular many years ago when discrete electrical components were large and widely spaced. They heated quickly to facilitate soldering. However, their tips do not possess sufficient size to contain sufficient heat to solder most connectors and larger items. The large size of their tips prohibits work inside most modern electronic items.

1.1.6 Ensure Proper Cleanliness

The single biggest factor that prohibits good soldering is contamination. That is, contamination of component leads and PCB pads and, yes, the tip of your soldering iron. Flux alone cannot clean all these in the instant that solder flows. You must do your part to ensure proper cleanliness.

Some soldering irons are equipped with a sponge that should be wetted for tip cleaning. Leave the sponge in its holder and wipe the iron's tip in a rotating motion after it is hot to remove contamination, old flux, and solder. Do this before starting to solder each joint.

A good way to clean component leads is to split an artist's eraser about an inch lengthwise. Place the component leads one at a time in the split, holding the sides of the eraser against the lead and draw them through the eraser.

1.1.7 Resoldering

Most PCB solderable pads and plated through-holes should be received clean and should require no further cleaning. If you are resoldering, use solder wick or a hand-operated solder sucker to remove solder. Flow a small amount of solder onto the pad afterward to prepare the surface for further soldering.

Before touching the soldering tip to the components to be soldered, carefully touch your solder to the tip to melt a small amount of solder onto the tip. Next, touch the tip to both the component lead and the PCB pad. Within

1 to 2 seconds, touch the junction of these three with the solder and allow a small amount of solder to flow. Remove the solder and tip within 3 to 4 seconds. The result should be a smooth, grayish-colored surface that is slightly concave. Solder lumps, and grainy and very dull surfaces should be cleaned and resoldered.

1.2 Lead-Free Solder Tip

The higher percentage of Sn and the higher melting temperature of Pb-free solders act more aggressively on the soldering tip and accelerate the reduction in tip life. In addition, Pb-free solders typically use a more aggressive flux formulation to compensate for the higher melting point alloys. For example, using a Sn–Cu alloy at 770°F versus 680°F decreases the durability by about 60%. The two most significant reasons for soldering tip failures are as follows:

1. Nonwettability of the iron layer due to oxidation or surface contamination.
2. Erosion of the iron layer due to flux activity level, mechanical degradation, cracks, voids, etc.

With proper care of the soldering tip, life expectancy can be increased to a reasonable level, even when using Pb-free solders. Eventually, the soldering tip is degraded by the soldering application. This means that the iron layer that protects the copper core is compromised. The exposed copper erodes quickly because of the extremely high temperature and corrosion rate of copper. The durability of the soldering tip is directly related to the iron layer thickness. As soon as the iron layer is compromised, tip life is over and the condition will be indicated by a noticeable hole in the copper core. The following steps can increase the life of solder tips:

- Choose the largest possible tip for the application. Larger tips provide better heat transfer. Larger-dimensioned tips have more iron plating, which helps to extend tip life.
- Do not exceed 725°F. Lead-free solder does not require a higher soldering temperature. Higher temperatures increase tip plating erosion. Fluxes degrade faster at high temperatures and black residues remain on the tip surface. Lowering the soldering temperature reduces both oxidation and flux splattering.
- Select the right solder alloy (SAC or Micro additive, if possible) and flux to reduce wear of soldering tips.

- High-powered soldering tools (80 to 150 watts) with optimum temperature control can in most cases do the job at lower temperatures.
- Dry cleaning keeps the tip wettable longer. Wet sponges cause thermal shock, remove the majority of the tinning, and do not properly remove flux residues.
- Always tin the tip to prevent oxidation and surface contamination. Apply a thin coating of solder to the tip after cleaning, and before placing in the soldering tool holder.
- Programmable station functions such as standby and auto-off, along with new stop-and-go stands, increase tip life by reducing the temperature while not in use.

Remember: a soldering tip is a consumable component (much like the tires on a car) and a degraded tip is not an actual tip defect.

1.3 Lead-Free Solder Bumps

There are five different types of Pb-free solder bump interconnections for flip-chip electronic packaging applications. Lead-free solder bumps are fabricated from

1. Pure tin (Sn)
2. Tin–bismuth (Sn–Bi)
3. Eutectic tin–copper (Sn–Cu)
4. Eutectic tin–silver (Sn–Ag)
5. Ternary tin–silver–copper (Sn–Ag–Cu, SAC) alloys

Tin-rich lead-free solders are currently the predominant interconnection material for the microelectronics industry. The resulting interconnects provide both electrical and mechanical connection between integrated circuit devices and their substrates. The emerging trends in the miniaturization of electronic devices demand a significant reduction in the size (pitch) of interconnects. As a consequence, the solder's reliability is of paramount importance in deciding the integrity of the device. In the past decade, researchers have demonstrated various approaches by which the properties of the solder can be improved. Among them is the composite approach, wherein suitable particles are introduced into the solder matrix; this approach seems to be one of the potential solutions to engineer a stabilized microstructure with improved mechanical and thermomechanical properties.

1.4 Flip-Chip Technology

Present means of connecting chips to substrates include

- Wire bonding
- Flip chip: C4 interconnect (i.e., "controlled collapse chip connection")

At present, "flip chip" is the strategic industry interconnection direction. When compared with wire bonding, flip chip offers the following advantages:

- Smaller overall size
- Higher I/O capacity/density
- Lower cost
- Improved electrical performance

Flip-chip attachment to organic carriers presently relies on Pb solders. High Pb content (>85 wt%), high melting temperature (>300°C) solders are used on the chip-side bumps and lower Pb content, lower melting temperature (eutectic; 37 wt% Pb; 185°C melting temperature) solder is used on the substrate side. In this scheme, the low-melt solder component is melted, fusing it to the unmelted high-melt solder and forming a viable electrical joint. This allows for attachment processing at temperatures that organic carriers can sustain, both in their materials of construction and in their fabricated form. It is not possible to safely attach chips with high-melt solder to current organic substrates without the use of a lower temperature melting solder because temperatures above the degradation temperature of the organic materials (e.g., greater than 300°C) are used to melt high-Pb solder.

For chips that will be assembled through "flip-chip" means, high Pb content solder is presently the standard connection finish. Other interconnection finishes and means are under development; those include copper–tin, tin–silver–copper, gold stub, and electrically conductive adhesive. None of these technologies has progressed to the point where it can currently be considered an available alternative to Pb solder in flip-chip applications.

More and more electronics manufacturers are adopting the latest flip-chip packaging technology in their designs. To incorporate this technology successfully, manufacturers must make some modifications to their surface-mount technology (SMT) assembly equipment, materials, and processes. A number of manufacturing-related issues must be addressed, including fluxing and underfill, and additional attention must be paid to yield- and quality-related issues. Since its development, and with the appropriate tweaks to assembly equipment and processes, flip-chip technology has been

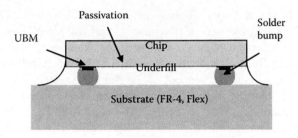

FIGURE 1.1
Sketch of flip chip on substrate.

successfully integrated into many electronics products. A sketch of flip-chip technology is shown in Figure 1.1.

As shown in Figure 1.1, the structure of flip chip involves the following:

- I/O pads
- Passivation layer
- Under-bump metallurgy (UBM)
- Solder bumps

By implementing flip-chip technology, it is possible to reduce cost, increase yield, and reduce the number of process steps overall. All these benefits can be achieved if a company converts to a flip-chip design and selects appropriate equipment and materials for the assembly process.

1.5 Flip-Chip Assembly Process

To incorporate flip-chip technology into an existing product design, a number of changes to standard SMT processes are required. Typically, placement of the flip-chip device can be achieved using existing placement machines; in contrast, the underfill process required for flip-chip assembly does require special equipment. A typical flip-chip production line may consist of a fluxing, placement, and reflow segment, and a separate section with underfill and curing.

1.5.1 Significant Impacts on Reliability

The material and method of flux application have significant impacts on reliability and yield. Dip fluxing and spray fluxing are the most commonly used methods. Most of the fluxes used for spray fluxing are so-called "no-residue" fluxes with an alcohol content of 98% to 99%. The fluxes are sprayed on the substrate before placement, and the alcohol evaporates rapidly at room temperature. The remaining flux does not provide sufficient tackiness to hold the flip chip in place during board handling and reflow, and frequently results in movement of the die. As a result, a dip-flux process may be adopted to apply a tacky flux to the flip-chip components. Typically, no-clean flux may be applied in a thick film on a rotating plate using a doctor blade. The die bumps can then be dipped in the flux and placed onto the substrate.

Compared to liquid fluxes, which provide no tackiness for the die, tacky paste fluxes will hold the die in place during handling. As a result, problems due to handling are unlikely. The choice of flux and encapsulant is influenced by die passivation, bump metallurgy, substrate, solder mask, and pad metallurgy. A good knowledge of bump height distributions and possible bump defects is essential in determining the process window for the dip fluxing process.

The minimum flux film thickness depends on the bump height variations within a die. To ensure good soldering of the eutectic bumps on the die, all the bumps must be dipped in the flux. To establish a suitable process window for the fluxing, experiments may be performed using bare laminate.

1.5.2 Placement Stage

The flip-chip components themselves can be handled out of trays. There are several configuration choices that should be made when considering flip-chip assembly on a pick-and-place machine. They include optical resolution, lighting geometry, processing capability, pick-and-place tooling, and substrate lifter tooling. Because the interconnect medium of a flip chip is a solder bump, simple die edge techniques are inadequate for locating and placing flip chips.

To avoid the possibility of placing flip chips in the wrong orientation, the programming bump pattern used for recognition should be asymmetric. The pick-and-place machine must have the optical resolution, lighting geometry, and processing capability to locate a pattern of bumps on the bottom side of the flip chip. The pick-and-place machine must then possess the intrinsic accuracy to place the die precisely on a substrate. In addition, pickup tooling must be properly sized and of the correct material such that when a die is dipped in flux (thin-film flux application), it is not dropped in the flux applicator and there is a clean release when it is placed.

1.5.3 Protection against Deformation

The under-board support must rigidly capture the substrate so that there is no movement in the substrates due to die placement contact. A well-chosen support mechanism will simplify the handling while supporting the substrate during printing, and will protect it from warpage and deformation due to sagging in the reflow and curing ovens. As always, there will be other issues that are specific to a given process that must be considered when choosing a pick-and-place configuration.

1.5.4 Yield of the Placement Process

The yield of the placement process depends on a number of parameters, including board tolerance (soldermask, copper layer), pad design, and placement accuracy. Soldermask tolerance is one of the major factors affecting flip-chip assembly yields; because of this, it may be necessary to experiment with different parameters to minimize the soldermask effect. For example, smaller pitches may introduce the effect of soldermask registration due to different required pad designs and therefore require drastically better placement accuracy and soldermask tolerances.

1.5.5 Underfill Stage

Figure 1.2 summarizes electromigration which is influenced by multiple interactions during the underfill stage. The multiple interactions between the various different materials involved clearly show that the materials used must be treated as a system; this should be reevaluated as soon as any one of the materials or the application changes.

1.5.6 Need for Process Optimization

Specifically, the process windows for flow temperatures and dispense pattern and time must be established for the individual application and materials. Even more importantly, these parameters, as well as the choice of materials, have a tremendous impact on the reliability performance.

1.5.7 Prevention of Contamination and Moisture

Contamination of board and die during handling can have a major impact on the underfill process and overall reliability. Because of the robustness

---

Electromigration is caused by high current density stress in metallization patterns and is a major source of breakdown in electronic devices. Voids can be formed due to electromigration in a Flip-Chip Solder Joint. It is therefore an important reliability issue to verify current densities within all stressed metallization patterns.

FIGURE 1.2
Formation of voids due to electromigration in a flip-chip solder joint.

of the soldering process, a clean room is not required, although handling boards with gloves and protection of the boards from contamination, dust particles, and chemicals is mandatory before underfill curing.

Board cleaning after misprints or contamination of the board during repair must be avoided. Exposure of assemblies to moisture before underfill will impact the reliability performance. Unfortunately, a tolerable exposure to moisture depends on substrate design and materials, as well as the reliability requirements.

1.5.8 Inspection and Testing

Inspection of materials and testing in the production is necessary to ensure high yield in production and acceptable reliability. The assembly yields and necessary process parameters depend, as discussed, on the substrate and die quality. Consequently, the quality of these materials must be monitored.

1.5.9 Process Capability

Specifically, the impact on the process of any changes such as new cleaning procedures or process flows for the substrate, as well as any changes in the bumping technique and underfill material, must be considered to evaluate the impact upon the process.

Additionally, substrate tolerance must be monitored closely to ensure high yields in assembly. To ensure quality, it is necessary to understand the bumping process and monitor the bump height distributions. In addition, shear tests of the bumps can disclose systematic interface weaknesses of the bump. Depending on bumping technology, a change in failure mode or shear force can indicate systematic problems with the wafer lot. A change could indicate a weakness and possible interface problem of the under-bump metallization (UBM) with the die or solder bump.

1.5.10 Quality Control

Lot-to-lot changes in the encapsulants' properties are often difficult to detect. Underfill needs storage temperatures below −40°C and rapidly ages at room temperature. Temperature variations that occur during shipping or storage can substantially alter the performance of the underfill materials. Simple standardized tests can be performed to monitor changes in the encapsulant batches. Possible tests include a flow test and a test measuring the wetting angle to the substrate.

Material selection becomes one of the most essential parameters for reliability as well as processability. For example, materials with low thermal mass will make uniform heating in reflow, during underfill and curing, easier.

1.5.11 Testability

Off-line x-ray test equipment is being used to do regular testing of the soldering quality. Because of the circular pad design with attached traces, a two-dimensional x-ray is, in this case, sufficient to determine solder joint quality. Non-wetted pads on the substrate are easy to detect because the bumps will not deform during reflow.

Other observable failures are solder voids and solder bridging. Solder bridging can be caused by solder extrusions filling the voids touching the soldered bumps. Depending on the bump pitch, these voids can link two solder joints and cause bridging during thermo-cycling or additional reflow. Due to the wide pitch of 400 μm, the risk of failures like these is very small for this particular application.

The underfill quality of the process was monitored with acoustic microscopy. Because the submersion of all assemblies in water for testing is undesirable, spot-checks were performed in order to detect voids and incomplete flow. Alternatively, a destructive test can be performed by grinding the die off the assembly in order to inspect the underfill layer.

1.5.12 Integrate Flip Chip into Standard Surface-Mount Technology Process

Various issues must be addressed for the integration of flip chip into a standard SMT process flow. The material selection of flux and encapsulant is influenced by die passivation, bump metallurgy, substrate, solder mask, and pad metallurgy. Process windows and robustness make a dip fluxing process preferable. Board design and component pitches will significantly impact the assembly yield. For the underfill process, the process windows for flow temperatures, dispense patterns, and time should be established for the individual application and materials. Monitoring of the incoming materials such as bump height distribution and underfill flow performance, as well as board quality, is important because of the number of interactions in the process, specifically the underfill process. Contamination of the board and die during handling can have a major impact on the underfill process and overall reliability. Additional board baking might be required to ensure sufficient reliability and good underfill performance.

New materials and processes make it very difficult to maintain high yields, acceptable quality, and a "competitive edge." Ongoing support and partnership with a large, application-relevant R&D organization is required. Nevertheless, standard solutions for conservative applications in terms of pitch and reliability requirements are becoming possible.

The selection of the material system is the key to required reliability as well as assembly yield. Different from other technologies, it is possible to implement flip-chip assembly in a step-by-step fashion as far as investments in facilities, equipment, and training are concerned. The decision to go with flip chip in a specific application may then be made on a case-by-case basis.

1.6 Mechanical Stress and Electromigration

1.6.1 Coupling of Two Trends

The mechanical behavior of the solder joint is influenced by factors such as geometry and plastic constraint. Therefore, constraining effects on the mechanical response of the solder joint must be removed so as to determine the constitutive properties of the solder material in most realistic conditions. Additionally, these constitutive properties are of utmost importance for realistic modeling/simulation of advanced technology such as System-On-Chip or stacked 3-D packaging of electronic devices. As the size of microelectronic devices continues to decrease, interconnects in the devices are scaling down correspondingly. Meanwhile, the demand for performance in terms of current capacity continues to increase. The coupling of these two trends leads to a drastically increasing current density in the devices during operation. One of the most significant problems associated with this large current density is a process known as *electromigration* that is observed in the solder bumps that connect the integrated circuit chips to external circuitry.

1.6.2 Electromigration

Electromigration (EM) is the current-induced transport of the conducting material. In the presence of high current stresses, electron momentum is transferred to atoms in the conducting material, yielding a net atomic flux. This net flux causes the conducting material to be depleted "upwind" and accumulated "downwind." Regions where the interconnect material has been depleted will form a void, leading to interconnect open-circuit failure. Likewise, interconnect material can also accumulate and extrude to make electrical contact with neighboring interconnect segments, potentially leading to circuit failure due to the formation of a short-circuit. Either outcome can contribute to the gradual "wearing out" of a current-stressed interconnect over time.

1.6.3 Wind Force Equation

Electromigration is mass transport due to atomic displacement resulting from an electric field. The formation of voids and accumulations depends on the underlying microstructure of the metal film from which the interconnect has been patterned. Once deposited, the metal film has a distribution of grain sizes. This metal film is then etched to produce the desired interconnect layer. As electrons move through a metal lattice, they tend to scatter at imperfections or by interactions due to phonon vibrations. This scattering results in a change in the momentum of the electrons and hence exerts a force on the metal ions known as the wind force (F_{wind}) and given by Equation (1.1):

$$F_{wind} = Z^* e\rho j \qquad (1.1)$$

where
  Z^* is the effective charge number
  e is the charge of the electron
  ρ is the resistivity
  j is the current density

Hence, one can see that this force is directly proportional to the current density; and above some threshold, this force is large enough to motivate atoms to diffuse from their original positions in the direction of the electron flow. The threshold current density is much smaller in the solder bump than in the Al and Cu interconnects that the solder joins; thus, electromigration in the solder bump can be significant. Furthermore, it has been suggested that the diffusion of atoms caused by electromigration creates tension upstream and compression downstream. Experimental evidence verifies this idea as voids (created under tensile stress) are seen upstream in the solder bumps.

The accumulation and propagation of these voids lead to increased electrical resistance and eventual failure of the solder bump. However, electromigration is only one factor in the creation of the voids; a wide range of others—including chemical potential, temperature, creep, and mechanical stress—may play additional roles.

Electromigration estimation is separated into two steps. The first step checks for violations of the current density limits, and the second step assesses the mean-time-to-failure (MTTF) for all wire segments. While most interconnect segments exhibit AC current behavior, almost every signal interconnect line on a chip includes interconnect segments that exhibit DC current behavior. Therefore, signal wires must be checked for both peak and RMS (root-mean-square) current density violations. Hence, the cumulative probability of failure for a projected lifetime must be determined.

Electromigration failure of contacts and vias in deep sub-micron integrated circuit (IC) technologies is the key concern for interconnect reliability. Electromigration failure of contacts and vias in advanced interconnect systems, where the low-resistivity conductor such as Al is clad by refractory layers of Ti or TiN, occurs by the drift of conductor away from the contact, leaving the refractory materials intact. The reliability of a VLSI (very-large-scale integration) chip is ultimately limited by the failure characteristics of its basic building materials, under the stress imposed by the operating environment. Among all the IC technological trends, scaling is an important method for reducing die size and thus increasing circuit performance and complexity. Common wear-out processes in ICs are highly influenced by the scaling of device dimensions as this usually leads to increased electrical stress.

If the reliability of a system can be expressed in terms of a failure parameter, then it should be possible to express it as a numerical index so that it could be seen as a fitness of the design created. The MTTF of a system is the expected time a system will operate before the first failure occurs. It turns out that the presence of dormant faults can drastically reduce the MTTF of a system. The effect of electromigration on the time-to-failure has been investigated. The MTTF of a conductor under a constant current stress is expressed by the following equation:

$$MTTF = A \cdot J^{-n} e^{\left\{\frac{E_a}{k \cdot T}\right\}} \qquad (1.2)$$

where
E_a is the activation energy
J is the current density
T is the temperature, in Kelvin
A is a constant depending on geometry and material parameters—scaling factor
k is the Boltzmann constant
n is a constant ranging from 1 to 7

The electromigration, Vm, can be expressed as

$$Vm = G \cdot J \cdot e^{-\frac{E_a}{k \cdot T}} \qquad (1.3)$$

where
G is a proportionality constant

Equation (1.2) and Equation (1.3) show that the following parameters are involved in the calculation of MTTF:

- Activation energy
- Temperature
- Electromigration effects

One problem caused by electromigration is a reduction in the effective operating dimensions of interconnects to a micron or a sub-micron. The other kind is material related, which is basically caused by high current densities. Three associated problems in electromigration are referred to as joule heating, current crowding, and material reactions. The effect of each of these parameters is required to arrive at a reliability parameter in terms of the MTTF, with which the fitness of the design can be evaluated.

1.7 Residual Mechanical Stress

In addition to electromigration, another factor that we should take into consideration is a residual mechanical stress, as it has been observed experimentally that an applied mechanical stress decreases the MTTF of the solder bump (see Figure 1.3). Such a stress is introduced naturally during operation due to the thermal mismatch between the Si die on one side of the solder bump and the substrate (a printed circuit board, PCB) on the other (Figure 1.4).

To permanently connect the solder to the remainder of the system, the solder is heated to its melting point to produce an electrical connection and is subsequently cooled to room temperature. The temperature of the system is then increased again during operation of the device. This results in warpage (increased curvature) of the substrate during operation, which can lead to various mechanical stresses being placed on the solder bumps. Accurate estimation of these naturally occurring mechanical stresses will allow us to predict how much of a role they have in the overall stress state of the solder bump compared to the other factors such as creep coupled with electromigration.

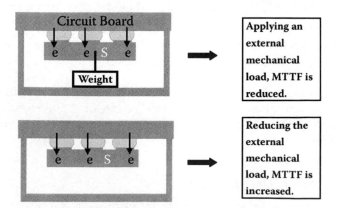

FIGURE 1.3
Applying an external mechanical load to the flip chip, the mean-time-to-failure is reduced.

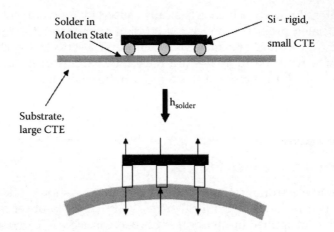

FIGURE 1.4
Predicted curvature.

1.8 Mitigate Deterioration of Lead-Free Tin Solder at Low Temperatures

1.8.1 Tin Plague

Tin plague is a characteristic of high-tin-content lead-free solders that may not be familiar to many people. It is the disintegration of pure tin into powder as it loses its crystalline structure at low temperatures. It is called a "plague" because it appears to spread like a disease. The phenomenon is also known as "tin disease" or "tin pest." Tin has the following two allotropes:

1. One is the familiar gray metal, called beta-tin (tetragonal crystalline structure).
2. The other is a crumbly, white, nonmetallic powder known as alpha-tin or white tin (cubic crystalline structure).

At temperatures below 13°C (56°F), a gradual change occurs in tin's crystalline structure from tetragonal to cubic. The rate of change reaches a maximum around −30°C (−22°F) and results in the disintegration of metallic tin into a powder.

1.8.2 How to Mitigate Tin Plague

Tin plague has been noted many times by historians. Napoleon's army that besieged Moscow wore trousers and tunics fastened with shiny white tin buttons that in the severe Russian winter turned into brittle gray tin (so called "tin plague") that crumbled away. Tin plague was noticed by Aristotle (384 to 322 BC) and Plutarch (46 to 120 AD). In the severe European winter of 1850, the tin organ pipes of the Zeit church in Germany were destroyed by the cold.

Tin plague can be suppressed or even eliminated by alloying tin with metals such as bismuth (Bi), antimony (Sb), or lead (Pb) that are readily soluble in tin. Bismuth and antimony are effective at levels as low as 0.5%. However, the level of Pb should be at least 5%. Traditional eutectic Sn–Pb solder is 37% Pb so tin plague has not been a problem in the past. The alloying elements found in the most widely used Pb-free solders are copper (Cu) and silver (Ag). These form intermetallics with Sn but do not prevent tin plague. Research has shown tin plague in some Sn–Ag–Cu alloys. The potential impact on Pb-free electronics should be assessed for risk mitigation.

Electronics are often exposed to extended storage and operation at temperatures that are conducive to the growth of tin plague. Spending extended periods of time at these temperatures could create a definite threat to long-term reliability. Failure of the solder could cause open circuits, intermittent opens, or even units falling apart. Lead-free solders have a vulnerability to tin plague: the disintegration of solid Sn into powder after prolonged exposure to temperatures below 13°C (56°F). Alloying Sn with at least 5% Pb or 0.5% or more of Sb or Bi can mitigate tin plague. Risk mitigation should be emphasized in applications where reliability and long life are important.

1.9 Able to "Take the Heat?": Capability to Withstand High Temperature

When soldering with Pb-free materials (e.g., flux and solder pastes), heat is the most important parameter to consider because it affects everything from component packages, to board materials, to the flux. It is fairly well understood that along with the generally higher temperatures requisite of Pb-free alloys, the risk of thermally induced damage to components and board materials is also higher. Another concern, however, is the effect of higher temperatures on the stability and activity of the flux, and the potential for oven maintenance issues associated with the burn-off of solvents and other flux components. It is not just about peak temperature, however. The challenge for the flux chemist is to develop flux systems that provide greater thermal stability during heat exposure dwell time. Establishing this stability is more important than focusing on absolute temperature levels.

The process engineer implementing a Pb-free process needs to establish sound metallurgy and long-term joint reliability for all soldered connections on a product or products. The achievement of such ensures long-time product reliability and reduces the potential for failure in the field. The second concern is the changing process window associated with the production of Pb-free electronics. When we begin looking at the materials used in Pb-free soldering, the compatibility of the different components becomes critical.

Lead-free alloys (such as Sn–Ag–Cu [SAC] alloys) require higher process temperatures, and thus more heat affecting every part of an assembly. The materials that contain both metallic and organic components will accommodate this additional thermal load in different ways. Impacts on solder paste and flux associated with these higher temperatures include

- Early evaporation of solvents
- Different melting characteristics
- Activation range
- Thermal decomposition
- Recrystallization of certain constituents

These are some of the various chemical and physical changes in the materials on their journey along the temperature/time line of the process. One often-overlooked issue, however, involves the implications of heat in the production of solder powder—an essential component in soldering paste. Heat has its own effect on the topography of the solder particle during solidification of the droplet. The topography is influenced by parameters such as the cooling rate and the atmosphere in which solidification takes place. In turn, these parameters impact the distribution of the alloying elements on the surface and the formation of passive films (such as oxidation).

When Pb is removed from a solder alloy, even at ambient temperatures, the Pb-free material will oxidize significantly faster. When the temperature rises, oxidation accelerates. In turn, oxidation impacts both the topography of the solder particles as well as the surface tension of the solder. The topography of the particles is a parameter in the rheologic system and therefore affects the printing properties of the solder paste. The changes in surface tension affect the wetting of the surfaces to be joined, ultimately impacting soldering performance.

It is generally considered common knowledge that most defects in a surface-mount assembly process have their origins in the printing process; consequently, the printing properties of a solder paste are of critical importance in maximizing yields.

Heat affects flux systems; thus, three key parameters should be observed:

1. Volatilization of both the solvent systems as well the volatile fractions of other materials
2. Melting, melt viscosity, and spread rate of the solid substances
3. Decomposition of all organic materials

These parameters directly impact issues such as SIR (surface insulation resistance)/electromigration, IC testing, and condensation of volatile fractions on the electronic circuit and in the reflow equipment. However, when the organic system breaks down prematurely in the temperature/time line, the metallic parts in the solder joint, lacking their protective blanket, may exhibit early and more intense oxidation.

Qualification studies and field experience by major end users of Pb-free solder paste have uncovered significant issues with the material, such as a disappointingly short shelf life of several types of Pb-free solder paste and significantly different results regarding voiding. Chemists and metallurgists have concluded that both problems have a potentially common root, that is, oxidation of the solder powder during production. It is commonly understood that oxidation appears to be self-propagating; thus, when solder paste is manufactured with powder that is partially oxidized, it will further deteriorate once it is in suspension with specific flux systems. A QC technician opens a jar of solder paste only to discover that the material has acquired the hardness and consistency of concrete.

During manufacture of the powder, the cooling rate during the transformation of the droplet into a particle, as well as the atmosphere in which this solidification takes place, have an effect on the oxidation and distribution of the alloying elements on the surface of the particle. Generally, a rougher particle surface may be an indication of differences in oxide levels, but it may also result in less particle mobility, which impacts the paste's printing qualities. When it appears that particle characteristics and other properties change from batch to batch, it is not a good sign.

Chemists also maintain that increased rate voiding in Pb-free solder connections is related to oxidation of the pads, the component metallization, and/or the powder in the solder paste, rather than the organic material in the flux system. Why? Because the flux becomes extremely mobile when the paste is still in the pre-heat and soak zone. Consequently, it flows to the boundaries of the solder joint. Metal oxides break down at higher temperatures, releasing gaseous decomposition products when the joint is exposed to peak zone temperatures. This happens so close to the cool-down of the exterior of the solder joint that some of these gas bubbles are trapped inside the solder mass, creating voids.

Oven contamination and flux management schemes were common discussion topics well before the advent of Pb-free solder. Studies have shown that when oven contamination is a problem with Pb-bearing solder paste, it will be even more of a problem with Pb-free SAC alloys. Higher process temperatures will decompose condensed flux even more, making residues more difficult to remove. If maintenance schedules are not rigorously adhered to, clogged filters may disrupt gas flow and cause problems such as higher defect rates. A key advantage associated with the use of synthetic resins in flux is that their decomposition products are not only significantly lower in volume, but are also much easier to remove from reflow oven interiors and filters.

A nitrogen blanket on the Pb-free solder pot is considered an essential precaution to avoid dross formation and to reduce solder defects. The same approach can be used in Pb-free reflow in cases where longer soak times are required to minimize Delta-Ts (temperature difference) between large and small components.

Tombstoning appears to occur less frequently in Pb-free soldering than with traditional Pb-containing solder paste, but it still occurs. In such cases, the problem can be minimized using a special powder system consisting of 50% Sn95.5/Ag4.0/Cu0.5 with a eutectic temperature of 217°C, and the balance of the powder Sn96.5/Ag3.5 with a eutectic temperature of 221°C provides a T of 4°C between the initial and final melting of the solder mass on each pad. So, before the solder on one of the pads of a bipolar component is completely molten, at least 50% of the solder on the adjacent pad is liquid as well, thereby restoring equilibrium surface tension forces keeping the component in place.

Demand for higher-quality electronic assemblies is the driving force toward more consistent material performance. The introduction of Pb-free technology has presented additional challenges to deliver more batch-to-batch consistency of solder paste. It is not only the consistency of the flux system with a higher thermal stability that is critical to this end, but also the surface properties of the powder impacting the interaction with the flux system. Surface roughness and differences in the alloy between the mass of the solder particle and its specific surface area not only impact the wetting properties of a solder paste, but also its rheology and printing properties.

1.10 Solder Joint Fatigue

Solder joint fatigue is one of the predominant failure mechanisms in electronic assemblies exposed to thermal cycling. Reliable, consistent, and comprehensive solder constitutive equations and material properties are needed for use in mechanical design, reliability assessment, and process optimization. Mechanical characterization of solder materials has typically been hampered by the difficulties involved in preparing test specimens that reflect the same true material making up the actual solder joints (e.g., match the solder microstructure). In addition, the microstructure, mechanical response, and failure behavior of solder materials are constantly evolving when exposed to isothermal aging and/or thermal cycling environments. Such aging effects are universally detrimental to reliability and include reductions in stiffness, yield stress, ultimate strength, and strain to failure, as well as highly accelerated creep.

Changes due to aging are greatly exacerbated at higher temperatures typical of thermal cycling qualification tests. However, significant changes occur even with aging at room temperature. As shown in Figure 1.4, the low cycle fatigue-induced failure of solder balls in surface-mounted electronic devices has become one of the most critical reliability issues in ball grid array type packages. Solder ball reliability performance was found to be highly dependent on the configuration of the package, such as the combination of substrates and geometry/material properties of die, etc., which in turn is governed by bond pad geometry, solder ball configuration, thermal behaviors of each component, moisture conditions, as well as solder reflow characteristics, etc. Using finite element methods with multiphysics capabilities to predict the package reliability and to reduce the time-to-market is imperative in the area of the electronic packaging industry. As a microelectronic package increasingly demands more I/O counts, smaller pitch, smaller solder bumps, and higher electric current per solder bump, flip-chip solder joints are expected to be exposed to very high thermal stress and temperature. At high temperature, the microstructure of Sn-based solders would significantly change, and consequently their mechanical properties would be affected.

The high-temperature stability of the microstructure and mechanical properties of Sn-low Ag (0.5 wt% and 1.0 wt%) and Sn-0.7Cu (wt-%) alloys has been investigated during aging at 200°C and 150°C. The microstructural investigation of the solders includes the changes in Sn grain size and their crystal orientation, and the coarsening rate of IMC (intermetallic compound) particles (Cu_6Sn_5 versus Ag_3Sn) as a function of aging time. Fine-pitch copper wirebond has been introduced into the industry mainstream. Replacement of Au wire by Cu is the last frontier for packaging materials cost savings. At the same time, the introduction of advanced nodes and low-k materials will demand finer-diameter wires for Au as well as Cu below the 18 µm being practiced today. While Cu wirebond has been in use for power devices with 50-µ diameter wires and low I/O counts, fine-pitch Cu wire-bond is a recent development. Fine-pitch applications with Cu wire diameters of 25 µm and below require improvements in the understanding of wire properties, IMC formation and evolution, wire bonding processes, and equipment development and control for wire oxidation. Palladium-coated wire has been introduced to eliminate the need for forming gas in production.

1.11 Finite Element Analysis

Finite element analysis (FEA) can be employed to predict mechanical stress levels in these solder bumps due to temperature changes during assembly and operation. For stress in flip-chip solder bumps due to package warpage, the FEA model takes into account the

- Si die
- PCB substrate
- Solder bumps
- Underfill

The FEA model includes the geometry and mechanical properties for each component above. FEA originated in the early 1940s with Richard Courant, a German mathematician, to analyze vibration systems. The complex systems were able to be solved using numerical approximations. Currently, manual FEA is still taught in-depth at many universities, and is the basis for many computer programs. FEA exists to give approximate solutions to complicated problems. For example, its can be used to find how much a ball grid array (BGA) would deform for a given thermal cycling.

For electromigration, scaling down the geometry of integrated circuits increases the current density and associated Joule heating in interconnect tracks and vias, leading to a greater incidence of thermal stress and electromigration failure mechanisms. The current density peaks for geometries containing sharp corners, such as those found at a track–via junction. The peak current densities will therefore be scaled according to the finite-element discretization. Given the track and via dimensions, an approximate value for the current density can be determined by supposing it is constant across the cross-sectional area of the track. At bends in the track and at contact vias, the current density often exceeds these estimates by a significant amount and can lead to preferential failure at these points. The finite element method is used here for one-dimensional approximation. The temperature profile and peak temperature depend on the aspect ratio of the via and track widths. For the case where the track and via widths are equal, the peak temperature coincides with the volume of current crowding. This suggests that such structures are most susceptible to failure.

Bibliography

Arora, N.D., Raol, K.V., Schumann, R., and Richardson, L.M. (1996). Modeling and extraction of interconnect capacitances for multilayer VLSI circuits, *IEEE Trans. Computer Aided Design of Integrated Circuits and Systems*, 15(1), 58–66.

Bilotti, A.A., (1974). Static temperature distribution in IC chips with isothermal heat sources, *IEEE Trans. Electron Devices*, ED-21 (March), 217–226.

Black, J.R., (1969). Electromigration failure models in aluminium metallization for semiconductor devices, *Proc. IEEE*, 57(9), 1587–1594.

Blech, I.A., and Herring, C. (1976). Stress generation by electromigration, *Appl. Phys. Lett.*, 29, 131–133.

Chen, C., and Liang, S.W. (2007). Electromigration issues in lead-free solder joints, *J. Mater. Sci.*, 18, 259–268.

Hunter, W.R. (1997). Self-consistent solutions for allowed interconnect current density. I. Implication for technology evolution, *IEEE Trans. Electron Devices*, 44(2), 304–309.

Hunter, W.R. (1997). Self-consistent solutions for allowed interconnect current density. II. Application to design guidelines, *IEEE Trans. Electron Devices*, 44(2), 310–316.

Suo, Z. (2004). A continuum theory that couples creep and self-diffusion, *J. Appl. Mechanics*, 71, 646–651.

Teng, C.C., Cheng, Y.K., Rosenbaum, E., and Kang S.M. (1997). iTEM: A temperature-dependent electromigration reliability diagnosis tool, *IEEE Trans. Computer-Aided Design of Integrated Circuits and Systems*, 16(8), 882–893.

Tu, K.N. (2003). Recent advances on electromigration in very-large-scale integration of interconnects, *J. Appl. Phys.*, 94, 5451–5473.

Yeh, E.C.C., Choi, W.J., Tu, K.N., Elenius, P., and Balkan, H. (2002). Current-crowding-induced electromigration failure in flip chip solder joints, *Appl. Phys. Lett.*, 80, 580.

2

Tin Whiskers: New Challenge for Long-Term RoHS Reliability

With present circuit geometries so much smaller than they have been in the recent past, adjacent whiskers can easily bridge spaces between leads, and/or whiskers from adjacent areas can touch each other causing short-circuits. Loose whiskers can bridge board traces, foul optics, or jam microelectromechanical machines (MEMs).

Metallic whiskers were first documented in 1946. It has been established that tin (Sn), zinc (Zn), cadmium (Cd), and silver (Ag) can grow whiskers. This growth can begin any time from within hours to after years of dormancy. It has been known for more than 50 years that an Sn layer plated on a surface will spontaneously grow hair-like single-crystal filaments known as "tin whiskers." These growths are of the same character as those known as "whiskers" and which developed between the leaves of Cd-plated variable air condensers, causing considerable trouble in military equipment during the early part of World War II. These whiskers are electrically conductive and physically strong.

Tin whiskers grow in the absence of lead (Pb) in solder and pose a serious reliability risk to electronic assemblies. Tin whiskers have caused system failures in earth- and space-based applications. At least three tin whisker-induced short-circuits resulted in complete failure of in-orbit commercial satellites. Electroplating an object's surface with a thin layer of corrosion-resistant material (e.g., Sn, Zn) is a standard method of protecting mechanical parts against corrosion. Platings (e.g., Sn–Pb solder, Sn, gold (Au), palladium (Pd)) are also used on electrical parts to protect against corrosion and improve their solderability.

2.1 Tin Whisker Growth in Lead-Free Electronics

An expedient transition to Pb-free electronics has become necessary for most electronics industry sectors, considering the European directives, other possible legislative requirements, and market forces. In fact, the consequences of not meeting the European July 2006 deadline for transition to Pb-free electronic products may translate into global market losses.

The elimination of Pb in many electronic systems by the European Union 2006 Restriction of Hazardous Substances (RoHS) legislation could potentially result in reduced reliability and reduced predictability of high-performance electronic systems that use commercial off-the-shelf (COTS) parts.

One significant challenge is the assessment of the tin whisker risk resulting from the pervasive use of high-Sn-content Pb-free finishes by COTS manufacturers. Considering that Pb-based electronics have been in use for more than 40 years, the adoption of Pb-free technology represents a dramatic change. In less than 10 years, the industry needs to adopt different electronic soldering materials and surface finishes for both components and printed circuit boards. This challenge is accompanied by the need to requalify component-board assembly and rework processes, as well as test, inspection, and documentation procedures. In addition, Pb-free technology is associated with increased materials, design, and manufacturing costs. The use of Pb-free materials and processes has also prompted new reliability concerns, as a result of different alloy metallurgies and higher assembly process temperatures relative to Sn–Pb soldering.

As shown in Figure 2.1, tin whiskers are filamentary growths that can grow spontaneously from both electroplated Sn surfaces and certain Pb-free solders. Typical whisker diameters generally range in size between 1 and 5 μm while whisker lengths may exceed 5 mm. Whiskers are highly conductive and can result in short-circuits in electronic equipment by bridging gaps between closely spaced electrical conductors. Incidents of high-profile failures have been documented, some very recently. Tin and tin alloys have been widely used in the electronics industry for more than 60 years, and whisker problems have often been a cause for concern. The addition of Pb to Sn in quantities exceeding 3% by weight (wt%) has been recognized as a mitigation path.

However, the European Union has enforced the RoHS directive, which requires Pb removal from electronic assemblies. While certain hardware has been exempted from this directive, particularly high-reliability electronics, the widespread use of commercial electronic components in this hardware

FIGURE 2.1
Tin-plated connector pins after 10 years. (*Source:* Courtesy of NASA Goddard Space Flight Center (GSFC).)

has thereby exposed a new generation of electronic equipment to this failure mechanism. This has led to a significant amount of research in the area, with more papers written on the topic since the year 2000 than in the prior 50+ years that the occurrence of tin whiskers has been recognized.

The following sections of this chapter provide a brief review of the history of tin whiskers and present the various theories that have been formulated to date, including some of the latest research in the area of energy methods. An examination of the tin whisker mechanisms that aggravate growth, as well as an overview of the preventative measures, is made along with recommendations of areas for further exploration. Some modeling is performed to show the effect of diffusion on the formation of intermetallic compounds and Kirkendall voiding, which has been shown to be a significant mechanism in the generation of compressive stress in tin deposits, which has been shown to be a cause for filamentary growth. The effect of the corrosion/oxidation mechanism is also examined.

2.1.1 Nomenclature

a Basal lattice parameter in bct (Å), radius of tin whisker (m)

A_r Chemical affinity of a reaction r

b One half of the distance of whisker separation

C Concentration gradient

c Height lattice parameter in bct (Å)

D Diffusion constant

$d(\ )$ Differential of ( )

G Gibbs' free energy

i Index (dimensionless)

k Index (dimensionless), Boltzmann constant

n_k Amount of species k in a reaction

p Pressure (Pascal)

S Entropy (kcal/mol-K)

T Temperature (K)

V Volume (m³)

X_1 Chemical driving force of reaction

x,y Mole fraction (dimensionless)

α FCC (face-centered cubic) form of Sn (dimensionless)

β BCT (body-centered tetragonal) form of Sn (dimensionless)

ε Cu_3Sn (dimensionless)

η, η' Cu_6Sn_5 (dimensionless)

μ_k Chemical potential of species k

υ   Poisson's ratio (dimensionless)

ξ_r Extent of reaction

ψ_{kr} Stoichiometric number of species k in reaction r

Ω   Chemical potential

2.2 Variability with Tin Whisker Mechanisms

It has been widely documented in the literature that the growth of tin whiskers is a result of compressive stress. The various sources of this stress have been studied extensively but no holistic model has been presented to date. One of the major hurdles associated with an in-depth understanding of the problem is that there are a multitude of variables that may affect this compressive stress. These stresses can originate from intermetallic formation, CTE (coefficient of thermal expansion) mismatches, operational stress, corrosion, and residual stress from tin electroplating processes.

The actual mechanism of tin whisker growth is not fully understood, even after decades of research. There is no test that can accurately determine when whiskers will form, the number of whiskers, or their length.

- 1942–1943: Aircraft Radio Corporation Electrical Problems—The first recognition of electrical problems caused by metal whiskering appears to have occurred in 1942–1943 in aircraft radios made by Aircraft Radio Corporation in Boonton, New Jersey.

- Air-spaced variable capacitors were cadmium plated to retard corrosion; the cadmium plating whiskered, and these whiskers dropped the Q of the tuned circuits to unusable low values. (Q, the quality factor of the circuit, is determined by taking the bandwidth and dividing it by the resonant frequency of the circuit.) This company's radios included those used to land under conditions of zero visibility.

The growth of needle-like crystals on Cd deposits has caused considerable annoyance in the radio industry. These crystals are known as "whiskers." They grow between condenser plates of variable condensers, and, being electrical conductors, actually short-circuit the plates, thereby putting the radio set out of operation. Not much is known about the cause of the growth of these crystals. It seems to have taken the special circumstances of World War II to identify a systematic problem, and the cause of this problem.

That the growth of whiskers is not a new phenomenon may be concluded from the examination of undisturbed old equipment. For example, a number of Zn-plated details installed in a telephone central office in 1912 were recently removed for study. Surfaces that had been protected from cleaning operations and from excessive air circulation had numerous whiskers

- 1986 – Pacemaker FDA Class 1 Recall - Total Failure Crystal Oscillator Short
- 1998 – Galaxy IV & VII: on-orbit failures of satellite control processors (SCP) due to tin whisker induced short circuits.
- 2002 – Northrop Grumman Relay Failures - Military Aircraft - approximately 10 years old - failed.
- 2005 – Millstone Unit 3 Nuclear Reactor Shutdown due to "tin whiskers". In response to this event, Millstone implemented a procedure to inspect for these whiskers at every refueling outage, or 18 months.
- 2006 – Galaxy IIIR: on-orbit failures of satellite control processors (SCP) due to tin whisker induced short circuits.

FIGURE 2.2
Failures attributed to tin whiskers.

present. Bell Labs learned during the early part of 1948 that "channel filters," used in carrier telephone systems, were failing, and Bell eventually traced the problem to whiskers growing from Zn-plated steel. There is a high degree of difficulty in identifying the cause of the shorting: In some cases, the whisker that caused the short disappeared, and the fault could not be reproduced in the lab. Even when the whisker was still present and shorting, its diameter was less than that of a human hair, and could easily escape attention, even to a careful inspector. But they were good, and eventually "nailed" this problem.

There are many plating formulations that have been used through the years. Each of these may perform differently with regard to the generation of tin whiskers. In addition, the variation in current density that is used for deposition can also result in a wide variation of the stress states in the plating, ranging from compressive to tensile. Organic additives that result in a bright tin deposit (the manifestation of fine grain size) have also been considered to play a role, but the influence is poorly understood. Contamination of plating baths, particularly from copper (Cu) ions, have been linked to the formation of tin whiskers. Figure 2.2 summarizes failures attributed to tin whiskers.

2.3 Tin Whisker Risk: Lesson from the Nuclear Industry

A nuclear reactor shutdown is a particularly disturbing event. On April 17, 2005, Millstone Unit 3 in Waterford, Connecticut, experienced a reactor trip from full power, and one of two trains of the Safety Injection (SI) and Main Steam Isolation actuated when low steam line pressure was sensed on one of four steam generators.

Within a few hours, under a high-powered microscope, engineers spotted a thin filament of metal, barely visible to the naked eye, spanning the card's surface and bridging a line of conductive material, called a trace. That metal fragment, they soon learned, had single-handedly caused the electrical short that gave a false low-pressure reading and forced an unplanned shutdown. The cause was the growth of a tin whisker between two components on a Westinghouse Solid State Protection System (SSPS) universal logic card, causing a short-circuit to ground and triggering the single train of SI and the subsequent automatic reactor trip. A Westinghouse SSPS engineer revealed that the component where the whisker grew was a diode with a blue jacket.

The tin whisker that shorted out at Millstone's Unit 3 reactor on April 17th triggered an automatic shutdown designed to protect the reactor, but that is not what worries the Nuclear Regulatory Commission (NRC), responsible for overseeing safe operations in the industry in the United States. Rather, it is that the tin whisker could prevent a safety system from working properly.

On August 25, 2005, the NRC issued Information Notice 2005-25.1878, which specifically discussed the inadvertent trip at Millstone Unit 3 caused by tin whiskers, notified licensees about recent operating experience related to the growth of tin whiskers in electronic circuits in nuclear power stations, and informed licensees to consider appropriate actions to avoid similar problems.

2.4 What Are Tin Whiskers?

Tin whiskers are single-crystal, hair-like growths that grow spontaneously from surfaces that use tin, especially electroplated tin, as a surface finish. These whiskers are electrically conductive and physically strong.

Whiskers are typically 1 to 2 mm in length and 1 to 3 μm in diameter. Lengths up to 400 mils (10 mm) and densities of 200 whiskers per square millimeter have been observed. The mechanism of their growth is still not clearly understood, even after decades of observation and experimentation. This has promoted much scientific debate, contradictory experimental results, and conflicting evidence. Here is a summary about the findings with tin whiskers:

- Passive and active electronic components with Pb-free tin plating are subject to whiskering. Tin-plated mechanical parts (e.g., nuts, bolts, fasteners, panels) may also grow whiskers. Tin whiskers can pierce through thin or soft conformal coatings.
- Eutectic tin-lead solder (63% Sn, 37% Pb) has been a standard material since the beginning of the electronics industry. Solder's characteristics are well understood. Manufacturing methods and parts have been designed around and optimized for the use of Sn–Pb solder.

- Tin alloys with more than 3% Pb have a greatly reduced tendency to whisker, and any whiskers that may form are very small. Pure Sn or high Sn alloys that do not contain Pb are very likely to whisker.

Tin whiskers are electrically conductive, crystalline structures of Sn that sometimes grow from surfaces where Sn (especially electroplated tin) is used as a final finish. Tin whiskers have been observed to grow to lengths of several millimeters (mm) and, in rare instances, to lengths in excess of 10 mm. Numerous electronic system failures have been attributed to short-circuits caused by tin whiskers that bridge closely spaced circuit elements maintained at different electrical potentials. NASA Goddard Space Flight Center (GSFC) has rules prohibiting the use of pure Sn coatings, and also Zn and Cd coatings.

2.5 What Factors Influence Whisker Growth?

There are many factors that can influence whisker growth. The number and diversity of factors has made the development of accelerated whiskering susceptibility tests very difficult. Some theories suggest that tin whiskers may grow in response to a mechanism of stress relief (especially "compressive" stress) within the tin plating.

Other theories contend that growth may be attributable to recrystallization and abnormal grain growth processes affecting the tin grain structure, which may or may not be affected by residual stress in the Sn-plated film. Figure 2.3 summarizes how those advocating "stress" as crucial for metal whisker formation point to some commonly accepted factors that can impart additional residual stresses.

2.5.1 Residual Stresses within the Tin Plating

Residual stresses within the tin plating are caused by factors such as the plating chemistry and process. Electroplated finishes (especially "bright" finishes) appear to be most susceptible to whisker formation, reportedly because bright tin plating processes can introduce greater residual stresses than other plating processes.

2.5.2 Intermetallic Formation

It is generally accepted, however, that the primary mechanism for the growth of tin whiskers is through the formation of Cu6Sn5 intermetallic at the copper/tin interface, which is commonly encountered in electronics.

Residual Stresses:

1. Residual stresses within the tin plating
2. Intermetallic formation
3. Externally applied compressive stresses
4. Scratches or nicks
5. Coefficient of thermal expansion (CTE) mismatches

Influencing Factors:

- Plating chemistry
- Plating process
- Deposit characteristics
- Substrate
- Environment

FIGURE 2.3
Factors that may influence tin whisker growth.

The diffusivity of Cu in Sn is higher than Sn in Cu. This results in intermetallic growth in the direction of the tin plating, with resultant Kirkendall voiding of the Cu near the prior plating interface. The kinetics of this reaction are sufficient for it to occur at room temperature, and indeed, the growth of this intermetallic has been well characterized in the literature. With a net motion of Cu into the Sn layer, a compressive stress results. The compressive stress is then relieved by diffusion of Sn, mostly along the grain boundaries. This has been studied with dual Sn isotope tracing of a layered tin coating. Whiskers then form in preferential areas; these whiskers serve to relax the compressive stress in the coating by accommodating the excess atoms, but the mechanism is self-sustaining as more intermetallic continues to form.

2.5.3 Diffusion of Substrate Material into Tin Plating

The diffusion of the substrate material into the tin plating (or vice versa) can lead to the formation of intermetallic compounds (such as Cu_6Sn_5 for an Sn-over-Cu system) that alter the lattice spacing in the tin plating. The change in lattice spacing may impart stresses to the tin plating that can be relieved through the formation of tin whiskers. Similarly, it is known that the oxidation of Sn will occur in air, particularly air that is laden with humidity. The resulting oxidation of Sn results in the formation of SnO_2. The net result is localized compressive stress in the near-surface oxidized layer. This, in turn, results in compressive stress, which is similarly relaxed via Sn diffusion to unstressed whisker grains.

2.5.4 Coefficient of Thermal Expansion Mismatches

The coefficient of thermal expansion (CTE) causes mismatches between the plating material and substrate, a mechanism for growth related to CTE. This is particularly evident for tin platings deposited on low-CTE materials such as Kovar or Alloy 42. CTE mismatches during thermal cycling serve to provide tensile and compressive stresses that can relax by similar diffusion mechanisms, but in this case the overall stress in the tin plating may exceed the yield stress of tin, with dislocation glide resulting.

2.5.5 External Factors Induce Residual Stresses

The following external factors can induce residual stresses:

- Externally applied compressive stresses, such as those introduced by torquing a nut or a screw or clamping against a tin-coated surface can sometimes produce regions of whisker growth
- Bending or stretching of the surface after plating (such as during lead formation prior to mounting an electronic component)
- Scratches or nicks in the plating and/or the substrate material introduced by handling, probing, etc.

There are several efforts underway to develop an effective test method for assessing the propensity of tin plating to initiate and grow whiskers. At this time, there is no test that can accurately determine when whiskers will form or the extent of any whisker growth.

2.6 Why Whiskers Are a Serious Reliability Risk to Electronic Assemblies

Tin whiskers pose a serious reliability risk for electronic assemblies. Several instances have been reported wherein tin whiskers have caused system failures in both earth- and space-based applications. There are reports of tin whisker-induced short-circuits that resulted in the complete failure of on-orbit commercial satellites. There have also been whisker-induced failures in medical devices, weapon systems, power plants, and consumer products.

Manufacturers are rapidly switching to pure tin or lead-free tin-based alloys to comply with "green" initiatives. These initiatives eliminate lead as an alloying option. According to an October 2003 University of Maryland Computer Aided Life Cycle Engineering (CALCE) survey, approximately 70% of seventy-one major semiconductor suppliers currently or will shortly offer pure tin or lead-free tin-based alloys as final plating. Suppliers are rapidly converting product lines to lead-free materials.

Pure Sn and Pb-free Sn-based alloys allow whiskers to grow. With present circuit geometries so much smaller than in the recent past, adjacent whiskers can easily bridge spaces between leads, and/or whiskers from adjacent areas can touch each other, causing short-circuits. Loose whiskers can bridge board traces, foul optics, or jam microelectromechanical machines (MEMs).

In previous technologies, circuit voltage and current levels would almost instantaneously vaporize any whiskers, and the circuit would not even notice the event. However, low voltage and current levels in modern circuits do not have the energy needed to melt the whiskers so circuits stay shorted by the whiskers. The general risks fall into four categories as described in the following (see Figure 2.4).

2.6.1 Stable Short-Circuits in Low-Voltage, High-Impedance Circuits

Tin whisker results in stable short-circuits in low-voltage, high-impedance circuits where current is insufficient to fuse the whisker open. In such circuits there may be insufficient current available to fuse the whisker open and a stable short-circuit results. Depending on a variety of factors, including the diameter and length of the whisker, it can take more than 50 milliamps (mA) to fuse open a tin whisker.

2.6.2 Transient Short-Circuits

Tin whisker results in transient short-circuits until the whisker fuses open. At atmospheric pressure, if the available current exceeds the fusing current of the whisker, the circuit may only experience a transient glitch as the whisker fuses open.

- Stable short circuits in low voltage, high impedance circuits where current is insufficient to fuse the whisker open
- Transient short circuits until the whisker fuses open
- Plasma arcing in a vacuum where a whisker fuses open, but the vaporized tin initiates plasma that conducts over 200 amps. This phenomenon is reported to have occurred on at least three commercial satellites and rendered the spacecraft non-operational
- Whiskers or pieces of whiskers that break loose and bridge conductors or interfere with optical surfaces or jam MEMs

---

FIGURE 2.4
Tin whisker failure mechanisms.

2.6.3 Metal Vapor Arc

Plasma arcing occurs in a vacuum where a whisker fuses open, but the vaporized tin initiates plasma that conducts over 200 A. This phenomenon is reported to have occurred on at least three commercial satellites and rendered the spacecraft nonoperational.

If a tin whisker initiates a short in an application environment possessing high levels of current and voltage, then a very destructive phenomenon known as a *metal vapor arc* can occur. The ambient pressure and temperature, and the presence of arc-suppressing materials, also affect metal vapor arc formation.

In a metal vapor arc, the solid metal whisker is vaporized into a plasma of highly conductive metal ions, which are more conductive than the solid whisker itself. This plasma can form an arc capable of carrying hundreds of amperes. Such arcs can be sustained for a long time (several seconds) until interrupted by circuit protection devices (e.g., fuses, circuit breakers) or until other arc-extinguishing processes occur. This kind of arcing is happening in the metal vapor.

When an arc-quenching agent (e.g., air) is present, more power must be installed into the event to replace power lost to the uninteresting processes happening in the quenching agent. Therefore, as air pressure is reduced, less power is required to initiate and sustain a whisker-induced metal vapor arc.

Past experiments have demonstrated that at atmospheric pressures of about 150 torr, a tin whisker could initiate a sustained metal vapor arc where the supply voltage was approximately 13 volts (or greater) and supply current was 15 A (or greater). Tin (or other materials) from the adjacent surfaces can help sustain the arc until the available material is consumed or the supply current is interrupted.

2.6.4 Debris/Contamination

Whiskers or parts of whiskers may break loose and bridge isolated conductors, interfere with optical surfaces, or jam MEMs.

There is no test that can accurately determine a plating's propensity to whisker. Suppliers cannot detect differences in the whiskering susceptibility of different platings. A new plating material will appear to have the same reliability as the one it is replacing. Because there is no detectable change in form, fit, function, quality, or reliability, some suppliers see no need to issue a product change notice.

Many suppliers and distributors are unaware of the tin whisker issue. The stratification of electronics manufacturing gives systems integrators little insight into the materials that are actually being incorporated into their products.

Just-in-Time (JIT) manufacturing and the use of COTS parts are common practices. JIT allows parts to be directly incorporated into systems with little or no inspection. COTS parts typically have very few restrictions on materials. Suppliers may not even be required to provide any prior notice for changes in materials or processes. This means that there is a high likelihood of lead-free tin plating in a part or an assembly.

There have been a number of failures because of tin whiskers in high-reliability applications. Systems experiencing tin whisker-caused failures during the past 3 years have included commercial satellites, power management modules, nuclear and conventional power plants, and military/aerospace systems.

The vast majority of these items that failed were manufactured before the major shift to tin plating and contained relatively few tin-plated items. Reported failure rates ranged from 1% to 10%. Systems with higher numbers of tin-plated items will probably experience higher failure rates.

2.7 How to Mitigate Tin Whisker Risk

The uncertainties associated with tin whisker growth make it extremely difficult to predict if and when tin whiskers might appear. With the exception of the total avoidance of Pb-free Sn, none of the following commonly available mitigation techniques has been proven to provide the degree of protection required by high-reliability equipment. With present circuit geometries so much smaller than in the recent past, adjacent whiskers can easily bridge spaces between leads, and/or whiskers from adjacent areas can touch each other, causing short-circuits. Loose whiskers can bridge board traces, foul optics, or jam MEMs.

In an early endeavor to mitigate tin whisker risk, Bell Labs carried out a research program, alloying Sn with each member of the periodic table that they could figure out how to get into a plating bath. In the mid-1950s, they showed that the addition of 1% to 5% Pb to plated Sn quenched whiskering. Other studies showed that as little as 0.5% Pb was effective. And these have been repeated with the same findings. Because many plating shops do not hit the target of Pb concentration with high precision, specifications often call for 2% or even 3%, in order to increase confidence that one will get at least 0.5%.

For decades the common method of whisker control and lowering the melting point of Sn alloys has been the addition of Pb to the Sn. At least 3% Pb in Sn alloys effectively eliminates the growth of whiskers that could cause harm. The worldwide movement to eliminate Pb from all products has caused the tin whiskers threat to reemerge. There are serious potential consequences to manufacturers and users of high-reliability electronics equipment.

Commercial manufacturers are rapidly adopting tin plating to advance their economic interests and to comply with European and Japanese legislation on Pb recycling and prevention. By the end of summer 2004, it may be difficult to find anything other than Pb-free Sn finishes for some parts. It seems impossible that any user of electronic components will be able to totally avoid Pb-free tin plating.

There is no Pb-free mitigation technique for pure Sn or Sn alloys that will guarantee whiskerless hardware. None of the commonly available mitigation techniques have been proven to provide the degree of protection required by high-reliability applications.

The transition by suppliers responding to the worldwide Pb-free movement poses a major reliability risk to manufacturers of high-reliability systems. System manufacturers must take immediate steps to understand their current exposure, provide strong direction to their suppliers, and aggressively support development and implementation of effective mitigation techniques.

Alloys of Sn and Pb are generally considered acceptable where the alloy contains a minimum of 3% Pb by weight. Although some experimenters have reported whisker growth from Sn–Pb alloys, such whiskers have also been reported to be dramatically smaller than those from pure tin plated surfaces and are believed to be sufficiently small so as not to pose a significant risk for the geometries of today's microelectronics.

Avoidance of pure Sn is the first and best choice. However, suppliers are converting to Pb-free Sn at a rapid rate, and system integrators are receiving pure Sn despite contractual prohibitions. As shown in Figure 2.5, the common tin whisker mitigation techniques and limitations are summarized as follows:

1. Avoid the use of *pure tin*-plated components, if possible. Utilization of procurement specifications that have clear restrictions against the use of pure tin plating is highly recommended. Most (but not all) of the commonly used military specifications currently have prohibitions against pure tin plating. Studies have shown that alloying tin with a second metal reduces the propensity for whisker growth.

- Avoid the use of PURE TIN plated components if possible.
- Alloys of tin and lead are generally considered to be acceptable.
- Matte tin (tin with a dull low gloss finish and larger grain size) is more resistant to whiskering than bright tin.
- Annealing tin can reduce the stresses in plating that contribute to whisker growth.
- Solder dip the plated surfaces sufficiently using a tin-lead solder to completely reflow and alloy the tin plating.
- Conformal Coat or foam encapsulation over the whisker-prone surface can significantly reduce the risk of electrical short circuits caused by whiskers.
- Replate the whisker prone areas.
- Post procurement verification.
- Evaluate application specific risks.

FIGURE 2.5
How to mitigate tin whisker risk.

2. Matte tin (tin with a dull low gloss finish and larger grain size) is more resistant to whiskering than bright tin.

3. Annealing tin can reduce the stresses in plating that contribute to whisker growth. However, the benefits are limited and only short term.

4. Solder dip the plated surfaces sufficiently using a Sn–Pb solder to completely reflow and alloy the tin plating. Obviously, special precautions are required to prevent thermal shock-induced damage, to prevent loss of hermeticity and to avoid thermal degradation. This approach may have limited success because it may be difficult to ensure that the entire surface is properly reflowed. Therefore, robotic solder dipping with Sn–Pb solder is a solution for some, but not all, components. Components must be handled carefully to avoid damage during the process.

5. Conformal coat or foam encapsulation over the whisker-prone surface can significantly reduce the risk of electrical short-circuits caused by whiskers. The choice of coating material, thickness, and possible degradation with time/environmental exposure can impact the effectiveness of the coating. NASA GSFC experiments have shown that the use of Arathane 5750 (formerly Uralane 5750) conformal, when applied uniformly to a nominal 2- to 3-mil thickness, can provide significant benefit by containing whisker growth outward through the coating. This coating is also resistant to being penetrated by whiskers attempting to puncture the coating from the outside. Therefore, conformal coatings can be applied, but their success is very dependent on the coating material, thickness, and application process. In summary, conformal coating is a promising technique, but long-term success in whisker control has not been established. This complex topic requires further investigation.

6. Replate the whisker-prone areas. Some manufacturers may be willing to strip the pure tin plate from finished products and re-plate using a suitable alternate plating material such as Sn–Pb or nickel (Ni). Caution is advised if considering use of an external plated finish (e.g., Sn–Pb or Cu) on top of an existing pure tin deposit. There is some evidence that whiskers may still form from the pure tin layer and protrude through the thin external deposit. Thus, stripping the finishes and replating with Sn–Pb solder is possible but requires extra handling and exposure of finished parts to corrosive materials. This sets the stage for corrosion-related issues.

7. Post procurement: It can be dangerous to rely on the part manufacturer's certification that pure tin plating was not used in the production of the product supplied. NASA GSFC is aware of several instances where the procurement specification required "No Pure Tin" but the product supplied was later determined to be pure tin. In some of these instances, tin whisker growths were also discovered.

Users are advised to analyze the plating composition of the products received as an independent verification.

8. Evaluate application-specific risks. A variety of application-specific considerations may be used to assess the risk of whisker-induced failures and assist in making "use as-is" or "repair/replace" decisions. These factors include circuit geometries that are sufficiently large to preclude the risk of a tin whisker short, mission criticality, mission duration, collateral risk of rework, schedule, and cost.

2.8 Use Finite Element Modeling to Assess Tin Whisker Risk

For finite element modeling (FEM), tin whisker mechanisms are to be modeled and a resulting compressive stress is to be calculated. FEM helps to assess the following regarding tin whisker risk mitigation:

- Identify typical densities of tin whisker occurrences.
- Estimate the average whisker growth length.

Such FEMs are useful, in that the length of a tin whisker is directly proportional to the risk that it may result in a short-circuit.

Several research studies have found that the growth of tin whiskers occurs from the surface. That is, the diffusion of Sn to the regions of favorable formation result in the whisker being pushed upward from the surface.

2.8.1 Allotropic Transformation of Tin

At room temperature, Sn has a BCT crystal structure, β-tin. Tin experiences an allotropic transformation at temperatures below 13.2°C to α-tin, which has an FCC structure. This transformation ($\beta \rightarrow \alpha$) is accompanied by a 27% volume expansion. As a result of the expansion, this can cause a phenomenon referred to as tin pest, which manifests as a crumbling gray powder when disturbed. The phenomenon has been studied, but is not often seen in experience as it is considered that the Sn must exist in a pure form. Often, Sn is electroplated from alkaline baths, which tend to co-deposit impurities that may suppress the transformation.

2.8.2 Diffusion and the Formation of Copper–Tin Intermetallics

Copper (Cu) and tin form two distinct intermetallics: Cu_6Sn_5 and Cu_3Sn. Cu_3Sn has been found to occur at temperatures in excess of 60°C (Black, 1969) and is generally of minor importance in the study of tin whiskers as at these temperatures, rate effects dominated the formation of hillocks in lieu of whiskers.

Hillocks are eruptions of the surface that occur as a result of stress relaxation, but as will be discussed later, these are much shorter in length than whiskers. Hence, their relative risk is minimal. Tin can also form an intermetallic with nickel (Ni). The formation of an intermetallic is limited by diffusion of species, which is a function of temperature according to Fick's First Law:

$$D = D_O * \exp\left(-\frac{C}{R*T}\right) \qquad (2.1)$$

2.8.3 Oxidation of Tin

Tin is known to form two distinct oxides: SnO and SnO_2. This is speculated to occur in humid air with the oxidation and reduction reactions starting at the aqueous/tin interface:

$$\text{Anodic:} \quad Sn \rightarrow Sn^{2+} + 2\,e^-$$
$$\text{Cathodic:} \quad O_2 + 2\,H_2O + 4\,e^- \rightarrow 4\,OH^-$$

These hydroxide ions react with the tin ions to form tin(II) hydroxide:

$$Sn^{2+} + 2\,OH^- \rightarrow 2\,Sn(OH)_2$$

This corrosion product is manifested as a "tin hydroxide." Also occurring is the oxidation of the tin(II) ion by

$$2\,Sn^{2+} + O_2 + 6\,H_2O \rightarrow 2\,Sn(OH)_4 + 4\,H^+$$

The corrosion product is manifested as a white tin hydroxide. Dehydration of the hydroxides occurs by

$$Sn(OH)_2 \rightarrow SnO \cdot xH_2O + (1-x)H_2O$$
$$Sn(OH)_4 \rightarrow SnO_2 \cdot yH_2O + (2-y)H_2O$$

SnO and SnO_2 are very different from β-tin in volume, as shown in Table 2.1.

Hence, with the increase in the molar volume, one would expect that compressive stress in the surface layers of the Sn would result. With the relative hardness of the oxide as compared to β-tin, it would be expected that the oxide would not deform as significantly as the tin. The mobility of Sn atoms in pure Sn is much higher than that in bound Sn oxide.

2.8.4 Excessive Energy Theory

This model considers that in a microstructure such as a whisker, the surface free energy of the formation for the new whisker surface is a determining factor as to whether or not they can form. The model further elaborates that

TABLE 2.1

SnO and SnO2 Are Very Different from β-Tin in Volume

Property	β tin	SnO (romarchite)	SnO$_2$ (cassiterite)
Molar mass (g/mol)	118.69	134.69	150.69
Density (g/cm³)	7.17	6.45	6.9
Molar volume (g/cm³)	16.3	21.04	21.8
Mohs hardness	1.5	2–2.5	6.5
Crystal structure	BCT	BCT	BCT
	a = 3.1817 Å	a = 3.79 Å	a = 4.737 Å
	c = 5.8311 Å	c = 4.83 Å	c = 3.186 Å
Elastic modulus (GPa)	41.6–44.3		
Melting temperature (°C)	231.9		
Poisson's ratio	0.33		

there is a critical dimension for whisker growth. The model starts by calculating the excessive chemical potential of the tin atoms that form a whisker.

2.8.5 Cracked Oxide Theory

In this model, it is assumed that a crack in the oxide layer results in a driving force for whisker growth. The oxide that forms on the tin surface is known to be stable and protective. However, for the model, isolated areas are assumed to crack. Now, because of the loss of load-bearing capability of the film, the compressive stress that forms as a result of the Cu_6Sn_5 intermetallic is released at these areas. In this case, a lateral flux of Sn atoms (and vacancies) is now free to develop. The oxide continues to form on both the whisker and the free surface, but a crack persists due to the growth of the whisker from the surface as the tin has no time to develop oxide as the whisker continues to grow. This results in the ability to form vacancies via a Nabarro–Herring creep model for stressed elastic solids. The growth of the intermetallic is assumed to be linear. The reaction at the Cu_6Sn_5/Cu interface is assumed to be limited by the interfacial reaction. For most applications, we only are concerned with the single reaction:

$$6\,Cu + 5\,Sn \rightarrow Cu_6Sn_5$$

The Gibbs' free energy change for this reaction is calculated from

$$dG = -S\,dT + V\,dp - A\,d\xi \tag{2.2}$$

For constant temperature and pressure, this results in the driving force of the reaction as

$$X_1 = -(dG/dy)|_{T,p} = A\,(d\xi/dy) \tag{2.3}$$

perpendicular to the interface. Because of the earlier assumption that the reaction rate was linear, this means that $(d\xi/dy)$ is constant or, stated differently, that the force is independent of the location of the reaction. The whisker field is modeled as a uniform field of defects laid out in hexagonal form. Here the gradient of the chemical potential in the Sn in cylindrical coordinates is given by

$$Xr = (\partial\sigma/\partial r)\Omega \tag{2.4}$$

The distance of separation is mechanical force. Now the Sn whisker growth rate can be found. The continuity equation in cylindrical coordinates is defined as

$$\nabla^2\sigma = (\partial^2\sigma/\partial r^2) + (1/r)\,(\partial\sigma/\partial r) \tag{2.5}$$

where the boundary conditions are $\sigma = \sigma_0$ at $r = b$, and $\sigma = 0$ at $r = a$. The solution of the equation is

$$\sigma = B\sigma_o\,\ln(r/a) \tag{2.5a}$$

with $B = [\ln(b/a)]^{-1}$ and σ_o is the stress in the Sn film, which is assumed to be constant while the reaction is occurring. The driving force can then be calculated and then the flux to grow the whisker at $r = a$ is calculated as

$$J = C\,(D/kT)\,X_r = (B\sigma_0 D)/(kTa) \tag{2.6}$$

According to the theory of elasticity, the elastic theory and compressive stresses and strains are related by the following relation:

$$\varepsilon_{ij} = (\upsilon/E)\,(\Sigma(k = 1,3)\,\sigma_{kk}\,\delta_{ij} + [(1+\upsilon)/E]\,\sigma_{ij} \tag{2.7}$$

For the problem of a thin film, particularly a soft thin film such as tin plating, it is reasonable to assume plane stress conditions may exist. Taking the "3" direction to be through the plating thickness,

$$\sigma_{33} = 0 \tag{2.8}$$

Also, the symmetry of the thermal expansion problem results in a biaxial stress state:

$$\sigma_{11} = \sigma_{22} \tag{2.9}$$

FEM and tin whisker risk mitigation are shown in Figure 2.6.

- Provide formal FEM training to suppliers and users on tin whiskers and related lead-free issues.

- Require that contractors audit their manufacturing operations to define their present and future exposure through FEM.

- Require an audit of existing hardware to understand level of exposure and degree of risk through FEM.

- Instruct contractors to establish incoming inspection procedures using FEM, X-ray fluorescence (XRF), or Energy Dispersive X-ray (EDX) to positively verify composition of incoming materials.

- Issue strong policy guidelines regarding tin use and mitigation techniques based on FEM.

- Support conformal coating whisker mitigation using FEM.

FIGURE 2.6
FEM and tin whisker risk mitigation.

2.9 How to Evaluate Tin Whisker Impact on High-Reliability Applications

High-reliability systems (e.g., military, aerospace, servers, medical equipment) require long operational lives, high availability, and extraordinary reliability. The longer the application life of the system with Pb-free tin plating, the greater the probability that tin whiskers will form. Based on the failure rates of past systems having few tin-plated parts, unless aggressive mitigation and control efforts are implemented, it is expected that double-digit percentage failure rates will be experienced in high-reliability applications.

Many systems do not have internal diagnostics (e.g., BIT) that are able to detect whisker formation. Short-circuits will not be detected until power is applied, and this would be too late in most cases. The shock and vibration associated with missile launch or aircraft takeoff and landing may dislodge whiskers and move them to areas where they can create problems.

It appears that it will be impossible to totally avoid tin plating. Mitigation techniques are not very robust, and the ability to predict a plating's propensity to whisker will not be available for the foreseeable future. It is critical to establish mitigation efforts that address both tin avoidance and tin adaptation as soon as possible. The following actions should be taken to reduce the risk posed by tin whiskers.

- Establish a Lead-Free Council to measure exposure, coordinate avoidance and mitigation efforts, organize education, and provide senior management with progress reports and risk assessments.

- Develop and implement an ongoing audit program to monitor and report the status of hardware and contractor tin mitigation efforts.

- Provide support to develop effective manufacturing processes for tin whisker mitigation.

- Encourage federal legislation to protect critical manufacturing technologies.

- Build tin whiskers surveillance units to monitor critical hardware using tin plating for where normal inspection and/or testing are impossible.

Bibliography

Antolovich, S.D., and Antolovich, B.F. (1996). An introduction to fracture mechanics, in *ASM Handbook 19 Fatigue and Fracture*, ASM International®.

Arora, N.D., Raol, K.V., Schumann, R., and Richardson, L.M. (1996). Modeling and extraction of interconnect capacitances for multilayer VLSI circuits, *IEEE Trans. Computer Aided Design of Integrated Circuits and Systems*, 15(1), 58–66.

Bilotti, A.A. (1974). Static temperature distribution in IC chips with isothermal heat sources, *IEEE Trans. Electron Devices*, ED-21, 217–226.

Black, J.R. (1969). Electromigration failure models in aluminium metallization for semiconductor devices, *Proc. of the IEEE*, 57(9), 1587–1594.

Blech, I.A., and Herring, C. (1976). Stress generation by electromigration, *Appl. Phys. Lett.*, 29, 131–133.

Chen, C., and Liang, S.W. (2007). Electromigration issues in lead-free solder joints, *J. Mater. Sci.*, 18, 259–268.

Gale, W.F., and Totemeier, T.C. (2004). *Smithells Metals Reference Book, (8th edition)*. Maryland Heights, MO: Elsevier.

Galyon, G.T. (2003). *Annotated Tin Whisker Bibliography*, a NEMI publication, July.

Hunter, W.R. (1997). Self-consistent solutions for allowed interconnect current density. I. Implication for technology evolution, *IEEE Trans. Electron Devices*, 44(2), 304–309.

Hunter, W.R. (1997). Self-consistent solutions for allowed interconnect current density. II. Application to design guidelines, *IEEE Trans. Electron Devices*, 44(2), 310–316.

Suo, Z. (2004). A continuum theory that couples creep and self-diffusion, *J. Appl. Mechanics*, 71, 646–651.

Teng, C.C., Cheng, Y.K., Rosenbaum, E., and Kang, S.M. (1997). iTEM: A temperature-dependent electromigration reliability diagnosis tool, *IEEE Trans. Computer-Aided Design of Integrated Circuits and Systems*, 16(8), 882–893.

Tu, K.N. (2003). Recent advances on electromigration in very-large-scale integration of interconnects, *J. Appl. Phys.*, 94, 5451–5473.

Tu, K.N. (1994). Irreversible processes of spontaneous whisker growth in bimetallic Cu-Sn thin-film reactions, *Phys. Rev. B*, 49(3), 2030–2034.

Yeh, E.C.C., Choi, W.J., Tu, K.N., Elenius, P., and Balkan, H. (2002). Current-crowding-induced electromigration failure in flip chip solder joints, *Appl. Phys. Lett.*, 80, 580.

3

Fatigue Characterization of Lead-Free Solders

Fracture by crack opening mode prevails because nonuniformity in the shear deformation of a solder joint creates a body rotation that results in crack opening stress rather than shear. While the crack growth in shear fatigue is found to vary sensitively with variations in the mechanical constraints on the assembly, such as solder shape and elastic modulus of the chip mold, it is also sensitive to variations in solder microstructure. This sensitivity to the assembly constraints and solder microstructure makes it ideal for investigating fatigue properties of solder joints as well as identifying the structural and microstructural features responsible for reliability failure.

3.1 Surface-Mount Technology

Surface-mount technology (SMT) is a method for constructing electronic circuits in which the components are mounted directly onto the surface of printed circuit boards (PCBs); see Figure 3.1.

The SMT is characterized by the following features:

- Component terminals are short stamped "legs" or metallized ends of components.
- Component mounting (placement on PCB and soldering) is completely automated.
- PCB without mounting holes is used for interconnections.

In the early 1980s, the industry began to replace the traditional through-hole mounting technique (TMT) with SMT. Special surface-mounted devices (SMDs) replaced traditional wire-leaded components. The historical roots of SMT can be seen in the hybrid and microwave circuits that comprised "leadless components" on ceramic substrates (mid-1970s).

SMD components are placed by a pick-and-place machine (chip shooter), with placement rates up to 40,000 cph. Placement precision is 0.05 mm at 3σ. The outline dimensions of a smallest two-terminal chip component (resistor, capacitor) are 0.2×0.4 (mm). Fine-pitch multi-terminal components

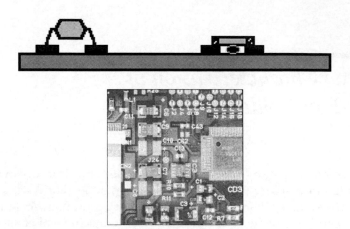

FIGURE 3.1
Surface-mounted components on PCBs.

(ICs) have a pitch down to 0.3 mm. SMT components may be leaded or lead-less (chips). Advantages of SMT components (when compared with TMT components) include

- Smaller dimensions
- Better HF (high frequency) performance (low parasitic inductance, low package propagation delay, low electromagnetic interferences)
- Lower cost

Problems related to SMT components include

- Rigid terminals of chips result in sensitivity of the solder joints to PCB bending and thermal fatigue
- Tombstone effect
- Labeling (marking) problems (the parts are sometimes too small to be marked)

3.2 Fatigue and Thermal Fatigue of Solder Joints

3.2.1 Solder Joint Fatigue

In materials science, fatigue is the progressive and localized structural damage that occurs when a material is subjected to cyclic loading. The nominal maximum stress values are less than the ultimate tensile stress limit, and may be below the yield stress limit of the material. Fatigue occurs when a material is subjected to repetitive loading and unloading. If the loads are above a certain threshold, microscopic cracks will begin to form at the surface. Eventually, a crack will reach a critical size, and the structure will suddenly fracture. The shape of the structure will significantly affect the fatigue life; square holes or sharp corners will lead to elevated local stresses where fatigue cracks can initiate. Round holes and smooth transitions or fillets are therefore important to increase the fatigue strength of the structure. Fatigue of solder joints in Leadless Chip Components (LCCs), such as chip resistors and capacitors, can influence the long-term reliability of electronic products operating in environments with temperature cycling.

3.2.2 Thermal Fatigue of Solder Joint

Mismatch of thermal expansion between an SMT component and the PCB may result in solder joint fracture after thermal cycling. Thermal fatigue of solder joints can be investigated by means of the finite element method (FEM). During the usual thermal cycle regime with relatively slow temperature ramping rates, the solder constitutive response is dominated by secondary creep.

Viscoplastic finite-element simulation methodologies were utilized to predict solder joint reliability for a same die size, stacked, chip scale, ball grid array package under accelerated temperature cycling conditions (–40°C to +125°C, 1-min ramps/15-min dwells). The effects of multiple die attach material configurations were investigated along with the thickness of the mold cap and spacer die. The solder structures accommodate the bulk of the plastic strain that is generated during accelerated temperature cycling due to the thermal expansion mismatch between the various materials that encompass the stacked die package. Because plastic strain is a dominant parameter that influences low-cycle fatigue, it was used as a basis for evaluating solder joint structural integrity. Thermal fatigue, shown in Figure 3.2, is a type of low-cycle fatigue.

Solder joints refer to the solder connections between a semiconductor package and the application board on which it is mounted. In unmounted

FIGURE 3.2
Thermal fatigue of surface-mounted components.

devices, it may also refer to the package's solder connection features themselves, for example solder balls, solder bumps, solder studs, etc., in the context of their attachment to the package body. Because solder joints are one of the most fragile elements of a package (mainly due to the fact that they are small in size and used at high temperatures relative to their melting points), their reliability is of utmost concern to assembly engineers. Solder joint failures occur for various reasons:

- Poor solder joint design
- Poor solder joint processing
- Solder material issues
- Excessive stresses applied to the solder joints, etc.

In general, however, solder joint failures are simply classified in terms of the nature of the stresses that caused them, as well as the manner in which the solder joints fail. Suppose that a chip resistor having length d is soldered to the surface of a PCB as shown in Figure 3.3. The solder layer thickness is δ; the ambient temperature is T_1.

Fatigue, or failure resulting from the application of cyclical stresses, is a major category of solder joint failures. It is often considered the largest and most critical failure category because it is encountered in many different situations that are difficult to control. Solder joint fatigue failure is attributed primarily to stresses brought about by temperature swings and mismatches between the coefficients of thermal expansion (CTEs) of the mounted devices' solder joints and the application board. Suppose that the ambient temperature change is ΔT:

$$T_2 = T_1 + \Delta T \qquad (3.1)$$

Prior to the actual fatigue fracture, solder joints first undergo cyclic deformation from the cyclic stresses as the temperature alternates between its high and low values. Improper design of the solder joint aggravates the effects of this cyclic deformation, which can occur in large steps (on the order of 1%), especially in cases of low cycle fatigue.

Low cycle fatigue, or failure from stress cycles that involve long periods wherein the cycle time is several hours, is a prevalent cause of solder joint

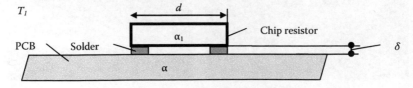

FIGURE 3.3
A chip resistor having length d is soldered to the surface of a PCB.

failures in the field. Even in the case when a chip resistor does not dissipate power, a shear strain in the solder joint arises because of different CTEs of chip expansion and PCB expansion (see Table 3.1).

Chip length after heating will be

$$d(1+\alpha_1\Delta T) \tag{3.2}$$

After the same heating, the respective part of the PCB will have length

$$d(1+\alpha\Delta T) \tag{3.3}$$

3.2.3 Shear Strain

As shown in Figure 3.4, shear strain is the tangent of an angle formed by the sheared line and its original orientation:

$$\Delta\gamma_p \tag{3.4}$$

We can define shear strain exactly the same way we do longitudinal strain: that is, the ratio of deformation to original dimensions. In the case of shear strain, however, it is the amount of deformation perpendicular to a given line rather than parallel to it. The ratio turns out to be tan A, where A is the angle the sheared line makes with its original orientation. Note that if A equals 90°, the shear strain is infinite. When scissors cut paper, they cause the paper to undergo a shear strain so large that the paper yields; it comes apart where it is strained. When a sample is subjected to a shear force, the stress is referred to as a shear stress, τ. The strain determined is referred to as shear strain γ.

TABLE 3.1

Different Rates of Chip Expansion and PCB Expansion

Material	Coefficient of Thermal Expansion (CTE)
Alumina (chip)	$\alpha_1 = 6$ ppm/°C
Epoxy glass (FR4) PCB	$\alpha = 18$ ppm/°C

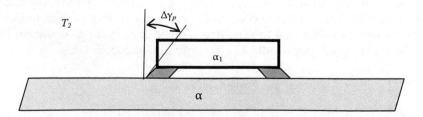

FIGURE 3.4
Shear strain is the tangent of an angle formed by sheared line and its original orientation.

With shear strain we are only concerned about the change in angles. Commonly, this angle is small and therefore the angle itself may be taken instead of its tangent:

$$\Delta\gamma_p \approx \tan\left(\Delta\gamma_p\right) = \frac{|\alpha_1 - \alpha| \cdot \Delta T \cdot d}{2\delta} \qquad (3.5)$$

Multiple deformations of the solder joint result in its cracking, followed by destruction (fatigue phenomenon).

3.2.4 Coffin–Manson Relationship for Thermal Fatigue

Where the stress is high enough for plastic deformation to occur, the account in terms of stress is less useful and the strain in the material offers a simpler description. Low-cycle fatigue is usually characterized by the Coffin–Manson relationship, the first-order approximation of the relationship between the strain and the number of cycles to failure.

$$N_f = \frac{1}{2}\left(\frac{\Delta\gamma_p}{2\varepsilon_f}\right)^{\frac{1}{c}} \qquad (3.6)$$

where
  $\Delta\gamma_p$ is the strain
  N_f is the number of cycles to failure
  ε_f is the fatigue ductility coefficient
  c is thefatigue ductility exponent, commonly ranging from -0.5 to -0.7 for
    metals in time-independent fatigue.

Slopes can be considerably steeper in the presence of creep or environmental interactions.

These two physical constants are influenced by temperature, dwell time, and other factors. The Coffin–Manson equation has been criticized as a means of estimating the thermal fatigue life of solders because it was developed for temperatures below $0.5T_m$, where T_m is the melting temperature in Kelvin. Solders generally operate at high homologous temperatures. "Modified Coffin–Manson" models have been used with more or less success to model crack growth in solder due to repeated temperature cycling. Equation (3.6) may be represented graphically as a straight line in coordinates:

$$\log(\Delta\lambda_p), \log(N_f)$$

In addition, Equation (3.6) may be used for extrapolation of accelerated test results to long-lasting failure processes. Plotted experimental data for 60Sn-40Pb solder are shown in Figure 3.5.

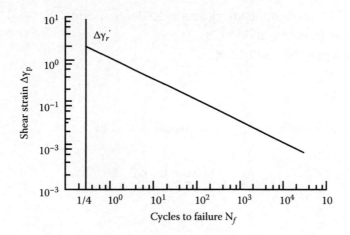

FIGURE 3.5
Plotted experimental data for 60Sn-40Pb solder.

The key temperature for predicting board-level solder joint reliability is the temperature of the solder balls at the board—it is not the device junction temperature. The examples given are for illustration purposes. The Coffin–Manson predictions are generally recognized as "conservative" or to pessimistically predict failures under most long-term use conditions. The design of the board to which the CBGA is assembled affects the solder joint reliability as well as attachment of heat sinks. Most life testing is done on an FR-4 2S, 2P board that is 0.062 inches thick, and the components are assembled on one side. Substituting $\Delta\gamma_p$ in Equation (3.6) by Equation (3.5) we get Equation (3.7):

$$N_f = \frac{1}{2}\left(\frac{|\alpha_1 - \alpha| \cdot \Delta T \cdot d}{4\varepsilon_f \cdot \delta}\right)^{\frac{1}{c}}$$ (3.7)

In the case of 60Sn-40Pb solder, it may be supposed roughly that

$$\varepsilon_f \approx 0.33$$

$$c \approx -0.5$$

3.2.4.1 Example 3.1

A chip resistor having length d is soldered to the surface of a PCB. Suppose that

- $d = 6.3 \cdot 10^{-3}$ m (the biggest size of chip component)
- $\Delta T = 35$ K,

- $\delta = 5\cdot10^{-5}$ m (common thickness of solder in a joint)
- $\alpha_1 = 18$ ppm/K $= 18\cdot10^{-6}$ 1/K
- $\alpha_2 = 6$ ppm/K $= 6\cdot10^{-6}$ 1/K;
- $\varepsilon_f \approx 0.33$;
- $c \approx -0.5$

The number of cycles to failure can be calculated by

$$N_f = \frac{1}{2}\left(\frac{|\alpha_1 - \alpha|\cdot\Delta T\cdot d}{4\varepsilon_f\cdot\delta}\right)^{\frac{1}{c}}$$

$$= \frac{1}{2}\left(\frac{1.3\cdot5\cdot10^{-5}}{|18-6|\cdot10^{-6}\cdot35\cdot6.3\cdot10^{-3}}\right)^2$$

$$\approx 300$$

The main alloys under consideration for surface-mount applications include Sn3.9Ag0.6Cu. For wave soldering applications, the main alloys include Sn0.7Cu and Sn3.5Ag. Thermal cycling tests on assemblies are being performed to compare the Sn3.9Ag0.6Cu alloy with the Sn37Pb control. In order to provide a basis for models of thermomechanical fatigue for Pb-free alloys, a comprehensive mechanical/materials property database is needed. Such data for Pb-free alloys are limited and must be collected using common measurement methods by the microelectronics industry, national laboratories, and academic institutions. Examples of real-world events that can lead to fatigue failures include:

- Powering up of equipment during the day and turning it off at night
- The frequently repeated cycle of driving a car and parking it, with the application board under the hood
- The orbiting of a satellite that exposes it to the alternating direct heat of the sun and cold vacuum of space.

Obtaining an understanding of how these alloys behave is required for the following reasons:

- It is impractical (and probably impossible) to rerun "standard" reliability qualifications, such as accelerated thermal cycling (ATC), for all packages and assemblies currently in use.
- Reliability estimation/extrapolation techniques using mechanical testing data and thermal cycling test data to predict the package/

assembly field performance do not exist for the new materials. These must be developed.

- Finite element model (FEM)-based projections require new valid constitutive equations (stress/strain rate/temperature relations) and new fatigue damage criteria based on thermomechanical loading history for the Pb-free solders.

- Engineering acceleration factor models, such as the frequency-modified Coffin–Manson criteria, are nonexistent or at least untested for the new materials in the limit of actual solder joints, such as BGA solder joints.

- Rules of thumb or engineering intuition relating such things as ATC test results or solder joint defects to field performance have not been established.

At this point it is not clear what criteria should be used for ranking and selecting new materials. Bulk solder mechanical properties must be characterized to provide a basis for FEM modeling, with this modeling used for the development of fatigue damage criteria based on ATC test results. Several temperature ranges and cycling frequencies are required to fully validate the fatigue criteria. ATC tests should include multiple temperature ramp rates and extremes for developing the damage criteria. Throughout the industry, different temperature cycling regimes are used, depending on end use and company tradition. The temperature cycling regimes include

- −55°C to 125°C cycling regime
- −40°C to 125°C cycling regime
- 0°C to 100°C cycling regime

These temperature cycling regimes are being used in the NEMI (National Electronics Manufacturing Initiative) lead-free thermal cycling trials.

3.3 Fatigue, Microstructure, and Microstructural Aging

3.3.1 Microstructure and Microstructural Aging

The test samples must have representative microstructures, that is, equivalent with the microstructure produced in actual solder joints. This cannot be over-emphasized. The plastic deformation behavior and the fatigue mechanisms depend critically on the microstructure. Microstructural aging processes coupled with deformation should be studied for selected alloys. An evolving microstructure will control the inelastic/plastic deformation behavior of the materials. Aged joints may be more indicative of actual field conditions. A PCB is used to mechanically support and electrically connect electronic components using conductive pathways, tracks, or signal traces etched from copper sheets laminated onto a nonconductive substrate. In the United States, PCB specifications include

- MIL-I-24768 Insulation, Plastics, Laminated, Thermosetting (military specification)
- IPC-2221 Generic Standard on Printed Board Design (IPC—Association Connecting Electronics Industries, deals with interconnection technologies)

3.3.2 Mechanical Characterization

Mechanical characterization of the strain rate, as a function of stress, temperature, and microstructure of the materials should be undertaken utilizing stress relaxation testing techniques along with some conventional stress/strain testing under isothermal, constant strain-rate controlled conditions. The strain rate range of interest is from approximately 10^{-3} to 10^{-7}/s. This strain rate range should be measured at temperatures from approximately $-55°C$ to $150°C$. Some cyclic stress versus strain responses should be measured to fully understand the monotonic and cyclic plastic hardening behavior as well as the elastic behavior. Boards are generally defined by

- Type of dielectric material
- Number of conductor layers
- Copper thickness, which is traditionally characterized by "copper weight"

Copper weight defines the number of ounces of copper in a 1 sq. ft. area, such as 0.5, 1.0, 2.0, 3.0 oz., etc. Usually this parameter is used to specify the thickness of the copper on each layer of the board. It is very easy to work out the thickness of 1.0 oz. copper on a 1 sq. ft. board based on the copper density (8.9 g/cm^3), which is 34.29 μm and rounded up to 35 μm. So we have the following list:

- 0.5 oz. means 17.5-μm copper thickness
- 1.0 oz. means 35-μm copper thickness
- 2.0 oz. means 70-μm copper thickness

Copper weight is measured in ounces of copper per square foot ($m = 1$ oz. $= 28.35$ g, $S = 1$ ft.$^2 = 0.305^2$ m$^2 = 0.093$ m^2). The common range of the parameter is 0.25 oz. to 3 oz.; a typical value is 1 oz. Higher numbers are used in high-current and high-frequency applications.

$$m = 1\,oz. = 28.35\,g = 0.02835\,kg \qquad m = \gamma S h$$
$$S = 1\,ft^2 = 0.3052\,m^2$$
$$\gamma = 8920\,kg/m^3 \qquad h = \dfrac{m}{\gamma S}$$
$$h = ?$$

$$h = \dfrac{0.02835}{8920 \cdot 0.305^2} = 34.2 \cdot 10^{-6}\,(m)$$

$$\dfrac{kg}{\left(kg/m^3\right) \cdot m^2} = m$$

FIGURE 3.6
Calculate 1-oz. copper layer thickness h.

Copper density is $\gamma = 8{,}920\,kg/m^3$. Let us now calculate 1 oz. copper layer thickness h.

Suppose that the copper weight number has unlimited precision (idealization) (Figure 3.6).

It is supposed that 1 oz. of copper per square foot corresponds to 35-μm copper thickness. Let us calculate the so-called sheet resistivity of a 1-oz. copper layer. Sheet resistivity is the resistivity of a square pattern of given thickness. The resistance of a rectangular pattern may be calculated as follows:

$$R = \rho \dfrac{l}{S}$$

$$= \rho \dfrac{l}{h \cdot b} \qquad (3.8)$$

$$= \dfrac{\rho}{h} \cdot \dfrac{l}{b}$$

where
  L is length [m]
  B is width [m]
  H is thickness [m]
  S is cross-sectional area [m²] (perpendicular to the length direction) of the
    pattern
  ρ is material resistivity [Ω·m]

Conclusion: The resistance of a PCB foil pattern depends on its length:width ratio and does not depend on the absolute values of its linear outline dimensions.

Suppose that $l = b$ (square pattern). Its resistance we shall call "sheet resistivity" and designate as in Figure 3.7. The sheet resistivity ω is equal to ρ/h and has [Ω/square] or [Ω/□] dimensions.

ω.

$h = 35 \cdot 10^{-6}\,m$

$\rho = 1.7 \cdot 10^{-8}\,\Omega \cdot m$

$l = b$

$\omega = ?$

$$\omega = R = \rho \frac{l}{S} = \rho \frac{l}{h \cdot b} = \frac{\rho}{h}$$

$\omega = 1.7 \cdot 10^{-8} \cdot m / 35 \cdot 10^{-6}$

$m \approx 0.50 \cdot 10^{-3}/\text{sq}$

FIGURE 3.7
Calculate the sheet resistivity ω of 1-oz. copper layer thickness h.

3.3.3 Stress Relaxation Testing

During accelerated testing, low cycle fatigue of the solder takes place. An accelerated testing regime is typically thermal cycling or power cycling, during which stress accumulates due to localized CTE material mismatches. The relatively weak solder joints are the most compliant materials in the assembly, and hence stresses accumulated due to the localized CTE mismatches (i.e., between components and substrate) are relieved by the solder. This scenario is different from that in creep testing where a constant load is applied. In this case, an instantaneous and fixed strain is applied, which is more representative of thermal cycling in which the CTE mismatch applies a fixed elongation. Layers of copper traces in PCBs are divided by dielectric layers.

- The number of copper layers may vary from 1 to 30.
- External layers (top and bottom) are covered by a screen-printed solder mask.
- Commonly, the board thickness is 15 to 250 mil (1 mil = 25.4 μm).
- The dielectric thickness is 1.3 to 2 mil.

Stress relaxation testing is quite general and equivalent to conventional stress versus strain testing over a spectrum of constant strain rates. However, the stress relaxation testing can be more economical and better suited to the low strain rate end of the desired strain rate range. If the microstructure is evolving, there are issues of how to capture such a microstructure evolution in a database, especially because the evolution (for Pb-Sn anyway) is not uniform. There are different evolution paths as a function of position and time, as the joints unzip and local stress state changes. The effect of pad finish (OSP [organic solderability preservatives], NiAu, palladium, silver) should be incorporated into the material/mechanical characterization of the specimens. Wherever possible, specimens should be prepared and tests conducted, which are either standardized or related to standards, and issued by relevant international and national standardization authorities. For example, recent work (Bradley, 2007) on shear strength testing of solder joints has been conducted to compare and correlate a particular test method with a standardized shear test method used on other materials in industry.

3.3.4 Finite Element Modeling

PCBs having unfavorable deflection have been considered for better mechanical design. During the reflow process, PCBs have experienced a range of temperatures between ambient temperature and peak temperature in the furnace. Out-of-plane deflection is generated due to the property mismatch of material composition. Through finite element modeling (FEM), a combination degree of board and copper plating can be researched for further investigation as to the cause of the trouble and design improvement. FEM can be used to assess the following design parameters' impact on thermal fatigue.

Line width and spacing between lines: down to 5 mil or 0.13 mm. Line (trace) thickness and width are selected depending on the expected current. Spacing between the traces depends on expected voltage. Both line width and spacing between lines are equally important in improving density.

Vias (metallized holes for interlayer connections): A via is a primitive design object. It is used to form an electrical connection between two signal layers of a PCB. Vias are like round pads, which are drilled and usually through-plated when the board is fabricated. They allow the electrical and thermal connection of conductors on opposite sides of the PCB.

A via is a vertical electrical connection between different layers of conductors in a PCB. It consists of two pads, in corresponding positions on different layers of the board, which are electrically connected by a hole through the board. The hole is made conductive by electroplating, or is lined with a tube or a rivet. High-density, multi-layer PCBs may have microvias, such as:

- Blind vias are exposed only on one side of the board, while buried vias connect internal layers without being exposed on either surface.

- A through-hole via is a plated hole made to extend completely through a circuit board hole for the sole purpose of connecting conductors on one or more layers.

- A buried via is like a blind via; however, it is not visible on either the top or bottom of the circuit board.

FEM-"ready" constitutive models must be developed from the mechanical testing data described above for the new materials and embedded in a commercial FEM code. These models can be calibrated using the test results discussed above. For instance, the conventional stress-versus-strain curves can be generated from the load relaxation data, using the FEM models. In addition, it is important that the FEM models be verified against the behavior of single solder ball joints or arrays of such solder joints in isothermal load versus displacement testing to assess any microstructural questions and to assess model accuracy.

3.3.5 Fatigue Criteria

Numerical methods have become a useful means to predict the thermomechanical reliability of solder interconnects in electronics, at least comparatively. In principle, each calculation finally rests on a creep–fatigue model or criterion. This section reviews some important failure models of solder found in literature, discusses their benefits and drawbacks, and finally compares them on the example of a semi-analytical assembly model. Cyclic fatigue criteria constitute the classical approach in creep–fatigue modeling. Strain range, strain range partitioning, inelastic energy, and energy partitioning are all approaches that have led to the successful development of failure models for solder in electronics (Massiot, 2004). In addition to cyclic criteria, continuous damage mechanical models have been developed and discussed recently in different publications. They are basically extended sets of constitutive equations, including a new variable called *damage*. In such approaches, damage may be integrated over any cyclic or noncyclic solicitation.

Fatigue criteria should be developed based on the ATC test data and FEM modeling for various types of solder joints and packages. The FEM analysis requires that valid constitutive (stress, strain, and strain rate) equations and damage criteria be developed. The ATC testing should include multiple strain ranges, temperatures, and cyclic frequencies with explicitly determined ramp-versus-dwell time. Possibly, a general fatigue model can be developed and used in conjunction with FEM analysis. If the testing outlined for the alloys were successfully undertaken, then the resulting data would serve as a benchmark for reliability modeling throughout the industry.

As shown in Figure 3.8, PCB dielectric materials can be divided into two major classes based on the type of reinforcement used. These are woven glass reinforced and nonwoven glass reinforced laminates. Woven glass reinforced laminates are lower in cost than nonwoven laminates and are cheaper to produce and process. Because of the amount of glass in the woven glass cloth, the dielectric constants of laminates based on it are higher than laminates based on other reinforcements. Increased packaging density, along with large-scale implementation of ball grid arrays (BGAs), chip-scale packages, and direct chip attachment (DCA) in portable products has resulted in the need for innovative techniques to increase the I/O density in PCBs. With the development of high-density interconnects (HDIs), via-in-pad has emerged as one of the key enabling technologies for increasing the I/O density. Via-in-pad permits the use of subsurface layers for fan-out and, consequently, smaller packages with higher I/O can be utilized in the design. Additionally, because traces no longer need to be routed between pads, the solder joint pitch can also be decreased. The reliability of via-in-pad under mechanical bend fatigue is examined in this section. Mechanical cycling

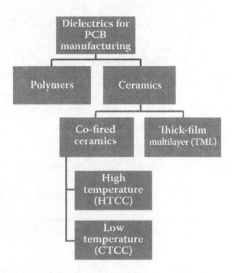

FIGURE 3.8
Dielectrics for PCB manufacturing.

fatigue reliability is especially critical for portable products where keypad actuation often induces repeated bending in the PCB.

3.3.5.1 Polymer PCBs: FR4, FR5

For a resin system, Tg is a certain temperature (different for each polymer) called the glass transition temperature. When the polymer is cooled below this temperature, it becomes hard and brittle, like glass. Some polymers are used above their glass transition temperatures, and some are used below it.

All common laminate resins exhibit changing temperature coefficients of expansion as temperature increases. The *glass transition temperature Tg* is the temperature at which the temperature coefficient of expansion makes a significant change from a low value to a much higher value. This corresponds to a phase change in the resin system. As shown in Table 3.2, hard plastics such as polystyrene are used below their glass transition temperatures; that is, they are in their glassy state.

As shown in Table 3.2, their T_g values are well above room temperature, both at around 100°C. Rubber elastomers such as polyisoprene are used above their T_g values, that is, in the rubbery state where they are soft and flexible. Properties that change around T_g are density, specific heat, dielectric coefficient, rates of gas/liquid diffusion through the polymer, and conductivity or charge mobility.

All resin systems absorb some moisture or water when exposed to high humidity environments. This *moisture absorption* affects the PCB in two ways. Water has a relative dielectric constant of approximately 73. If a laminate

TABLE 3.2

Resin System

Resin System	Epoxy	Modified Epoxy	Epoxy/ Blend	BT	Polyimide	Rogers 4350
T_g *min,* (°C)	140°	170°	210°	185°	260°	280°
Permittivity (Dk @ 1 MHz)	4.3	4.3	4.0	4.0	4.0	3.2
Permittivity (Dk @ 1 GHz)	4.0	4.0	3.8	3.8	3.7	3.5[a]
Loss Tangent (DF @ 1 MHz)	0.02	0.018	0.009	0.01	0.005	0.005
Loss Tangent (DF @ 1 GHz)	0.01	0.009	0.008	0.008	0.004	0.004[a]

[a] Tested at 10 GHz.

absorbs a significant amount of water, the resulting relative dielectric constant of the combination will be higher than the 4.1 used to calculate impedance and can cause impedance mismatches.

A more important effect of moisture absorption is increased leakage current. Materials with high moisture absorption may exhibit leakages in excess of what the circuits housed on them can withstand. In order to use high-absorption materials in such applications, it is often necessary to seal them with a special coating after first baking them dry. This represents an added cost as well as a problem when rework must be done, as the coating must be removed to do the rework and then reapplied. Two materials that have this problem are polyamide and cyanate ester. The moisture absorption levels of the FR-4 derivatives are satisfactory for all digital applications.

3.3.6 Ceramic PCBs

The shear strain experienced depends on the CTE mismatch between the materials and the length:height ratio of the joint. As CTE mismatch increases, so does the strain, and thus the thermal cycling life decreases. If rigid solder joints are to survive cycling during the specified life, the component size may have to be limited or the stand-off height increased to withstand large temperature fluctuations and CTE mismatch. The Column Grid Array is an example of a package where the stand-off height is deliberately made higher than a normal BGA (using columns of high-melting solder) in order to accommodate CTE differences between its ceramic body and a PCB substrate. Three ceramic-based technologies are used to manufacture ceramic PCBs:

1. Thick-Film Multilayer (TFM)
2. High-Temperature Co-fired Ceramic (HTCC)
3. Low-Temperature Co-fired Ceramic (LTCC)

The same technologies are used for manufacturing of Multi-Chip Modules (MCMs).

Thick-film multilayer (TFM). Multilayer circuits formed by sequential screen printing. Additional thick-film compositions may be applied to the same substrate by repeating the screen printing, drying, and firing processes. In this way, complex, interconnected conductive, resistive, and insulating films can be generated. Using this sequential process, double-sided and multilayer circuits with two to four circuit layers are manufactured in high volume for commercial applications such as those in automotive and telecommunications. For military and aerospace applications, circuits with six to eight circuit layers are manufactured in moderate volume.

Co-fired ceramics. In the case of co-fired ceramics, sheet dielectric materials are cast as a tape, and then all the layers are processed in parallel. Ready screen-printed layers are laminated (stacked and pressed together) and then co-fired.

- *High-Temperature Co-fired Ceramic (HTCC).* High-temperature co-fired ceramic (HTCC) is a multilayer packaging technology for the electronics industry, used in military electronics, MEMS, microprocessor, and RF applications. HTCC packages generally consist of multilayers of alumina oxide with tungsten and moly-manganese metallization. The ceramic is fired at around 1600°C, which is significantly higher than a similar-technology low-temperature co-fired ceramic, or LTCC. Compared to LTCC, HTCC has higher resistance conductive layers.

  - Ceramic material is alumina. It is fired at 1600°C to 1800°C in a hydrogen atmosphere. Only tungsten (W) and molybdenum (Mo) can be used as conductors because of the high firing temperature. However, their conductivity is significantly lower when compared with silver.

  - Advantages of HTCC include mechanical rigidity and hermeticity, both of which are important in high-reliability and environmentally stressful applications. Another advantage is their thermal dissipation capability, which makes this a microprocessor packaging choice, especially for higher-performance processors.

- *Low Temperature Co-fired Ceramic (LTCC).* Multilayer ceramics based on LTCC are gaining increasing interest in the manufacturing of highly integrated devices for microelectronics applications. In many applications, the parts are exposed to mechanical stresses, which is an important issue regarding the reliability of the device.

To predict the lifetime of LTCC multilayer devices, and to extend the application range of LTCC, basic mechanical data of this material is needed. Sintered LTCC laminates have been investigated concerning their flexural strength, crack growth rate, and lifetime prediction. The flexural inert strength of the investigated LTCC material is about 450 MPa. In applications with a stress level of 100 MPa, like mass flow sensors for the measurement of injected fuel quantities, acceptable lifetimes are achieved. This means that LTCC is an interesting material to fabricate devices, in which LTCC fulfills the requirements of a functional and structural material. An example is the DuPont Green Tape™ Low Temperature Co-fire Ceramic System. This system consists of a glass/ceramic dielectric tape. The tape is blanked to size, and registration holes are punched in it. Vias are formed in the dielectric tape by punching or drilling, and conductor lines are screen-printed on the tape. High-conductivity silver metallization is used. When all layers have been punched and printed, they are registered, laminated, and co-fired. Process parameters include:

- Lamination:
  - 50–130 kg/cm²
  - 0.5–3 min
  - 30–80°C
- Firing:
  - 1–4°C/min (rise)
  - 880–920°C (peak)
  - 10–40 min (dwell)
  - 1–6°C/min (cooling)

The advantage of a co-firing process is the possibility of inspection of all printed layers prior to lamination. This ensures better capability of the process compared to sequential screen printing of multilayer circuits. Multilayer PCBs with fifty or more layers have been manufactured using LTCC technology. This multilayer technology is well-suited for constructing RF modules for wireless applications with capacitors and inductors integrated into the substrate.

For PCB thermal management, thermal resistance primarily depends on the thermal conductivity of the PCB and the thermal connection between the

TABLE 3.3

Thermal Conductivity of PCB Materials

Material	Thermal Conductivity(W/m·K)
FR4	0.25
LTCC	2.5–4
HTCC	16–30 (Al_2O_3)
	180 (AlN)
	260 (BeO)
Copper	401

package and the PCB (see Table 3.3). The thermal management of PCBs is of increasing importance as the power density of components and circuits continues to rise. The situation is complicated by the use of boards with multiple sheets of copper embedded in the electrically insulating boards to provide electromagnetic shielding and to allow more three-dimensional connectivity of the circuitry by the use of vias. Because of the much higher thermal conductivity of copper, the thermal properties are expected to be drastically altered by the embedded layers, possibly introducing overall anisotropy as well as thermal resistance at the interfaces.

3.3.7 Soldering

Permanent connection of mechanical parts may be done using the following techniques:

- Welding
- Soldering
- Gluing

Their essence is the introduction of a liquid phase between the parts to be connected (molten base metal in welding, molten solder in soldering, glue in gluing). The liquid wets the parts, and gets in intimate interfacial contact with them due to intermolecular attractive forces. Then the liquid solidifies, keeping the parts connected.

As shown in Figure 3.9, soldering is a process by which the parts are joined together using a substance (solder) that has a lower melting point than the materials of the parts. Welding is based on the melting of parts material. Glues are liquid materials that solidify as the result of chemical reaction.

Solder joints are widely used in the electronics packaging industry to provide electrical, thermal, and mechanical connections between the package and the PCB. One of the common solder failures is the formation of fatigue crack at the solder/package interface under cyclic thermal loading. The growth of the crack will eventually cause the failure of the package. How to accurately predict solder joint fatigue failure without resorting to the rather lengthy and costly tests has always been the concern of every IC company.

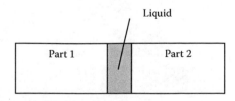

FIGURE 3.9
Soldering: A process by which the parts are joined together using a substance (solder).

It remains a formidable task even today. This is largely due to the complicated deformation behavior of the solder material under elevated temperatures.

The solder in most cases is a metallic element or alloy. Sometimes it is glass. Glass solder is used, for example, to attach optic cable to metal connectors or to assemble ceramic and metallic parts of an IC package. Most solders that are used in electronic applications are based on tin. Tin is a unique metal that wets most of the metals. Six metals not wetted by tin (and therefore not solderable by tin alloys) are

- Cast iron
- Chromium (Cr)
- Titanium (Ti)
- Tantalum (Ta)
- Magnesium (Mg)
- Beryllium (Be)

The fact that some metals are not wetted by solder is important. These metals are used in soldering machinery for manufacturing machine parts that get in touch with molten solder.

There are many tin-based alloys. The real choice depends on material properties, including:

- Wetting
- Ductility
- Electrical conductivity
- Thermal conductivity
- Thermal expansion
- Tensile strength
- Toxicity

Figure 3.10 shows families of solders.

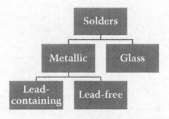

FIGURE 3.10
Families of solders.

3.3.8 Issues with Lead-Free Solder

The most frequently used lead-free solders are composed of silver (3% to 4%), copper (0.5% to 0.7%), and tin (the remainder). This family of solders is commonly called SAC (Sn–Ag–Cu). The melting point of the respective eutectic ternary alloy is 217°C (34° higher than the melting point of eutectic tin-lead alloy).

When compared with standard Sn63Pb37 solder, Pb-free solder has a higher melting point (217°C versus 183°C) and pushes peak reflow temperatures from 220°C to 260°C. This results in a(n)

- Energy consumption increase and impact on the ambiance
- Total product cost increase (more expensive solder, higher energy spending)
- More rigorous requirements to soldered components (they are exposed to higher temperature)
- More rigorous requirements to fluxes
- Higher probability of fatigue cracking in solder joints

Finite element analysis (FEA) is the most commonly used tool to study solder behavior under thermal loading. Many different FEA-based solder models exist in the literature. They differ primarily on the fundamental mechanism viewed as being responsible for inducing crack damage. Some such models include strain based, stress based, energy based, or the recently introduced damage model based on fracture mechanics.

Bibliography

Antolovich, S.D., and Antolovich, B.F. (1996). *An Introduction to Fracture Mechanics in ASM Handbook 19. Fatigue and Fracture*, Materials Park, OH: ASM International®, 1996.

Arora, N.D., Raol, K.V., Schumann, R., and Richardson, L.M. (1996). Modeling and extraction of interconnect capacitances for multilayer VLSI circuits, *IEEE Trans. Computer Aided Design of Integrated Circuits and Systems*, 15(1), 58–66.

Bilotti, A.A., (1974). Static temperature distribution in IC chips with isothermal heat sources, *IEEE Trans. Electron Devices*, ED-21(March), 217–226.

Black, J.R. (1969). Electromigration failure models in aluminium metallization for semiconductor devices, *Proc. IEEE*, 57(9), 1587–1594.

Blech, I.A., and Herring, C. (1976). Stress generation by electromigration, *Appl. Phys. Lett.*, 29, 131–133.

Bradley, E., Handwerker, C.A., Bath, J., Parker, R.D., and Gedney, R.W. (2007). *Lead-free electronics: iNEMI projects lead to successful manufacturing*. Hoboken, NJ: John Wiley & Sons, Inc.

Chen, C., and Liang, S.W. (2007). Electromigration issues in lead-free solder joints, *J. Mater. Sci.*, 18, 259–268.

Dreezen, G., Deckx, E., and Luyckx, G. (2003). Solder alternative: Electrically conductive adhesives with stable contact resistance in combination with non-noble metallization, *CARTS Europe 2003*, pp. 223–227.

Gale, W.F., and Totemeier, T.C. (2004). *Smithells Metals Reference Book, (8th edition).* Maryland Heights, MO: Elsevier.

Galyon, G.T. (2003). *Annotated Tin Whisker Bibliography,* a NEMI Publication, July.

Hunter, W.R. (1997). Self-consistent solutions for allowed interconnect current density. I. Implication for technology evolution, *IEEE Trans. Electron Devices*, 44(2), 304–309.

Hunter, W.R. (1997). Self-consistent solutions for allowed interconnect current density. II. Application to design guidelines, *IEEE Trans. Electron Devices*, 44(2), 310–316.

Massiot, G. and Munier, C. (2004). A review of creep–fatigue failure models in solder material: Simplified use of a continuous damage mechanical approach. *Proceedings of the 5th International Conference on Thermal and Mechanical Simulation and Experiments in Microelectronics and Microsystems*, EuroSimE 2004, Brussels, Belgium, pp. 465–472.

Strauss, R. (1998). *SMT Soldering Handbook, (Second edition).* Maryland Heights, MO: Elsevier/Newnes.

Suo, Z. (2004). A continuum theory that couples creep and self-diffusion, *J. Appl. Mechanics*, 71, 646–651.

Teng, C.C., Cheng, Y.K., Rosenbaum, E., and Kang, S.M. (1997). iTEM: A temperature-dependent electromigration reliability diagnosis tool, *IEEE Trans. Computer-Aided Design of Integrated Circuits and Systems*, 16(8), 882–893.

Tu, K.N. (2003). *Solder Joint Technology: Materials, Properties, and Reliability.* Berlin, Germany: Springer.

Tu, K.N. (2003). Recent advances on electromigration in very-large-scale integration of interconnects, *J. Appl. Phys.*, 94, 5451–5473.

Tu, K.N. (1994). Irreversible processes of spontaneous whisker growth in bimetallic Cu–Sn thin-film reactions, *Phys. Rev. B*, 49(3), 2030–2034.

Yeh, E.C.C., Choi, W.J., Tu, K.N., Elenius, P., and Balkan, H. (2002). Current-crowding-induced electromigration failure in flip chip solder joints, *Appl. Phys. Lett.*, 80, 580.

4

Lead-Free Electronic Reliability: Finite Element Modeling

> Due to its brittle nature and lattice mismatch, solder cracks tend to be generated near compounds, and these cracks affect the mechanical integrity of solder joints. Therefore, it is important to study the effect of intermetallic compound development on the mechanical properties of solder joints.

Insight into the thermomechanical response of solder joints is critical to the design and deployment of reliable electronic circuit board assemblies. Understanding the mechanics of lead-free (Pb-free) soldered assemblies is also essential to the development of accelerated test plans, predictive reliability models, and to their use as effective tools for product reliability assessment. Lead-free electronic transition has been successfully completed for computing and consumer electronics applications. The solder joint reliability performance of Pb-free solders (that is, Sn–Ag–Cu [SAC] solders) has been extensively studied for various package types, and most of the studies to date have shown improved Pb-free electronic reliability performance compared to Sn–Pb solder. However, mixed reliability results were also observed for high cyclic strain components such as large ceramic ball grid arrays (BGAs) and quad flat no (QFN) leads packages. There is also renewed interest in understanding the reliability performance of Pb-free solders in other applications such as telecom, automotive electronics, and aerospace industries.

The life of Sn–Pb or Pb-free solder joints is limited by the fatigue damage that accumulates in solder materials. Accelerated testing provides distributions of failure times whose relevance to service life is determined by extrapolation to use conditions based on the appropriate *acceleration factors* (AFs). Because AFs are defined as the ratio of life under test and use conditions, their determination requires up-front predictions of solder joint lives under both sets of conditions. Such models are also of use for reliability analysis of circuit boards at the design stage. While a variety of life prediction models have been developed for near-eutectic Sn–Pb assemblies, to this author's knowledge, such models are not currently available for Pb-free soldered assemblies. The development of life prediction models requires a detailed understanding of failure modes, an adequate constitutive model that captures the thermomechanical behavior of Pb-free solders in electronic assemblies, and a reliability database that is needed for the empirical correlation of failure times under test and/or field conditions. This chapter reviews the reliability performance of Pb-free electronic solder and addresses key reliability challenges, which include:

- Material compatibility issues
- Material quality attributes affecting reliability
- Reliability test conditions
- Fatigue creep concerns
- Pb-free electronic acceleration model

This chapter also provides an overview of Pb-free electronic solder fatigue models, as well as issues and recommendations for Pb-free electronic reliability performance assessment.

4.1 Finite Element Modeling and Inelastic Strain Energy Density

Finite element modeling (FEM) is the dominant discretization technique in structural mechanics. The basic concept in the physical interpretation of FEM is the subdivision of the mathematical model into disjoint (nonoverlapping) components of simple geometry called finite elements (or elements, for short). The response of each element is expressed in terms of a finite number of degrees of freedom characterized as the value of an unknown function, or functions, at a set of nodal points. FEM has been widely used for the reliability assessment and analysis of various electronic components in the electronics packaging industry. Several modeling methodologies have been developed and used over a period of time, each having its own merits and demerits. Both 2-D (two-dimensional) and 3-D (three-dimensional) models have been used for the analysis. Usually, 3-D models have been shown to give better accuracy and more realistic results as compared to the 2-D models. Depending on the geometric and material symmetry of the actual package, there are different configurations of 3-D models, such as the full model, 1/4th or 1/8th symmetry model, and diagonal slice model. Use of the 3-D diagonal slice model configuration is preferred over the other configurations for the fully symmetric packages due to its computational efficiency and ability to capture true boundary conditions.

In this chapter, the 3-D diagonal slice model has been used for all the FEM models. Various other parameters, such as material nonlinearity, element type, shape, and size, also affect the simulation accuracy. The finite element models used in the current work have been checked for their element shapes (aspect ratio and angles), and a finer mesh in the region of solder bump has been employed for better accuracy. To capture the true nature of a solder bump, a time- and temperature-dependent nonlinear constitutive model has been employed for the simulation.

Finite element analysis obtains the temperatures, stresses, flows, or other desired unknown parameters through minimizing an energy function. The energy functional consists of all the energies associated with the particular finite element model. Based on the law of conservation of energy, the finite element energy function must equal zero. Exhaustive research has been done on the solder joint reliability of various electronic packages subjected to thermal cycling.

It has been established by many researchers that the fatigue life of the solder interconnect depends on the amount of *inelastic strain energy density* (ISED) accumulated by the solder joint during each thermal cycle. Some reliability data have been acquired or are in the process of being gathered by several Pb-free consortia. Efforts are also underway to characterize the mechanical behavior of Pb-free solders, and numerous studies have been published with emphasis on the secondary creep of solder. The derived creep rate equations are an important ingredient in constitutive models; however, their applications to circuit board assemblies are still the subject of validation studies. A correlation between solder joint reliability and the ISED output from FEM has also been developed for both leaded and lead-free interconnects.

As shown in Figure 4.1, the ISED calculations done from the simulations run for different configurations and various thermal cycles have been used in the current work for the solder joint reliability prediction using life prediction correlations. Several researchers have established ISED to be the damage proxy for the solder joint thermomechanical reliability of electronic packages. Damage relationships correlating the solder joint life

- The inelastic strain energy accumulated per unit volume per cycle, also referred to as inelastic strain energy density (ISED)

- ISED is given by the area enclosed within the hysteresis loop of the solder joint

- The ISED calculation for any given package depends on various factors such as:

  - the geometry and architecture of the package,

  - material properties,

  - assembly stiffness,

  - local and global thermal mismatch,

  - hysteresis loop approximation methodology, and

  - creep and other material constitutive relations.

FIGURE 4.1
Inelastic strain energy density (ISED).

with the ISED accumulated per thermal cycle by the solder joint have also been developed. These relationships can be used for the life prediction of the electronic packages with solder joint cracking as the failure mode. The ISED can be calculated from the simulation. In the current work, a 3-D diagonal slice model for an encapsulated flip-chip package has been developed. Both linear and nonlinear material properties have been used for analyzing the effects of several parametric variations on the reliability of the package. The parametric variations include both design parameters (geometry and material) and thermal cycling parameters. The two failure modes analyzed for the relative comparison of reliability due to the various parametric variations include solder joint failure and copper trace cracking.

4.2 FEM Model Description

A 3-D diagonal slice model incorporated with time- and temperature-dependent nonlinear and linear material properties has been developed in FEM. Several different geometries were modeled in order to generate a total of eighteen different configurations for the analysis.

4.2.1 FEM Variables

Eighteen different configurations were developed for the analysis of flip-chip solder joint integrity and three configurations for the copper trace integrity. The parametric variations in the model include both geometric and material property variables. Copper trace failure mode is another critical failure mode in flip-chip devices and other electronic packages. Because the copper trace has negligible effect on solder joint reliability, the copper trace was ignored in solder joint integrity analysis.

4.2.1.1 Geometry

For the FEM, the variations in geometry include three bump heights (low, mid, and high) and two bump sizes, or diameters. The dimensions of the fixed parameters are given in Table 4.1, and the dimensions of the variable parameters for all eighteen configurations are listed in Table 4.2.

4.2.1.2 Material

Both eutectic (62Sn-36Pb-2Ag) and Pb-free (95.5Sn-4.0Ag-0.5Cu) solder material were considered for the analysis. Also, three different underfill materials

TABLE 4.1

Dimensions of Fixed Parameters used for the
Finite Element Model

Parameter	Dimension (mm)
Pad thickness	0.0007
Slice length (Si die)	0.096
Slice length (PCB)	0.192
Slice width	0.0056
Height (Si die)	0.028
Height (PCB)	0.0212

TABLE 4.2

Material Variations and Dimensions of the Geometric Parameters

Test Case	Solder	Underfill	Bump Height (mils)	Bump Size (mils)
1	Eutectic	Kester	1.0 (Low)	4
2	Eutectic	FP 4526	2.5 (High)	4
3	Eutectic	Kester	2.5 (High)	4
4	Eutectic	Kester	2.0 (Mid)	4
5	Eutectic	FP 4526	1.0 (Low)	4
6	Eutectic	3M 3667	2.0 (Mid)	4
7	Eutectic	FP 4549	2.5 (High)	4
8	Eutectic	FP 4549	2.0 (Mid)	3
9	Eutectic	3M 3667	2.0 (Mid)	3
10	Pb-free	FP 4549	2.5 (High)	4
11	Pb-free	FP 4549	2.0 (Mid)	3
12	Pb-free	Kester	2.5 (High)	4
13	Pb-free	Kester	2.0 (Mid)	3
14	Pb-free	3M 3667	2.5 (High)	4
15	Pb-free	3M 3667	2.0 (Mid)	3
16	Pb-free	FP 4526	1.0 (Low)	4
17	Pb-free	FP 4526	2.0 (Mid)	4
18	Pb-free	FP 4526	2.5 (High)	4

were considered. The material variations used in all eighteen configurations
for the simulation are listed in Table 4.2.

4.2.1.3 Elements Used

The model was map meshed using brick-shaped eight-noded hexahedral
isoparameteric elements (Figure 4.2). Since the eight-noded hexahedral element provides higher accuracy as compared to tetrahedral elements, the
hexahedral elements were used for meshing the entire model. The 2x2x2
Gaussian quadrature integration option was used for the solution.

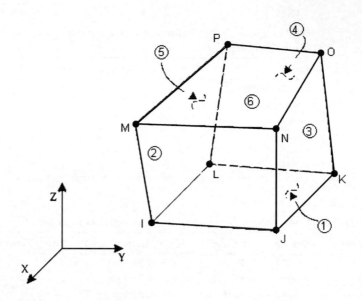

FIGURE 4.2
Eight-noded hexahedral isoparametric element.

- VISCO107 elements were used for the solder bump. VISCO107 is used for 3-D modeling of solid structures. It is defined by eight nodes having three degrees of freedom at each node: translations in the nodal x-, y-, and z-directions. The element is designed to solve both isochoric (volume preserving) rate-independent and rate-dependent large strain plasticity problems. Iterative solution procedures must be used with VISCO107 because it is used to represent highly nonlinear behavior. VISCO107 has rate-dependent large strain plasticity capabilities and is used to represent the nonlinear behavior of viscoplastic materials. Another advantage of using VISCO107 for solder is to facilitate the constitutive modeling of the solder material using Anand's model. The SOLID45 element also has large strain capabilities.
- SOLID45 elements were used for the rest of the materials.
- SOLID45 was used for package and board elements.

4.2.1.4 PCB Material Layers

Various material layers were modeled with the geometric dimensions provided by Seagate. For simplification, the underfill fillet was not considered for the solder joint integrity simulation but it was considered for copper trace integrity simulation as the fillet has significant impact on the stresses developed in the trace. The thickness and the material properties of each layer of the PCB are listed in Table 4.3.

TABLE 4.3

Material Properties and Thickness of PCB Layers

Layer Material	Thickness (mils)	Material Properties	
Cover film	1.0	E(Kpsi)	500.38
		α	18 ppm/°C
Adhesive 1	0.8	E(Kpsi)	105.88
		α	118 ppm/°C
Copper	0.7	E	9717.53
		α	17 ppm/°C
Adhesive 2	0.7	E(Kpsi)	105.88
		α	118 ppm/°C
Base polyimide	1.0	E(Kpsi)	500.38
		α	18 ppm/°C
Adhesive 3	1.0	E(Kpsi)	159.54
		α	218 ppm/°C
Aluminum	16.0	E(Kpsi)	10007.6
		α	10 ppm/°C

4.2.2 Material Properties for FEM

4.2.2.1 Linear Material Properties

Linear isotropic material properties were assigned to all the elements except for solder bump elements. The solder bump elements were modeled with both linear and nonlinear constitutive relationships. The linear isotropic material properties assigned in the model are listed in Table 4.4. The nonlinear constitutive relationships for the solder bump are discussed in the next section.

4.2.2.2 Nonlinear Solder Properties

Solder alloy is a viscoplastic material, which means that it exhibits both time-dependent creep and plasticity. Thus, it is essential to use a time- and temperature-dependent nonlinear material model for the solder bump. Viscoplasticity is a theory in continuum mechanics that describes the rate-dependent inelastic behavior of solids. Rate dependence in this context means that the deformation of the material depends on the rate at which loads are applied. The inelastic behavior that is the subject of viscoplasticity is plastic deformation, which means that the material undergoes unrecoverable deformations when a load level is reached.

Rate-dependent plasticity is important for transient plasticity calculations. The main difference between rate-independent plastic and viscoplastic material models is that the latter exhibit not only permanent deformations after the application of loads, but also continue to undergo a creep flow as a function of time under the influence of the applied load.

TABLE 4.4

Linear Isotropic Material Properties for Underfill Materials

Material	E(psi)	Poisson's Ratio	CTE (per °C)
Silicon	23641000	0.278	2.50E-06
FP 4526	1232800	0.25	3.30E-05
FP 4549	812211	0.25	4.50E-05
3M 3667	391602	0.25	6.10E-05
Kester	449617	0.25	6.70E-05
Pb-free solder	5729100	0.3	2.00E-05
Eutectic solder	4163600	0.35	2.45E-05
Copper pad	18710000	0.34	1.63E-05
Cover film	500400	0.3	1.80E-05
Adhesive 1	105900	0.35	1.18E-04
Copper	9717500	0.34	1.70E-05
Adhesive 2	105900	0.35	1.18E-04
Base polyimide	500400	0.35	1.80E-05
Adhesive 3	159500	0.35	2.18E-04
Aluminum	10008000	0.3	2.30E-05

Viscoplasticity is characterized by the irreversible straining that occurs in a material over time. A unified viscoplastic constitutive law, the Anand model, has been applied to represent the inelastic deformation behavior for solders used in electronic packaging. The material parameters of the constitutive relations for 62Sn-36Pb-2Ag, 60Sn-40Pb, 96.5Sn-3.5Ag, and 97.5Pb-2.5Sn solders were determined from separate constitutive relationships and experimental results.

The Anand model was tested for constant strain rate testing, steady-state plastic flow, and stress/strain responses under cyclic loading. It was concluded that the Anand model can be applied for representing the inelastic deformation behavior of solders at high homologous temperatures and can be recommended for finite element simulation of the stress/strain responses of solder joints in service. The widely used and established Anand model, which is also a standard feature in ANSYS™ has been used to model both eutectic (62Sn-36Pb-2Ag) and Pb-free solder material. In the Anand's model, both creep and plasticity were coupled together with the help of flow and evolutionary equations:

Flow equation:
$$\frac{d\varepsilon_p}{dt} = A(\sinh(\varsigma\sigma/s_o))^{\frac{1}{m}} \exp\left(\frac{-Q}{kT}\right) \tag{4.1}$$

Evolutionary equation:
$$\frac{ds_o}{dt} = \left\{h_o\left(|B|\right)^a \frac{B}{|B|}\right\} \frac{d\varepsilon_p}{dt} \tag{4.2}$$

where

$$B = 1 - \frac{s_o}{s^*}$$

$$s^* = s^{\wedge} \left[\frac{d\varepsilon_p/dt}{A} \exp\left(\frac{Q}{kT} \right) \right]^n$$

There are two basic features in Anand's model applicable to isotropic rate-dependent constitutive models for metals. First, there is no explicit yield surface; rather, the instantaneous response of the material depends on its current state. Second, a single scalar internal variable "s," called the deformation resistance, is used to represent the isotropic resistance to inelastic flow of the material. The specifics of this constitutive equation are the flow equation and the evolution equation. The constants of Anand's model for the simulation have been assigned through the standard input for VISCO107 element in ANSYS™. The value of the constants used for modeling both the eutectic and Pb-free solder material are listed in Table 4.5.

TABLE 4.5

Values of the Constants of Anand's Viscoplastic Model

Pb-free Solder Sn-3.5Ag-0.75Cu			Eutectic Solder 62Sn-36Pb-2Ag	
Value	ANSYS Input	Model Parameter	Value	Definition
135.72	C1	S_o (psi)	1800	Initial value of deformation resistance
8400	C2	Q/k (K^{-1})	9400	Activation energy/ Boltzmann's constant
4610000	C3	A (sec^{-1})	4000000	Pre-exponential factor
0.038	C4	ξ	1.5	Multiplier of stress
0.162	C5	m	0.303	Stain rate sensitivity of stress
448000	C6	h_o (psi)	200000	Hardening constant
150.8	C7	$s^{\wedge}$ (psi)	2000	Coefficient for deformation resistance saturation value
0.0046	C8	n	0.07	Strain rate sensitivity of saturation (deformation resistance) value
1.56	C9	a	1.3	Strain rate sensitivity of hardening

4.3 Inelastic Strain Energy Density

Several researchers have established inelastic strain energy density (ISED) as the damage proxy for the solder joint thermomechanical reliability of electronic packages. Damage relationships correlating the solder joint life with the ISED accumulated per thermal cycle by the solder joint have also been developed. These relationships can be used for the life prediction of electronic packages with solder joint cracking as the failure mode. The ISED can be calculated from the simulation.

Hysteresis loops provide useful information for engineering evaluations of solder joint reliability. For example, the width of the loop gives an estimate of the inelastic strain range that solder joints experience. The inelastic strain range is used in a Coffin–Manson-type of fatigue law. Another, more general approach consists of using the hysteresis loop area, which is a measure of the amount of cyclic strain energy density that is imparted to solder joints. Strain energy density is used in Morrow's type of fatigue law (Morrow, 1964; Obtani et al., 1985) where cycles to failure are given as a function of the cyclic inelastic strain energy density, ΔW_{in}:

$$N = \frac{C}{\Delta W_{in}}. \tag{4.3}$$

where
  C is a material constant
  N is an exponent found to be in the range 0.7 to 1.6 for several engineering metals, including soft solders.

In ANSYS™, the VISCO107 element has plastic work (PLWK) as a standard output; the ISED has been calculated by volume averaging PLWK over the whole solder bump volume. Some researchers volume average PLWK over a few layers in the vicinity of the interface instead of the whole solder volume. Plastic work accumulated in the solder joint over the complete thermal cycle has been found to be stabilized after the first cycle. Consequently the simulation is run for only two cycles, and the ISED is calculated based on the plastic work accumulated during the second thermal cycle. The ISED for the solder joint is given by

$$\Delta W_i(ISED) = \frac{\left(\sum_{n=1}^{total\_elements} \Delta W^{(n)} \times V_n \right)}{\left(\sum_{n=1}^{total\_elements} V_n \right)} \tag{4.4}$$

where $\Delta W^{(n)}$ is plastic work accumulated by n-th element during the second thermal cycle of the simulation, and it is standard output in ANSYS™ for the VISCO107 element in the form of PLWK and V_n volume of the n-th element.

4.4 Material Characterizaton of Underfill Materials

Electronic packaging designs are moving toward fewer levels of packaging to enable miniaturization and to increase the performance of electronic products. One such package design is flip chip on board (FCOB). In this method, the chip is attached face-down directly to a printed wiring board (PWB). Because the package is comprised of dissimilar materials, the mechanical integrity of the flip chip during assembly and operation becomes an issue due to the coefficient of thermal expansion (CTE) mismatch between the chip, the PWB, and interconnect materials. To overcome this problem, a rigid encapsulant (underfill) is introduced between the chip and the substrate. This reduces the effective CTE mismatch and reduces the effective stresses experienced by the solder interconnects. The presence of the underfill significantly improves long-term reliability. The underfill material, however, does introduce a high level of mechanical stress in the silicon die. The stress in the assembly is a function of the assembly process, the underfill material, and the underfill cure process. Therefore, selection and processing of underfill material is critical to achieving the desired performance and reliability. The effect of underfill material on the mechanical stress induced in flip-chip assembly during cure was presented in Tu (2003) and Bradley (2007). This section studies the effect of the cure parameters on a selected commercial underfill and correlates these properties with the stress induced in flip-chip assemblies during processing.

The material characterization of the underfill materials is provided in this section. The stress–strain curve for five different materials has been plotted from uniaxial tensile testing of the samples. The elastic modulus for the linear range has also been calculated. The simulations for the 0°C to 70°C ATC (accurate thermal cycle) were run with both the vendor data and the measured properties in order to see the impact of the difference in the material properties on the plastic work accumulated in the solder joint during the thermal cycle.

4.4.1 FP 4526 as an Underfill Material

A key step in the manufacture of direct-chip attachment underfill packaging is the capillary flow of an underfill material into the chip-to-substrate stand-off created by an array of solder bumps. FP 4526 is a capillary flow

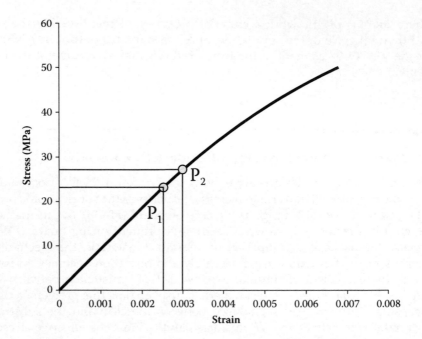

FIGURE 4.3
Stress–strain plot from the uniaxial tensile test for FP 4526 underfill material.

underfill material with highest elastic modulus value and the lowest CTE value among all the underfill materials tested. The stress–strain curve from the test data is plotted in Figure 4.3. The samples were tested at the strain rate of 0.001 sec^{-1}, as the strain rates encountered during the ATC of flip-chip fall in this range. The maximum strains encountered by the underfill during the simulation of both 0°C to 70°C and 0°C to 90°C thermal cycles had been measured in order to make sure that the elastic properties used for the simulations fall within the linear range.

The points P1 and P2 in Figure 4.3 correspond to the maximum strains encountered by the FP 4526 underfill during 0°C to 70°C and 0°C to 90°C ATC, respectively. It can be seen that the maximum strain for 0°C to 90°C is about 3,000 µm, and this point falls well within the linear range of the stress–strain curve. The elastic modulus of the FP 4526 underfill material for the linear region measured from the uniaxial tensile test was found to be 9.21 GPa, which is 8.35% higher than the value of 8.5 GPa provided by the vendor.

4.4.2 FP 4549 as an Underfill Material

The presence of an "underfill" encapsulant between a microelectronic device and the underlying substrate is known to substantially improve the thermal fatigue life of flip-chip solder joints, primarily due to load-transfer from the solder to the encapsulant. The primary effect of the underfill is to reduce the shear load carried by the solder joints, thereby reducing the inelastic strain sustained by the solder and thus enhancing solder life. It has been shown that the presence of an underfill results in substantial redistribution of the displacement fields within a joint, and thereby reduces the extreme local strain concentrations that occur in joints without underfill. It is generally thought that stiff underfills with a close CTE match with the solder are the most effective in enhancing package life.

FP 4549 is also a capillary flow underfill with an elastic modulus lower than that of FP 4526 underfill, while the CTE is higher. The stress–strain curve from uniaxial tensile test data is plotted in Figure 4.4, and the test again carried out at the strain rate of 0.001 sec⁻¹.

The points P_1 and P_2 plotted on the curve in Figure 4.4 indicate the maximum strain encountered during the 0°C to 70° and 0°C to 90°C test cycle simulation, respectively, and it is seen that the maximum strain during the 0°C to 90°C test cycle is about 5,000 μm, which is higher than that for FP 4526 underfill material but still lies well within the linear range of the material. The elastic modulus for the linear region measured from the uniaxial tensile test is 5.22 GPa, which is 6.78% lower than the value of 5.6 GPa provided by the vendor.

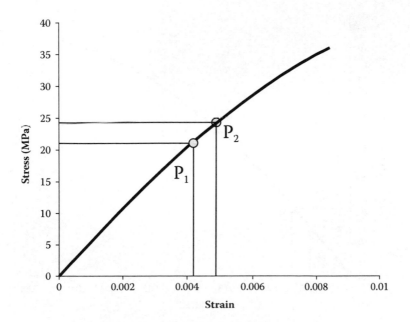

FIGURE 4.4
Stress–strain plot from the uniaxial tensile test for the FP 4549 underfill.

4.4.3 3M UF3667 as an Underfill Material

3M UF3667 is a reflow type of underfill material with elastic modulus lower and the CTE higher than capillary flow underfill materials. The relationship between the stress and strain that a material displays is known as a stress–strain curve. This curve is unique for each material and is found by recording the amount of deformation (strain) at distinct intervals of tensile or compressive loading. These curves reveal many of the properties of a material (including data to establish the modulus of elasticity, E). For 3M UF3667, the stress–strain curve from the uniaxial tensile test data is plotted in Figure 4.5.

It is visible from the plot in Figure 4.5 that the maximum strain during the 0°C to 90°C test cycle is around 5,500 µm, which lies well within the linear range of the material. The elastic modulus of the 3M UF3667 material for the linear region measured from the uniaxial tensile test is 3.11 GPa, which is 15.18% higher than the value of 2.7 GPa provided by the vendor.

4.4.4 Kester 9110S as an Underfill Material

Kester 9110S is also a reflow type of underfill material with elastic modulus lower and the CTE higher than the 3M UF3667 underfill material. The stress–strain curve from the uniaxial tensile test data is plotted in the Figure 4.6. It

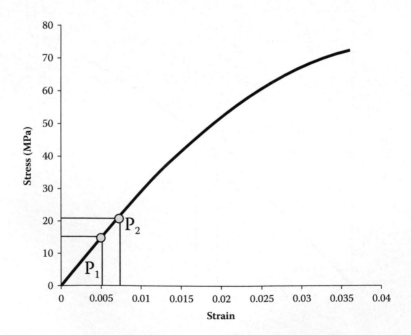

FIGURE 4.5
Stress–strain plot from uniaxial tensile test of 3M UF3667 underfill material.

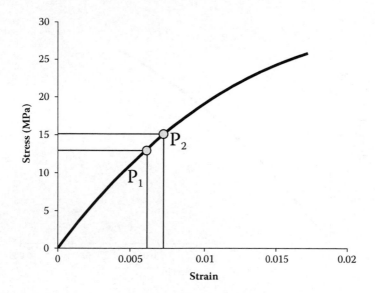

FIGURE 4.6
Stress–strain plot from uniaxial tensile test of Kester 9110S underfill material.

is visible from the plot that the maximum strain (5,500 μm) goes just beyond the linear region of the material. However, the linear isotropic material properties have been used in the simulation for the approximation. Figure 4.6 also shows the strain distribution in the Kester underfill material from the 0°C to 90°C thermal cycle simulation.

The maximum strain in the case of the Kester material lies at the solder bump and underfill interface edge, as compared to the die and underfill interface edge in the case of other underfill materials. The elastic modulus of the Kester material for the linear region measured from the uniaxial tensile test is 2.53 GPa, which is 18.4% lower than the value of 3.1 GPa provided by the vendor.

4.4.5 Kester 9110 SBB as an Underfill Material

Kester 9110 SBB is a reflow type of encapsulant with 45-μm sized beads. As shown in Figure 4.7, the elastic modulus for the linear region measured from the uniaxial tensile test for this material is 2.66 GPa.

4.4.6 Effect of Measured Properties on the Simulation Results

The correct input of material properties in the finite element analysis (FEA) model is vital for obtaining accurate results from the simulation. The simulations for 0°C to 70°C ATC were run with both the material properties measured in-house as well as with the material data provided by the vendor for

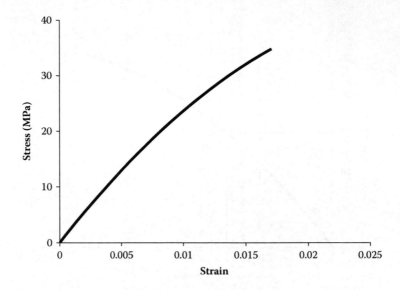

FIGURE 4.7
Stress–strain data for Kester 9110 SBB material from the uniaxial tensile test.

the fifteen different configurations provided by Seagate. Differences in the simulation results can be observed in Table 4.6.

Because a higher value of ISED for the same configuration implies lower reliability, then except for FP 4526 underfill, all three of the other underfills give higher reliability with the vendor data. So the models will under-predict the solder joint reliability in case of FP 4526 and over-predict in other cases if the vendor data is used in the simulation.

4.5 Solder Joint Integrity in Accelerated Thermal Cycling

The recent need to develop Pb-free electrical and electronic products has resulted in an increased demand to characterize solder joint integrity. Many standards exist to evaluate the long-term reliability of electronic components or the quality of an assembly process. However, no standards apply to custom PCB assemblies. For example, the IPC-9701 standard and the use of evaluation boards are directed at evaluating processes rather than products.

The most serious concern regarding flip-chip reliability is solder interconnect fatigue failure due to thermal cyclic loading. Due to the CTE mismatch within the various components of the flip-chip package, stresses develop in the component whenever thermal cycling takes place. These stresses lead to the fatigue failure of the component when the thermal cycling is continued

TABLE 4.6

Simulation Results for 0°C to 70°C Thermal Cycle with Vendor and Measured Data

Test Case	Solder Composition	Underfill	Bump Gap Height	Bump Size (mils)	Inelastic Strain Energy Density Per Cycle (psi)	
					Simulation 1	Simulation 2
1	Eutectic	Kester	Low	4	33.65	37.98
2	Eutectic	FP 4526	High	4	10.75	10.65
3	Eutectic	Kester	High	4	28.42	29.01
4	Eutectic	Kester	Mid	4	29.73	31.05
5	Eutectic	FP 4526	Low	4	12.49	12.13
6	Eutectic	3M 3667	Mid	4	25.95	26.89
7	Eutectic	FP 4549	High	4	16.52	16.65
8	Eutectic	3M 3667	Mid	3	16.89	17.11
9	Eutectic	3M 3667	Mid	3	28.22	27.19
10	Pb-free	3M 3667	High	4	14.09	14.06
11	Pb-free	3M 3667	Mid	3	15.80	15.86
12	Pb-free	Kester	High	4	25.52	25.67
13	Pb-free	Kester	Mid	3	29.40	30.05
14	Pb-free	3M 3667	High	4	22.66	22.31
15	Pb-free	3M 3667	Mid	3	26.48	25.82

for a long time. It has been established through various experimental studies and field failure analyses that the solder interconnect is the most susceptible to such fatigue failures. Due to this, FEA and simulation of solder joint integrity become extremely important.

4.6 Life Prediction and Field Life Correlation with ATC Life

Life prediction is a very essential part of product design and manufacturing for up-front reliability assessment in order to manufacture robust assemblies that will meet product life requirements. Modeling and ATC tests are employed for this purpose. Finite element modeling and analysis have been used here in order to analyze solder joint reliability and copper trace integrity.

4.6.1 Theory of Life Prediction

Thermal fatigue failure caused by CTE mismatches among silicon chip, substrate, and solder joint is the dominant failure mechanism in solder joint interconnections. In this chapter we evaluate the reliability of the produced solder joints for power chip interconnections. First, the current solder joint fatigue study approaches are introduced and the popular solder joint fatigue models of solder joint reliability assessment and life prediction are briefly reviewed. Accelerated temperature cycling tests as well as tensile and shear tests on solder joint assembly with different solder joint configurations are described, and the failure analysis methods are then elaborated.

Finally, we present the experimental results and discuss solder joint fatigue failure behaviors. The development of life prediction modeling involves several steps, including:

- Determination of an ATC in order to simulate the field condition and same failure mechanism as encountered in the field. An energy partitioning method has been used in the FEA for determining the correct ATC.
- Acquisition of actual test data on the components subjected to ATC with the various failure mechanisms encountered. Actual test failure data provided by Seagate have been used to determine the constants and the scaling factors for life prediction.
- Constitutive equations and material properties for the applicable range of stress-strain conditions. Constitutive equations from the published literature have been used here.
- Simulations for calculating the ISED for the life prediction of solder joints have been carried out in FEM using a 3-D diagonal slice model.

- Development of the correlation between the damage parameter and life of the package.
- Determination of the correct acceleration factor correlating the field life to the ATC life. Mapping based on statistical analysis has been used in order to determine the correct acceleration factors.

The fatigue failure of solder joints and their life prediction are one of the most important issues of solder joint reliability. In terms of the fatigue failure of solder joints, the current paradigm for assessing the in-service reliability of electronic packages is based on thermal/mechanical cycling and thermal shock tests with humidity, which is a time-consuming practice. Therefore, accelerated testing becomes more important and is the focus of a recent intensive research area, driven by short time to market and low cost. Rapid reliability assessment is highly desirable for electronic manufacturing industries. It has long been known that solder fatigue life under high homologous temperature cycling is difficult to predict due to the time-, rate-, and temperature-dependent viscoplastic behaviors of solder alloys. The complicated geometry of solder joints makes the task even more difficult. Moreover, the thermal fatigue life of solder joints under accelerated test conditions also depends on extreme temperatures (maximum and minimum temperatures), temperature ramping rates, dwell time and dwell temperature, evolution of solder microstructures, the presence of intermetallics, soldering defects, and residual strains due to processing.

4.6.2 Energy Partitioning Methodology

Determination of accurate ATC and correct acceleration factors for experimental testing for the thermomechanical reliability assessment of the electronic package is extremely important. The flowchart in Figure 4.8 describes the methodology used for energy partitioning in the determination of the correct ATC. Different ATCs induce different failure modes in the component; it is extremely critical to match the failure mode induced in the ATC to the failure mode encountered in the field application to obtain an accurate life prediction. In case the failure mode induced in the ATC is different from the one encountered in the field application, there is no way to predict the actual life of the component based on the ATC testing and it is not possible to establish a correlation between the field life and the life of the component subjected to ATC. The damage or the strain energy accumulated in the solder joint during its life, which is what causes it to crack and finally fail, can be divided into two portions: one due to time-independent plastic deformation and the second due to time-dependent creep deformation.

The philosophy behind the application of an energy partitioning methodology for the development of field life correlation and solder joint life prediction is that in order to induce the same failure mode in the ATC as encountered in the field, the ratio of the ISED accumulated due to both

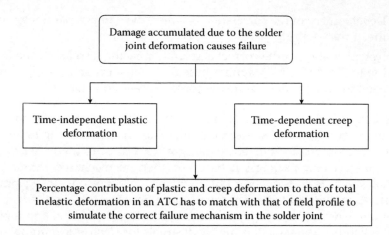

FIGURE 4.8
Energy partitioning methodology flowchart.

time-dependent plastic and time-independent creep deformation must remain the same. The energy partitioning methodology allows one to separate the contribution of both factors, thus enabling the user to determine the ratio of both contributions toward the total ISED. Thus, by matching the ratio, one can determine the correct ATC for the accelerated testing of a component.

The stress-strain analyses for solder joints in a thin single outline package (TSOP), a ball grid array (BGA) assembly, and a leadless ceramic chip carrier (LCCC) are carried out to investigate plastic-creep behavior and stress relaxation behavior due to the temperature cycling or isothermal cyclic loading. The temperature dependence of plastic behavior (yield stress) and creep behavior (creep properties) are taken into consideration in all numerical analyses. The results of FEA show that in an accelerated temperature cycling test, long high-temperature and low-temperature dwell times do not contribute to the increase in the cyclic inelastic equivalent strain range in solder joints (although the creep behavior occurring during the dwell times in an operating condition is important enough to be taken into consideration for estimating the fatigue life of solder joints).

Based upon the results of the strain analyses, some efficient testing processes of temperature cycling and isothermal fatigue tests for the microelectronic solder joints are proposed, and the cycling tests are carried out. The experimental results show good agreement with the analytical results. In order to separate the effect of plastic and creep deformation, separate creep and plasticity models must be used in FEM instead of Anand's viscoplasticity model. Anand's model combines both the effects in form of evolution and flow equations, preventing the user from separating the two effects.

4.6.3 Time-Dependent Creep Model

The constitutive modeling of creep has been extensively studied due to the importance of the creep failure mode in solder joints. However, there are very few studies that considered room-temperature aging contributions in their creep modeling studies. Researchers investigated constitutive modeling of creep of solders by taking into account the possible contribution of room-temperature aging. Lead-free solder (Sn-4.0Ag-0.5Cu) was found to have a higher creep resistance than Sn–Pb solder at the same stress level and testing temperature. The higher creep resistance was contributed by the second phase intermetallic compounds, Ag3Sn and Cu6Sn5. The precipitation of these intermetallic compounds can significantly block the movement of dislocations and increase the creep resistance of the material.

Constitutive models of creep for both Pb-free and Sn–Pb eutectic solders were constructed based on experimental data. The activation energy for SAC405 is much higher than that of Sn–Pb, which also indicates that SAC405 possesses higher creep resistance. The constitutive models can be used in the FEA of actual electronic packages to predict solder joint failure. The creep mechanisms of both Pb-free and Sn–Pb eutectic solders were also extensively discussed in Ma (2007). Dislocation gliding and climb is believed to be the major failure mode at high stresses, while lattice diffusion and grain boundary diffusion is believed to be the major failure mode at low stress levels. Grain boundary sliding is believed to contribute to creep deformation at both high stresses and low stresses. For eutectic Sn–Pb, superplastic deformation is a major part of the creep mechanism at low stresses and high temperatures. Garofalo's equation (4.5) has been used as a constitutive model for solder in the simulation for representing the time-dependent creep:

$$\dot{\varepsilon} = A[\sinh(\alpha\sigma)]^n \exp\left(\frac{-Q}{RT}\right) \tag{4.5}$$

The constants in Garofalo's equation used for both the SAC eutectic solder and the Pb-free solder are listed in Table 4.7.

Here, a method to separate plasticity and creep is discussed for a quantitative evaluation of the plastic, transient creep, and steady-state creep deformations of solder alloys. The method of separation employs an elasto-plastic-creep constitutive model comprised of the sum of the plastic, transient creep, and steady-state creep deformations. The plastic deformation is expressed by the Ramberg–Osgood law, the steady-state creep deformation by Garofalo's creep law, and the transient creep deformation by a model proposed here. A method to estimate the material constants in the elasto-plastic-creep constitutive model is also proposed. The method of separation of the various deformations is applied to the deformation of the Pb-free solder alloy Sn-3Ag-0.5Cu and the Pb-containing solder alloy Sn-37Pb to compare the differences in plastic, transient creep, and steady-state creep deformations.

TABLE 4.7

Constants Used for the Time-Dependent Creep Model of the Solder

Solder Alloy	A	α	N	Q
Eutectic [Zahn (2003) Liang, Chen, C., and Liang, S.W. (2007)]	10	0.001379	2.0	5401.2
Lead-free [Wiese (2001); Singh (2006)]	277984	1.688E-4	6.41	6500.0

The method of separation provides a powerful tool to select the optimum Pb-free solder alloys for solder joints in electronic devices.

A variety of Pb-free solder alloys have been studied for use as flip-chip interconnects, including Sn-3.5Ag, Sn-0.7Cu, Sn-3.8Ag-0.7Cu, and eutectic Sn-37Pb as a baseline. The reaction behavior and reliability of these solders have been determined in a flip-chip configuration using a variety of under-bump metallurgies (TiW/Cu, electrolytic nickel, and electroless Ni-P/Au). The solder microstructure and intermetallic reaction products and kinetics have been determined. The Sn-0.7Cu solder has a large grain structure, while the Sn-3.5Ag and Sn-3.8Ag-0.7Cu solders have a fine lamellar two-phase structure of Sn and Ag3Sn. The intermetallic compounds are similar for all the Pb-free alloys. On Ni, Ni3Sn4 formed and on copper, Cu6Sn5 Cu3Sn formed. During reflow, the intermetallic growth rate is faster for the Pb-free alloys, compared to eutectic Sn–Pb. In solid-state aging, however, the interfacial intermetallic compounds grow faster with the Sn–Pb solder than with the Pb-free alloys. The reliability tests performed included shear strength and thermomechanical fatigue. The lower-strength Sn-0.7Cu alloy also had the best thermomechanical fatigue behavior. Failures occurred near the solder/intermetallic interface for all the alloys except Sn-0.7Cu, which deformed by grain sliding and failed in the center of the joint. Based on this, the optimal solder alloy for flip-chip applications is identified as eutectic Sn-0.7Cu.

Bibliography

Antolovich, S. D., and Antolovich, B.F. (1996). An introduction to fracture mechanics, in *ASM Handbook. 19 Fatigue and Fracture*, ASM International®, 1996.

Arora, N. D., Raol, K.V., Schumann, R., and Richardson, L.M. (1996). Modeling and extraction of interconnect capacitances for multilayer VLSI circuits, *IEEE Trans. Computer Aided Design of Integrated Circuits and Systems*, 15(1), 58–66.

Bilotti, A. A. (1974). Static temperature distribution in IC chips with isothermal heat sources, *IEEE Trans. Electron Devices*, ED-21(March), 217–226.

Black, J.R. (1969). Electromigration failure models in aluminium metallization for semiconductor devices, *Proc. IEEE*, 57(9), 1587–1594.

Blech, I.A., and Herring, C. (1976). Stress generation by electromigration, *Appl. Phys. Lett.*, 29, 131–133.

Bradley, E., Handwerker, C.A., Bath, J., Parker, R.D., and Gedney, R.W. (2007), *Lead-free electronics: iNEMI projects lead to successful manufacturing*. Hoboken, NJ: John Wiley & Sons.

Chen, C., and Liang, S.W. (2007). Electromigration issues in lead-free solder joints, *J. Mater. Sci.*, 18, pp. 259–268.

Dreezen, G., Deckx, E., and Luyckx, G. (2003). Solder alternative: Electrically conductive adhesives with stable contact resistance in combination with non-noble metallization, *CARTS Europe 2003*, pp. 223–227.

Gale, W.F., and Totemeier, T.C. (2004). *Smithells Metals Reference Book, (8th edition)*. Maryland Heights, MO: Elsevier.

Galyon, G.T. (2003). *Annotated Tin Whisker Bibliography*, Herndon, VA: a NEMI Publication, July.

Hunter, W.R. (1997). Self-consistent solutions for allowed interconnect current density. I. Implication for technology evolution, *IEEE Trans. Electron Devices*, 44(2), 304–309.

Hunter, W.R. (1997). Self-consistent solutions for allowed interconnect current density. II. Application to design guidelines, *IEEE Trans. Electron Devices*, 44(2), 310–316.

Ma, H. (2007). "Characterization of lead-free solders for electronic packaging," Auburn University Dissertation (Doctor of Philosophy), Auburn, AL, May 2007.

Morrow, J. D. (1964). "Cyclic plastic strain energy and fatigue of metals," ASTM STP 378, ASTM, Philadelphia, PA, pp. 45–87.

Obtani and Kitarmura, T. (1985). "On the fatigue life law for smooth specimens at elevated temperature derived from fracture mechanics law of crack propagation," *J. Soc. Mater. Sci. Japan*, 34, 843–849 (in Japanese).

Singh, N. C. (2006). "Thermo-Mechanical Reliability Models for Life Prediction of Area Array Electronics in Extreme Environments," Auburn University Thesis (Master of Science), Auburn, AL, May.

Strauss, R. (1998). *SMT Soldering Handbook, (Second edition)*. Maryland Heights, MO: Elsevier/Newnes.

Suo, Z. (2004). A continuum theory that couples creep and self-diffusion, *J. Appl. Mechanics*, 71, 646–651.

Teng, C.C., Cheng, Y.K., Rosenbaum, E., and Kang, S.M. (1997). iTEM: A temperature-dependent electromigration reliability diagnosis tool, *IEEE Trans. Computer-Aided Design of Integrated Circuits and Systems*, 16(8), 882–893.

Tu, K.N. (2003). *Solder Joint Technology: Materials, Properties, and Reliability*. Berlin, Germany: Springer.

Tu, K.N. (2003). Recent advances on electromigration in very-large-scale integration of interconnects, *J. Appl. Phys.*, 94, 5451–5473.

Tu, K.N. (1994). Irreversible processes of spontaneous whisker growth in bimetallic Cu-Sn thin-film reactions, *Phys. Rev. B*, 49(3), 2030–2034.

Wiese, S., Schubert, A., Walter, H., Dudek, R., Feustel, F., Meusel, E., and Michel, B. (2001). "Constitutive behavior of lead-free solders vs. lead-containing solders: Experiments on bulk specimens and flip-chip joints," *51st Proceedings of the Electronic Components and Technology Conference*, pp. 890–902.

Yeh, E.C.C., Choi, W.J., Tu, K.N., Elenius, P., and Balkan, H. (2002). Current-crowding-induced electromigration failure in flip chip solder joints, *Appl. Phys. Lett.*, 80, 580.

Zahn, B.A. (2003). "Solder joint fatigue life model methodology for 63Sn37Pb and 95.5Sn4Ag0.5Cu Materials," *53rd Proceedings of the Electronic Components and Technology Conference*, pp. 83–94, May 27–30.

5

Lead-Free Electronic Reliability: Fatigue Life Model

With continuing miniaturization of packaging structure for electronic devices, the reliability failure at solder joints is becoming a major concern in the microelectronics industry. Among many mechanisms leading to solder joint failure, the fracture by cyclic bending, shear, and shock load is of particular concern. Conventionally, those load conditions rarely contribute to joint failure; however, with the expansion of mobile electronics, combined with compact packaging structure, they are becoming equally problematic to more conventional cyclic thermal load.

Insight into the thermomechanical response of solder joints is critical to the design and deployment of reliable electronic circuit board assemblies. Understanding the mechanics of lead-free (Pb-free) soldered assemblies is also essential to the development of accelerated test plans, predictive reliability models, and to their use as effective tools for product reliability assessment.

The life of tin–lead (Sn–Pb) or Pb-free solder joints is limited by the fatigue damage that accumulates in solder materials. Accelerated testing provides distributions of failure times whose relevance to service life is determined by extrapolation to use conditions based on the appropriate *acceleration factors* (AFs). Because AFs are defined as the ratio of life under test and use conditions, their determination requires up-front predictions of solder joint lives under both sets of conditions. Such models are also of use for reliability analysis of circuit boards at the design stage. While a variety of life prediction models have been developed for near-eutectic Sn–Pb assemblies, to this author's knowledge, such models are not currently available for Pb-free soldered assemblies. The development of life prediction models requires a detailed understanding of failure modes, an adequate constitutive model that captures the thermomechanical behavior of Pb-free solders in electronic assemblies, and a reliability database that is needed for the empirical correlation of failure times under test and field conditions.

Some reliability data has been acquired or is in the process of being gathered by several "lead-free consortia." Efforts are also underway to characterize the mechanical behavior of Pb-free solders, and numerous studies have been published with emphasis on secondary creep of the solder. The derived creep rate equations are an important ingredient of constitutive models; however, their applications to circuit board assemblies are still the subject of validation studies.

This chapter provides a quantitative review of the thermomechanical properties of Pb-free solders, with emphasis on tin-silver-copper (Sn–Ag–Cu, SAC) and Sn–Ag alloys. The Sn–Ag–Cu alloys of near-eutectic composition are labeled "SAC" throughout the report, including the NEMI-selected (National Electronics Manufacturing Initiative) Sn-3.9Ag-0.6Cu alloy. SAC solders are the alloys of choice for solder reflow assemblies. Eutectic Sn–Ag is also recommended for wavesoldering applications. The microstructure of both alloys consists of an Sn matrix with finely dispersed intermetallic precipitates. We thus expect similarities in the constitutive response of SAC and eutectic Sn–Ag solders. Given the more abundant literature on Sn–Ag, its properties are reviewed in detail in an attempt to better understand the qualitative behavior of precipitate-strengthened solder alloys.

To put things in perspective, this chapter starts with a review of Sn–Pb properties and lessons learned from the development of constitutive and life prediction models for near-eutectic Sn–Pb assemblies. This understanding of Sn–Pb behavior, although not fundamentally complete, has proven valuable to industry. It is also worthwhile observing that the characterization of Sn–Pb solder and the development of reliability models for Sn–Pb assemblies spans over three decades of research in industry and academia. For example, one of the earliest and well-known solder joint reliability models is the Norris–Landzberg (1969) model for non-underfilled flip-chip assemblies. While some of the lessons learned with Sn–Pb will transfer to Pb-free applications, a significant amount of research and development is needed for Pb-free reliability models to come up to par with their Sn–Pb equivalents. The chapter then proceeds with the results of an extensive, although not exhaustive, review of material properties for SAC and Sn–Ag alloys. Gaps in the material properties database are identified and suggestions are offered as to what additional testing and analyses are needed to develop constitutive models for engineering use.

5.1 Time-Independent Plasticity Model

Current techniques for nondestructive quality evaluation of solder bumps in electronics packages are either incapable of detecting solder bump cracks, or unsuitable for in-line inspection due to high cost and throughput. As an alternative, solder bump inspection systems are developed using laser ultrasound and interferometric techniques. Such a system

- Uses a pulsed Nd:YAG laser to induce ultrasound in electronics packages in the thermoelastic regime
- Measures the transient out-of-plane displacement responses on the surface of the chip or packages using laser interferometric technique

A systematic study on thermomechanical reliability of flip-chip solder bumps can be performed using a laser ultrasound-interferometric inspection system and finite element (FE) simulation. The correlation between the failure parameter of solder bumps extracted from FE simulation and the quality degradation results of solder bumps from laser ultrasound testing has also been studied. Accelerated thermal cycling (ATC) tests have been performed in two phases on flip-chip package (FCP) test vehicles with 63Sn-37Pb solder bumps.

The elastic-plastic-creep and Anand's viscoplastic constitutive models have been used in FE simulation to describe the inelastic deformation behavior of solder bumps. A three-dimensional FE model has been implemented using finite element modeling (FEM) on the same FCP as used in ATC testing. The stress-strain results have been extracted from FEM, and the inelastic strain energy density (ISED) per cycle calculated at the critical solder bump position was used as a failure parameter. A good correlation exists between ISED extracted from FE simulation and results from laser ultrasound testing.

5.2 Fatigue Life Prediction Models

The fatigue life prediction model, also known as the damage relationship, correlates the life of the solder joint to the strain energy density or the strain (plastic, creep, or total strain) accumulated during each ATC that the package is subjected to. Several strain energy and strain-based life prediction models for the eutectic solder have been used by researchers in the past and published in the literature. The life prediction models developed for the Pb-free solder are very few.

5.2.1 Eutectic (Sn–Pb–Ag) Solder

One lesson learned from Sn–Pb studies is that there is no unique constitutive model for surface-mount technology (SMT) solder joints, and thus the variety of models available throughout the literature. In the end, the applicability of a given model to real-life assemblies and a reasonable agreement between the ensuing life predictions and test results determine whether a constitutive model will be of use to design engineers and reliability analysts.

The mechanical behavior of solder depends on joint microstructure and is affected by many parameters, such as intermetallics, joint or specimen size, cooling rate of the assembly after soldering, aging in service, etc. Test factors such as specimen or load eccentricity, temperature variations, and measurement errors also contribute to the scatter in mechanical properties of solder, as is well known (e.g., for steady-state creep). Nevertheless, simplified constitutive models have been developed to help characterize the mechanical behavior of Sn–Pb solder and enable first-order stress/strain analysis of solder joints using methods of classical mechanics or numerical techniques such as the finite element method (FEM). Some of the fatigue life prediction models for eutectic solder published in the literature are listed below:

1. The shear strain in the solder is computed using equations developed by Werner Engelmaier for leadless components. This is then multiplied by the previously calculated shear stress to determine the cyclic strain energy. Engelmaier's model is based on total shear strain range:

$$N_f = \frac{1}{2}\left[\frac{\Delta\gamma_T}{0.65}\right]^{\left(\frac{1}{c}\right)} \tag{5.1}$$

where the ductility exponent $c = -0.442 - 6 \times 10\text{-}4(Tmean) + 1.72 \times 10\text{-}2$ $Ln(1+v)$, where $Tmean$ is the mean temperature and v is the cycling frequency.

The stresses on the solder joint are determined using a simplified structural model that accounts for the various stiffnesses of the structure. These strain and stress results are then used to determine the strain energy dissipated by the solder joint. The strain energy is then used to make life predictions using equations. While the model has been widely adopted for Sn–Pb solder joint reliability prediction, many issues that arise from simplifications in formulating input model parameters as well as from the complex physics of solder degradation challenge the model's ability to accurately estimate cycles to failure.

2. Solomon's low cycle fatigue model based on plastic shear strain range:

$$N_p = \left[\frac{1.36}{\Delta\gamma_p}\right]^{\left(\frac{1}{0.5}\right)} \tag{5.2}$$

3. Knecht and Fox's model based on creep shear strain range:

$$N_c = \frac{8.9}{\Delta\gamma_c} \tag{5.3}$$

4. Pang's model based on Miner's superposition rule of creep–fatigue interaction:

$$N_T = \left[\left(\frac{1}{N_c}\right) + \left(\frac{1}{N_p}\right)\right]^{-1} \tag{5.4}$$

5. Morrow's energy-based fatigue model modified by Spraul et al.:

$$N_f = C_1 (\Delta W)^{c_2} \tag{5.5}$$

where $C_1 = 537.15$, $C_2 = -1.0722$

6. Shi et al.'s frequency-modified model based on total strain range:

$$N_f v^{(h-1)} = \left[\frac{C}{\Delta \varepsilon_t} \right]^{\left(\frac{1}{m}\right)} \tag{5.6}$$

where
$\quad m = 0.731 - (1.63 \times 10\text{-}4)T + (1.392 \times 10\text{-}6)T2 - (1.15 \times 10\text{-}8)T3$
$\quad C = 2.122 - (3.57 \times 10\text{-}3)T + (1.329 \times 10\text{-}5)T2 - (2.502 \times 10\text{-}7)T3$
$\quad h = 0.919 - (1.765 \times 10\text{-}4)T - (8.634 \times 10\text{-}7)T2$
$\quad v = 0.0139$

7. Shi et al.'s frequency-modified model based on Morrow's energy model:

$$N_f v^{(h-1)} = \left[\frac{C \Delta \sigma}{\Delta W} \right]^{\left(\frac{1}{m}\right)} \tag{5.7}$$

where m = 0.7, C = 1.69, h = 0.9

8. The Darveaux model uses inelastic strain energy obtained from finite element analysis to correlate fatigue lives from over 100 experiments. Darveaux's energy-based model with CAVE-modified (Auburn University Center for Advanced Vehicle Electronics) constants is shown as follows:

$$N_e = N_0 + \frac{a}{da/dN} \tag{5.8}$$

where
$\quad$ a is the joint diameter at the interface
$\quad N_0 = K_1 (\Delta W)^{K_2}$ is the number of cycles for crack initiation
$\quad da/dN = K_3 (\Delta W)^{K_4}$ is the crack propagation rate

The values of the constants K_1, K_2, K_3, and K_4 used for calculation of characteristic life cycles are given in Table 5.1.

TABLE 5.1

Constants for the Damage Relationship

	K_1 (cycles/psi^{K_2})	K_2	K_3 (in/cycle/psi^{K_4})	K_4
Lall [2006, 2005, 2004, 2003]	28769	−1.53	6×10^{-7}	0.7684
Darveaux [2000]	48300	−1.64	3.8×10^{-7}	1.04

5.2.2 Lead-Free (Sn–Ag–Cu) Solder

At present, Sn–Ag–Cu appears to be the leading Pb-free solder in the electronics industry. Driven by miniaturization, decreasing the component size leads to a stronger influence of microstructure on the observed lifetime properties. Research has concentrated on the thermal fatigue response of a near-eutectic Sn–Ag–Cu solder alloy with the objective of correlating damage mechanisms with the underlying microstructure, on the basis of which a thermomechanical fatigue damage evolution model is characterized. Bulk Sn-4Ag-0.5Cu specimens are thermally cycled between 40°C and 125°C up to 4,000 cycles.

As a result of the intrinsic thermal anisotropy of the beta-Sn phase, thermal fatigue loading causes localized deformations, especially along Sn grain boundaries. Mechanical degradation of test specimens after temperature cycling is identified from a reduction in the global elasticity modulus measured at very low strains. Using orientation imaging microscopy (OIM) scans, the test specimens are modeled, including the local grain orientations and the detailed microstructure. Orientation imaging microscopy enables simultaneous determination of microstructural and crystallographic information of scanned samples. A traction-separation based cohesive zone formulation with a damage variable that traces the fatigue history is used to model interfacial interactions between grains. Damage evolution parameters are identified on the basis of the experimentally obtained global elastic moduli after a certain number of cycles. The resulting damage evolution law is applied to a number of numerical examples, and the mismatch factor is discussed in detail. Finally, the damage evolution law characterized in the research is exploited for fatigue life prediction of a 2-D microstructure-incorporated ball grid array (BGA) solder ball. Some of the fatigue life prediction models for Pb-free solder published in the literature are listed below:

1. Shi et al.'s frequency modified model based on total strain range:

$$N_f v^{(h-1)} = \left[\frac{C}{\Delta\varepsilon_t} \right]^{\left(\frac{1}{m}\right)} \tag{5.9}$$

where
$m = 0.731 - (1.63 \times 10\text{-}4)T + (1.392 \times 10\text{-}6)T2 - (1.15 \times 10\text{-}8)T3$
$C = 2.122 - (3.57 \times 10\text{-}3)T + (1.329 \times 10\text{-}5)T2 - (2.502 \times 10\text{-}7)T3$
$h = 0.919 - (1.765 \times 10\text{-}4)T - (8.634 \times 10\text{-}7)T2$
$v = 0.0139$

2. Shi et al.'s frequency modified model based on Morrow's energy model:

$$N_f v^{(h-1)} = \left[\frac{C\Delta\sigma}{\Delta W} \right]^{\left(\frac{1}{m}\right)} \tag{5.10}$$

where $m = 0.7$, $C = 1.69$, and $h = 0.9$.

3. Morrow's frequency-modified energy model by Pang et al.:

$$N_f v^{(h-1)} = \left[\frac{A}{\Delta W} \right]^{\left(\frac{1}{n}\right)} \qquad (5.11)$$

where $n = 0.877$, $A = 1487.3$, and $h = 0.82$.

The movement to Pb-free soldering will result in solder joints that are significantly stiffer than those made of Sn–Pb. This section presents the results from the first phase of a two-part study to understand and compare the isothermal mechanical fatigue behavior of SAC solder to that of Sn–Pb solder. A combination of experiments and finite element analysis (FEA) has been used to compare and predict the durability of Sn–Pb and SAC surface-mount solder joints. The experiments consist of cyclic four-point bend tests of printed wiring board (PWB) coupons populated with 2512 sized resistors at 5 Hz. This configuration has been chosen so the test would reflect actual electronic products and still be rapidly modeled using FEA.

This frequency should be sufficiently high to minimize solder creep during testing. The board-level strains have been verified with strain gauges, and the solder joint failures have been detected using a high-speed event detector. Tests have been conducted at two board-level strain values and then modeled in FEA to determine the strains and stresses developed in the solder joint. This information was then used to determine the appropriate cyclic fatigue relationship for both SAC and Sn–Pb solder. The results indicate that at high board-level strains, Sn–Pb solder outperforms SAC solder. However, at lower board-level strains, the SAC solder outperforms Sn–Pb. The second phase of the study involved bend testing at even lower board-level strains to characterize the high cycle fatigue behaviors of the solders.

5.3 Life Prediction Calculation Using Darveaux's Energy-Based Model

5.3.1 Motorola/Darveaux's Constitutive Model

Among all solder-fatigue life models, the energy-based method is arguably the most popular one. Over the years, the method has been applied to many types of packages. Excellent correlation with the actual test results has been reported. The energy-based method links the fatigue life to the inelastic strain energy dissipation of solder joints. Based on extensive testing of BGA solder joints and FE modeling, Darveaux proposed empirical equations to calculate the solder fatigue life. In Darveaux's model, the total fatigue life consists of the life before crack initiation and the life after it. The constant terms are derived by curve fitting the FEA prediction with the test data. Because the inelastic strain energy depends strongly on the finite element mesh of the solder joints, different constants values were given according to the element size. According to Darveaux, good correlation with the test data can be expected only when the solder joints are meshed such that the element size in the solder height direction has fixed values that fall in the range specified.

Darveaux and co-workers at Motorola conducted extensive mechanical testing of flip-chip and BGA solder joints and characterized the time-independent plastic flow and creep deformations of several solder alloys. Their constitutive model is described below for several alloys of electronic solder. Robert Darveaux implemented this model in two commercial finite element codes, ANSYS™ and ABAQUS™. His original publication included detailed recommendations on how to input material constants in the preprocessor of those two programs. One important feature of Darveaux's creep model is that it was found to apply consistently to several solder alloys over a wide range of temperatures and several orders of magnitude of strain rates. These solder alloys include

- 60Sn-40Pb
- 62Sn-36Pb-2Ag
- 96.5Sn-3.5Ag
- 97.5Pb-2.5Sn
- 100In
- 50In-50Pb

The initial, instantaneous strain that develops at the start of a creep test includes an elastic strain and an inelastic strain that represent time-independent plastic flow. The plastic strain is described by a plastic flow or strain hardening law of the form:

$$\gamma_p = C_6 (\tau / G)^m \tag{5.12}$$

where
γ_p is the plastic strain
C_6 is a material constant
M is a material constant
τ is the applied stress
$G = G(T)$ is the temperature-dependent shear modulus given as
$G(T) = G_0 - G_1 (T(^{\circ}k) - 273)$
G_0 is the shear modulus at 0°C
G_1 is the shear modulus temperature coefficient
$G_0 = 1.9$ Mpsi and $G_1 = 8.1$ kpsi/K for both alloys of 60Sn-40Pb and 62Sn-36Pb-2Ag

The elastic constants and the plastic flow parameters for several solder alloys, including Sn–Ag eutectic, are given in Table 5.2. Note that the elastic constants and the power-law exponent m are about the same for 60Sn-40Pb and 62Sn-36Pb-2Ag. However, the constant C_6 is about twice as low for 62Sn-36Pb-2Ag; that is under equal loads, Sn–Pb with 2% Ag will see half as much initial plastic strain as 60Sn-40Pb.

During primary or transient creep, the creep strain is given by

$$\gamma_c^1 = \frac{d\gamma_s}{dt}t + \gamma_\tau\left[1 - exp\left(-B\frac{d\gamma_s}{dt}t\right)\right] \tag{5.13}$$

where

γ_c^1 is the primary creep strain
γ_τ is the transient creep strain
B is the transient creep coefficient
$d\gamma_s/dt$ is the steady-state creep rate

The primary creep constants for several alloys are given in Table 5.3.

The primary creep rate is:

$$\gamma_c^1 = \frac{d\gamma_s}{dt}\left[1 + \gamma_\tau * B * exp\left(-B\frac{d\gamma_s}{dt}t\right)\right] \tag{5.14}$$

Initially, at time $t = 0$, the primary creep rate is a factor $(1 + \gamma_\tau B)$ times greater than the steady-state creep rate. For 60Sn-40Pb, this factor is

$$(1 + \gamma_\tau B) = +0.026 \times 403 = 11.48 \tag{5.15}$$

that is, the initial transient creep rate is over an order of magnitude higher than the steady-state creep rate. For Sn-3.5Ag, the rate factor is even larger:

$$(1 + \gamma_\tau B) = +0.167 \times 131 = 21.88 \tag{5.16}$$

Thus, primary creep may not be negligible in applications with high-temperature ramp rates or under thermal cycling conditions with short dwell times. More general relationships were found to apply to steady-state creep of solder in shear:

$$\dot{\gamma}_s = \frac{d\gamma_s}{dt} = C_4\frac{G(T)}{T}\left[sinh\left(\alpha\frac{\tau}{G}\right)\right]^n exp\left(\frac{-Q}{kT}\right) \tag{5.17}$$

TABLE 5.2

Solder Elastic Constants and Strain-Hardening Constants

	Symbol	Unit	Description	60Sn-40Pb	62Sn-36Pb-2Ag	96.5Sn-3.5Ag	97.5Pb-2.5Sn
Elastic Constants	G_0	Mpsi	Shear modulus at 0°C	1.9	1.9	2.8	1.3
	G_1	kpsi/K	Shear modulus temperature coefficient	8.1	8.1	10.0	1.5
Strain-Hardening Constants	C_6		Material constant	2.34E13	1.21E13	2.04E11	6.36E7
	M		Material constant	5.58	5.53	4.39	3.10

Source: From Darveaux, R. (1996), How to Use Finite Element Analysis to Predict Solder Joint Fatigue Life, *Proc. VIII Int. Congress on Experimental Mechanics*, Nashville, TN, pp. 41–42, June 10–13, 1996; Darveaux, R., Banerji, K., Mawer, A. and Dody, G. (1995), Reliability of plastic ball grid array assemblies, Chap. 13, in *Ball Grid Array Technology*, Ed. J.H. Lau, McGraw-Hill, New York.

TABLE 5.3

Solder Primary Creep Constants

Primary Creep Constants	Symbol	Description	60Sn-40Pb	62Sn-36Pb-2Ag	96.5Sn-3.5Ag	97.5Pb-2.5Sn
	γ_z	Transient creep strain	0.026	0.040	0.167	0.115
	$B.$	Shear modulus temperature coefficient	403	152	131	137

Source: From Darveaux, R. (1996), How to Use Finite Element Analysis to Predict Solder Joint Fatigue Life, *Proc. VIII Int. Congress on Experimental Mechanics*, Nashville, TN, pp. 41–42, June 10–13, 1996; Darveaux, R., Banerji, K., Mawer, A. and Dody, G. (1995), Reliability of plastic ball grid array assemblies, Chap. 13, in *Ball Grid Array Technology*, Ed. J.H. Lau, McGraw-Hill, New York.

or in a simplified form:

$$\dot{\gamma}_S = C_5 \left[sinh(\alpha_1 \tau) \right]^n exp\left(\frac{-Q_a}{kT} \right)$$ (5.18)

where
$\dot{\gamma}$s is the steady-state strain rate
$G(T)$ is the temperature-dependent shear modulus
T is the absolute temperature (in Kelvin)
τ is the applied stress
n is the constant exponent that depends on the controlling creep mechanism
k is Boltzmann's constant ($k = 8.620 \times 10\text{-}5$ eV/K)
Q is the creep activation energy
Q_a is the apparent activation energy
α, C_4, C_5 are constants

The above constants and activation energies are given for several solder alloys in Table 5.4.

For decades, BGA solder joint fatigue in thermal cycling has been studied by various researchers. Finite element modeling techniques have been used for years to calculate the inelastic strain energy accumulated during thermal cycling. Fatigue life prediction methods were then used to measure and predict accumulative fatigue. For years, different stress-based, strain-based, damage-accumulation-based, inelastic-energy-based theories were proposed and studied by researchers. Darveaux proposed a damage-accumulation-based, inelastic-energy-based theory in 2000 that is being commonly used in industry. Darveaux's method measures the number of cycles for cracks to initiate and propagate through the joint based on average inelastic energy density accumulated during thermal cycling, using the crack initiation and crack growth constants he calculated for BGA joints.

For thin-profile fine-pitch BGA (TFBGA) packages, board-level solder joint reliability during thermal cycling testing is a critical issue. In this section, both global and local parametric 3-D FEA fatigue models are established for TFBGA on board with considerations for the detailed pad design, realistic shape of the solder joint, and nonlinear material properties. The FEA fatigue models have the capability of predicting the fatigue life of a solder joint during the thermal cycling test within ±13% error. The fatigue model applied is based on a modified Darveaux approach with nonlinear viscoplastic analysis of solder joints. A solder joint damage model is used to establish a connection between the strain energy density (SED) per cycle obtained from the FEA model and the actual characteristic life during the thermal cycling test. For the test vehicles studied, the maximum SED was observed at the top corner of the outermost diagonal solder ball. The modeling predicted that fatigue life is first correlated to the thermal cycling test results using modified correlation constants, curve-fitted from in-house BGA thermal cycling test data.

TABLE 5.4

Solder Steady-State Creep Parameters for Common Solders

	Symbol	Unit	Description	60Sn-40Pb	62Sn-36Pb-2Ag	96.5Sn-3.5Ag	97.5Pb-2.5Sn
Steady-State Creep Parameters	$C_{4}.$	sec/psi	Constant	0.198	0.0989	3.13E-3	1.62E-7
	α		Constant	1300	1300	1500	1000
	n		Constant exponent that depends on the controlling creep mechanism	3.3	3.3	5.5	7.0
	Q	eV	Creep activation energy	0.548	0.548	0.50	1.10
	C_5	1/sec	Constant	2.778E-5	1.39E-5	2.46E-5	4.60E-11
	α_1		Constant	8.0E-4	8.0E-4	6.3E-4	8.0E-4
	Q_a	eV	Creep activation energy	0.70	0.70	0.75	1.15

Source: From Darveaux, R. (1996), How to Use Finite Element Analysis to Predict Solder Joint Fatigue Life, *Proc. VIII Int. Congress on Experimental Mechanics*, Nashville, TN, pp. 41–42, June 10–13, 1996; Darveaux, R., Banerji, K., Mawer, A. and Dody, G. (1995), Reliability of plastic ball grid array assemblies, Chap. 13, in *Ball Grid Array Technology*, Ed. J.H. Lau, McGraw-Hill, New York.

Subsequently, design analysis was performed to study the effects of fourteen key package dimensions, material properties, and thermal cycling test conditions. In general, smaller die size, higher solder ball stand-off, smaller maximum solder ball diameter, bigger solder mask opening, thinner board, higher mold compound coefficient of thermal expansion (CTE), smaller thermal cycling temperature range, and depopulated array type of ball layout pattern contribute to longer fatigue life.

5.4 Solder Joint Integrity in Accelerated Thermal Cycling

Experimental studies for creep and fatigue of solder-interconnects in microstructures have been performed. The strains were directly measured in the fillet area of solder-joints with a typical linear dimension of 50 μm. An analytical approach was developed for calculating shear stress based on the shear strain measurement and the established solder constitutive relationships. Also obtained was the strain-rate as well as the separate elastic, plastic, and creep components from the measured total strain. The data enables the determination of the strain energy density per temperature cycle for the characterization of the solder joint creep–fatigue behavior. Case studies provided evidence for the shear dominance and the creep–fatigue mechanism in thermally induced solder joint deformation in surface-mounted electronic assemblies. Although a similar trend of variation in stress–strain was found in the joints of different solders, the substantial differences in the hysteresis loop area and shape as well as in the creep rate suggested that the solder constitutive parameters should have a profound impact on the creep–fatigue endurance of the joints.

One of the most serious concerns in the flip-chip reliability is the solder interconnect fatigue failure due to thermal cyclic loading. Due to the CTE mismatch within the various components of the flip-chip package, stresses develop in the component whenever thermal cycling takes place. These stresses lead to the fatigue failure of the component when the thermal cycling is continued for long periods of time. It has been established through various experimental studies and field failure analysis that the solder interconnect is the most susceptible to such fatigue failures. Due to this, finite element analysis and simulation of solder joint integrity become extremely important. Simulation of 0°C to 70°C ATC along with three different field profiles has been done and is presented here.

The thermal profile shown in Figure 5.1 has been simulated in FEM for the analysis of solder joint integrity and the comparative study in order to see the effect of various variations on the solder joint reliability. The cycle starts at room temperature (25°C) and has a dwell time of 12 minutes at the extreme temperatures of 0°C and 70°C, and the ramp time between the extreme

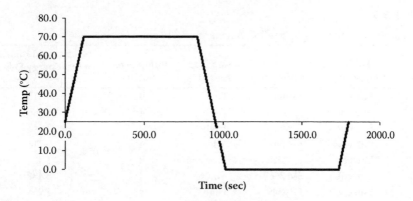

FIGURE 5.1
0°C to 70°C accelerated thermal cycle profile.

temperatures is 3 minutes; therefore, the total time period for the cycle is 30 minutes. Simulations for the eighteen different configurations listed previously have been run for this profile, and the simulation results are listed in Table 5.5. Inelastic strain energy density averaged over the complete solder volume has been calculated, as discussed earlier. Because ISED is a true indicator of the damage accumulated by the solder joint, which leads to the crack initiation and finally failure of the solder joint, it has been used as the basis for the comparison of reliability for different flip-chip configurations.

Finite element modeling simulates the distribution of the magnitude of the ISED accumulated in the solder joint during one complete thermal cycle for test case-1, which consists of low-gap height eutectic solder bumps with FP 4526 underfill. The magnitude of the ISED accumulated is maximum at the top-left corner of the bump, which indicates that the crack would initiate at this location and then propagate; also, there is a high ISED accumulation region at the bottom-right corner of the bump, which implies a possibility of secondary crack initiation in this region before the primary crack completely propagates.

Finite element modeling can simulate the distribution for the ISED accumulated and the Von Mises stress remains more or less the same, but in this case it can be observed that the region for the secondary crack has shifted from the bottom right (test case-1) to the top right (Figure 5.1). This shift in the secondary crack location may speed up the failure, as now the primary crack does not have to travel to the other end in order to cause a failure.

5.4.1 Effect of Solder Joint Material Composition

Figure 5.2 shows that the value of ISED accumulated per cycle for eutectic solder bumps is higher as compared to the Pb-free bump for the capillary flow FP 4526 underfill material for both high-gap and low-gap height bumps. For eutectic solder bumps, the difference in the ISED accumulated between

TABLE 5.5

Simulation Results for 0°C to 70°C Accelerated Thermal Cycle

Test Case	Solder Composition	Underfill	Bump Gap Height	Bump Size (mils)	Normalized Inelastic Strain Energy Density per Cycle
1	Eutectic	Kester	Low	4	4.415
2	Eutectic	FP 4526	High	4	1.208
3	Eutectic	Kester	High	4	3.543
4	Eutectic	Kester	Mid	4	3.716
5	Eutectic	FP 4526	Low	4	1.423
6	Eutectic	3M 3667	Mid	4	3.487
7	Eutectic	FP 4549	High	4	1.964
8	Eutectic	FP 4549	Mid	3	2.024
9	Eutectic	3M 3667	Mid	3	3.550
10	Pb-Free	FP 4549	High	4	1.795
11	Pb-Free	FP 4549	Mid	3	1.890
12	Pb-Free	Kester	High	4	3.444
13	Pb-Free	Kester	Mid	3	3.890
14	Pb-Free	3M 3667	High	4	3.047
15	Pb-Free	3M 3667	Mid	3	3.483
16	Pb-Free	FP 4526	Low	4	1.045
17	Pb-Free	FP 4526	Mid	4	1.012
18	Pb-Free	FP 4526	High	4	1.000

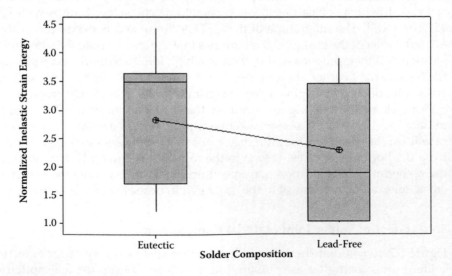

FIGURE 5.2

Boxplot of normalized inelastic strain energy density versus solder composition.

the high- and low-gap height bumps is higher as compared to Pb-free solder bumps.

Detailed analysis reveals the following:

- The eutectic solder bumps have higher values of ISED than the Pb-free bumps in the case of FP 4549 underfilled flip chip, and also for both the mid- and high-gap height bumps.
- However, the difference in the value of ISED between mid- and high-gap for FP 4549 is higher in the case of Pb-free bumps.
- For both reflow encapsulant materials, the ISED accumulated in the solder joint per cycle is less for the Pb-free bump.
- Kester is more sensitive to solder joint composition as far as solder joint reliability is concerned.

5.4.2 Effect of Underfill Composition

Underfill is today's polymer magic that enables the increasingly popular second-generation flip chip lower-cost assembly on organic substrate. This section provides a basic understanding of the chemistry of underfills and the physical properties that are important to their successful use. Underfills are a carefully formulated composite of organic polymers and inorganic fillers. Fillers are the most important single ingredient in the modern underfill. We must therefore examine in some detail the effects of filler characteristics on underfill pre-cured and post-cured properties. Finally, we investigate the total rheology of underfill in order to understand the material/machine interface and better define the underfill process.

Underfill was originally tested as a sealant to protect flip-chip (FC) joints from corrosion on IBM mainframe computer modules. This early work led to the unexpected discovery that thermomechanical fatigue could be reduced. The improved thermocycle performance, while of small value for already robust ceramic FC systems, is absolutely essential for reliable FC-organic systems with their significant thermomechanical mismatch. Solder joint reliability is also sensitive to the underfill material. The trend of higher reliability with the capillary flow encapsulants as compared to reflow encapsulants is shown in Figure 5.3, a boxplot of normalized inelastic strain energy density versus underfill composition.

Flip chip is becoming a mainstream micro-assembly method as the infrastructure rapidly builds and technical hurdles are overcome. Flip chip on organic substrates generally requires underfill to reduce joint fatigue. State-of-the-art underfills have achieved rapid flow, snap cure, and the capability of flowing under the smallest gaps. The science of underfill is a multidisciplinary realm embracing both chemistry and physics. Understanding the dynamics of underfill is a prerequisite to the efficient application of this class of materials on modern automated dispensing machines. Fortunately,

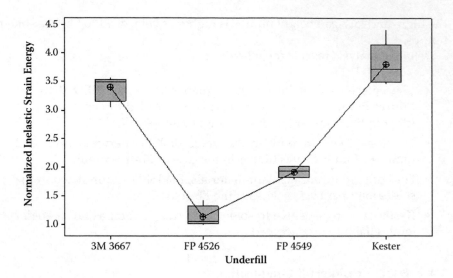

FIGURE 5.3
Boxplot of normalized inelastic strain energy density versus underfill composition.

underfill materials follow the laws of science and the most important relationships have been present here.

Underfills have come a long way since the days when flow could take minutes and curing required up to 6 hours. The best materials flow rapidly, up
to 3 or more centimeters per minute and cure in about 5 minutes. Advances
in filler technology allow underfills to be made that can flow under gaps of
well under 1 mil.

Underfills are well behaved in terms of their chemistry and physics.
Science can and must be applied to the use of underfills as the "art" content
continues to decrease. Simplified property and use rules can be applied that
have a firm basis in chemistry and physics as well as the specialized areas
of surface chemistry, rheology, and fluid dynamics. The optimum material/
machine interface requires that the science of underfills be applied to the
technology of dispensing and curing equipment.

5.4.3 Effect of Bump Gap Height

Figure 5.4 shows the normalized ISED accumulated by the eutectic solder
joint of three different gap heights for a flip-chip with Kester encapsulant.
A trend of increasing ISED with decreasing gap height is observed. The same
trend is also observed for the Pb-free solder bumps with both reflow and
capillary flow encapsulant.

A variety of Pb-free solder alloys were studied for use as flip-chip interconnects, including Sn-3.5Ag, Sn-0.7Cu, and Sn-3.8Ag-0.7Cu, with eutectic
Sn-37Pb as a baseline. The reaction behavior and reliability of these solders

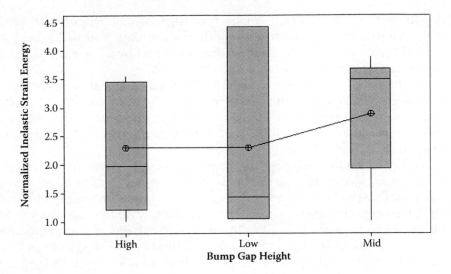

FIGURE 5.4
Boxplot of normalized inelastic strain energy density versus bump gap height.

were determined in a flip-chip configuration using a variety of under-bump metallurgies (TiW/Cu, electrolytic nickel, and electroless Ni-P/Au). The solder microstructure and intermetallic reaction products and kinetics were determined. The Sn-0.7Cu solder has a large grain structure and the Sn-3.5Ag and Sn-3.8Ag-0.7Cu solders have a fine lamellar two-phase structure of Sn and Ag_3Sn. The intermetallic compounds were similar for all the Pb-free alloys. On Ni, Ni_3Sn_4 formed; and on copper, $Cu_6Sn_5Cu_3Sn$ formed. During reflow, the intermetallic growth rate was faster for the Pb-free alloys, compared to eutectic Sn–Pb. In solid-state aging, however, the interfacial intermetallic compounds grew faster with the Sn–Pb solder than for the Pb-free alloys. The reliability tests performed included shear strength and thermomechanical fatigue. The lower-strength Sn-0.7Cu alloy also had the best thermomechanical fatigue behavior. Failures occurred near the solder/intermetallic interface for all alloys except Sn-0.7Cu, which deformed by grain sliding and failed in the center of the joint. Based on the research, the optimal solder alloy for flip-chip applications is identified as eutectic Sn-0.7Cu.

5.4.4 Effect of Bump Size

The influence of joint size on low cycle fatigue characteristics of Sn–Ag–Cu (SAC) has been investigated by a miniature joint specimen fabricated using micro-solder balls. The influence of the size on crack initiation life is not remarkable, while the life, which reaches complete failure, decreases greatly when the ball size is decreased to less than 150 µm. The reduction in life is due to the fact that the crack propagation stage does not appear and complete failure occurs simultaneously with crack initiation in ball sizes less than 150 μm. Failure in the smallest size is induced by subgrain boundary formation that occurs in the whole area of the joint.

Two different bump sizes were used for the simulation of the eighteen configurations. The solder bumps with the bigger size have lower ISED values and thus higher reliabilities. Flip chips with eutectic bumps and 3M 3667 reflow encapsulant were used (Figure 5.5) to show the effect of bump size on thermal reliability.

The motivation to enlarge the passivation opening is to reduce the severity of the stress concentration caused by the original design, and also to increase the contact area between the solder bump and the aluminum bump pad. It was confirmed in the thermal shock test that with the new design, package fatigue life improved by more than 70%. To numerically predict this improvement represents a unique challenge to modeling. This is because in order to capture the slightest geometrical difference on the order of a few microns between the two designs, the multilayer solder/die interface must be modeled using extremely fine mesh, while the overall dimensions of the package and the test board are on the order of millimeters. To bridge this tremendous gap in geometry, a single finite element model that incorporates all the necessary geometrical details is deemed computationally prohibitive and impractical. We applied a global–local modeling scheme that was also suggested by other research. The global model contains the complete package with a much-simplified solder/die interface, whereas the local model includes only one solder joint, but with detailed solder/die interface. Unlike most global–local models proposed

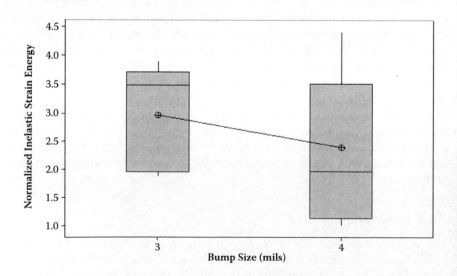

FIGURE 5.5
Boxplot of normalized inelastic strain energy density versus bump size.

by others, this one included time-independent plasticity and temperature-dependent materials in the global model. This greatly improved model correlation accuracy with only moderate increase in runtime. An energy-based solder fatigue model was used to correlate the inelastic strain energy with the package fatigue life. Researchers have found that Darveaux's equations tended to be conservative when applied to the micro-SMD (surface mount device), and hence new correlations based on curve-fitting the test data were derived. We used the newly derived equation and achieved less than 20% error in N_{50} life (number of thermal cycles where 50% of the parts are expected to fail due to fatigue) for both designs, which is on a par with Darveaux's equations when used for BGAs. The analysis also revealed two factors that may account for the life improvement: (1) a slight decrease in inelastic energy dissipation after enlarging the passivation opening, and (2) the shift in crack initiation location, which leads to a longer crack growth length for the new design. The second factor was also independently confirmed by failure analysis.

A more realistic and accurate prediction of fatigue life in the second-level solder interconnect was conducted through power cycling. Fatigue is the dominating failure mechanism of solder interconnects, and enhancement of solder life is one of the major concerns for package designers and users. Conventionally, fatigue life is obtained empirically through accelerated thermal cycling (ATC) with hundreds of parts. To reduce development time and cost, virtual qualification attempts were made using numerical simulation tools, such as FEA. Modeling of life prediction has been conducted for ATC conditions, which assumes uniform temperature throughout the assembly. In reality, an assembly is subjected to power cycling, that is, nonuniform temperature with chip as the only source of heat generation. This nonuniform temperature and the different CTEs of each component make the package deform differently than in the case of uniform temperature. In this work, a proper power cycling (PC) analysis scheme was proposed and conducted to predict solder fatigue life for a flip-chip plastic ball grid array (FC-PBGA) package. Numerical simulations were performed by a combination of computational fluid dynamics (CFD) and FEA. CFD analysis was used to extract transient heat transfer coefficients while subsequent thermal and structural FEA was performed with heat generation and heat transfer coefficients from CFD as thermal boundary conditions. It was found that for organic packages, PC was the more severe condition and caused solder interconnects to fail earlier than in ATC. In summary,

- Lead-free solders have lower ISEDs compared to eutectic Sn–Pb solder for all investigated underfills. However, this does not necessarily mean that Pb-free solder will have better reliability, as the damage relationships for the two different solder materials will be different.

- An increase in gap height lowers the ISED for both eutectic and Pb-free solders; this means that the flip chip with higher gap height will have higher reliability.

- For the underfills investigated, capillary flow underfills exhibit lower ISEDs as compared to the reflow encapsulant for both eutectic and Pb-free solders.

- An increase in the bump size lowers the ISED for both eutectic and Pb-free solders.

5.5 Effect of T_g of the Underfill Material

In flip-chip microelectronics packages, solder bumps are used to connect the silicon die and package substrate for electrical functionality. However, due to the large mismatch between the CTE of silicon and the organic package, the solder bumps undergo large viscoplastic deformations in temperature cycling tests and in field operations. The viscoplastic damage accumulates cycle by cycle, which leads to bump failure by fatigue cracking after hundreds or thousands of thermal cycles. Underfill plays a key role in the solder joint fatigue lifetime. In addition to the effects of underfill modulus, CTE, and the glass transition temperature (T_g), the slope of the temperature-dependent modulus curve in the range of glass transition is also critical. Because the solder deformation is very sensitive to underfill modulus, the viscoplastic work is much larger when the temperature is above T_g than when it is below T_g. Because the glass transition occurs within a range of temperature, there are various definitions of T_g in practice. Also, different materials could have the same T_g based on some definition, but the transition behaviors could be quite different. As long as the accurate temperature-dependent properties of underfill are not used, the predicted solder fatigue life based on a single value of T_g and a sharp transition is subject to skepticism. This research has studied the effect of glass transition slope on solder fatigue life under accelerated thermal cycling (ATC: 0°C ~ 100°C) and deep thermal cycling (DTC: −40°C ~ 125 °C) test conditions. The results show that lifetime variation can be 80% less for a steeper slope and doubled for a shallower slope.

Material properties such as CTE and elastic modulus change drastically once the working temperature of the material rises beyond the T_g of that material. The material properties of the underfill have a very significant impact on the solder joint reliability; thus, it is advisable to use the underfill material that has T_g above the working temperature of the component. Kester has a T_g in the range of 70°C to 75°C; therefore simulations for special test cases of the Kester with modified properties were run in order to understand the effect of T_g on solder joint reliability.

Four different material models (A, B, C, D; Table 5.6 and Table 5.7) were considered for the analysis. The material model D represents the actual behavior of the Kester underfill. Model C represents the behavior of Kester

TABLE 5.6

Simulation Results for the Analysis of the Effect of T_g on Solder Joint Reliability

Temp Cycle	Material Model	Normalized ISED
0–90°C	A	206.63
0–90°C	B	64.4
0–90°C	C	51.85
0–70°C	D	33.65
0–75°C	D	182.02
0–80°C	D	191.75
0–85°C	D	257.49
0–90°C	D	289.72

TABLE 5.7

Material Model Data Used for Simulation

Temp (°C)	A CTE (ppm)	A E (GPa)	B CTE (ppm)	B E (GPa)	C CTE (ppm)	C E (GPa)	D CTE (ppm)	D E (GPa)
25	67	3.1	67	3.1	67	3.1	67	3.1
50	67	3.1	67	3.1	67	3.1	67	3.1
70	67	3.1	67	3.1	67	3.1	67	3.1
75	281	3.1	100	3.1	67	3.1	281	0.74
80	281	3.1	100	3.1	67	3.1	281	0.74
90	281	3.1	100	3.1	67	3.1	281	0.74

in case the T_g of Kester is above 90°C. The results from the simulation are listed in Table 5.6, which shows that the ISED value for the Kester reduces to 0.18X if the T_g of the Kester material is above 90°C for the 0°C to 90°C ATC. The simulation runs for the various temperature ranges show that there is a drastic increase in ISED once the operating temperature goes above the T_g range. The increase in the maximum temperature from 70°C to 75°C leads to an increase in the ISED value by 5.4X.

Bibliography

Amagai, M., Watanabe, M., Omiya, M., Kishimoto, K., and Shibuya, T. (2002). Mechanical characterization of Sn–Ag based lead-free solders, *Micoelectronics Reliability*, 42, 951–966.

Antolovich, S.D., and Antolovich, B.F. (1996). *An Introduction to Fracture Mechanics in ASM Handbook. 19 Fatigue and Fracture*, Materials Park, OH: ASM International®, 1996.

Arora, N.D., Raol, K.V., Schumann, R., and Richardson, L.M. (1996), Modeling and extraction of interconnect capacitances for multilayer VLSI circuits, *IEEE Trans. Computer Aided Design of Integrated Circuits and Systems*, 15(1), 58–66.

Bilotti, A.A. (1974). Static temperature distribution in IC chips with isothermal heat sources, *IEEE Trans. Electron Devices*, ED-21(March), pp. 217–226.

Black, J.R. (1969). Electromigration failure models in aluminium metallization for semiconductor devices, *Proc. IEEE*, 57(9), 1587–1594.

Blech, I.A., and Herring, C. (1976). Stress generation by electromigration, *Appl. Phys. Lett.*, 29, 131–133.

Chen, C., and Liang, S.W. (2007). Electromigration issues in lead-free solder joints, *J. Mater. Sci.*, 18, 259–268.

Darveaux, R. (2000). Effect of simulation methodology on solder joint crack growth correlation, *Proc. 50th ECTC*, May 2000, pp. 1048–1058.

Darveaux, R. (1996). How to use finite element analysis to predict solder joint fatigue life. *Proc. VIII Int. Congress on Experimental Mechanics*, Nashville, TN, June 10–13, 1996, pp. 41–42.

Dreezen, G., Deckx, E., and Luyckx, G. (2003). Solder alternative: Electrically conductive adhesives with stable contact resistance in combination with non-noble metallization. *CARTS Europe 2003*, pp. 223–227.

Gale, W.F., and Totemeier, T.C. (2004). *Smithells Metals Reference Book, (8th edition)*. Maryland Heights, MO: Elsevier.

Galyon, G.T. (2003). Annotated Tin Whisker Bibliography, a NEMI Publication, July.

Hunter, W.R. (1997). Self-consistent solutions for allowed interconnect current density. I. Implication for technology evolution, *IEEE Trans. Electron Devices*, 44(2), 304–309.

Hunter, W.R. (1997). Self-consistent solutions for allowed interconnect current density. II. Application to design guidelines, *IEEE Trans. Electron Devices*, 44(2), 310–316.

Lall, P., Hariharan, G., Tian, G., Suhling, J., Strickland, M., and Blanche, J. (2006). "Risk Management models for flip-chip electronics in extreme environments." *ASME International Mechanical Engineering Congress and Exposition*. Chicago, IL, November, pp. 1–13.

Lall, P., Islam, N., Shete, T., Evans, J., Suhling, J., and Gale, S. (2004). "Damage mechanics of electronics on metal-backed substrates in harsh environments," *Proceedings of 54th Electronic Components & Technology Conference*, IEEE, Las Vegas, NV, June 1–4, pp. 704–711.

Lall, P., Islam, N., Suhling, J., and Darveaux, R. (2003). "Model for BGA and CSP reliability in automotive underhood applications," in *Proc. 53rd Electronic Components and Technology Conference*, May 27–30, 2003, pp. 189–196.

Lall P., Pecht, M., and Hakim, E. (1997). *Influence of Temperature on Microelectronic and System Reliability*. Boca Raton, FL: CRC Press.

Lall, P., Singh, N, Suhling, J., Strickland, M., and Blanche, J. (2005). "Thermomechanical reliability tradeoffs for deployment of area array packages in harsh environments," *IEEE Transactions on Components and Packaging Technologies*, 28, 3, September, pp. 457–466.

Ma, H. (2007). "Characterization of lead-free solders for electronic packaging," Auburn University Dissertation (Doctor of Philosophy), Auburn, AL, May 2007.

Norris, K.C., and Landzberg, A.H. (1969). "Reliability of Controlled Collapse Interconnections," *IBM Journal of Research Development*, 13, 266–271.

Pang, J.H.L., and Chong, D.Y.R. (2001). Flip chip on board solder joint reliability analysis using 2-D and 3-D FEA models, *IEEE Trans. Advanced Packaging*, 24(4), 499–506.

Pang, J.H.L., Xiong, B.S., and Low, H. (2004). Creep and fatigue characterization of lead free 95.5Sn-3.8Ag-0.7Cu solder, *Proc. 54th ECTC,* June 2004, pp. 1333–1337.

Shi, X.Q., Pang, H.L.J., Zhou, W., and Wang, Z.P. (1999). A modified energy-based low cycle fatigue model for eutectic solder alloy, *Scripts Material,* 41(3), 289–296.

Singh, N. C. (2006). "Thermo-Mechanical Reliability Models for Life Prediction of Area Array Electronics in Extreme Environments," Auburn University Thesis (Master of Science), Auburn, AL, May.

Strauss, R. (1998). *SMT Soldering Handbook, (Second edition).* Maryland Heights, MO: Elsevier/Newnes.

Suo, Z. (2004). A continuum theory that couples creep and self-diffusion, *J. Appl. Mechanics,* 71, 646–651.

Syed, A.R. (2004). Accumulated creep strain and energy density based thermal fatigue life prediction models for SnAgCu solder joints, *Proc. 54th ECTC,* June 2004, pp. 737–746.

Syed, A.R. (1995). Creep crack growth prediction of solder joints during temperature cycling: An engineering approach, *Trans. ASME,* 117 (June), pp. 116–122.

Teng, C.C., Cheng, Y.K., Rosenbaum, E., and Kang, S.M. (1997). iTEM: A temperature-dependent electromigration reliability diagnosis tool, *IEEE Trans. Computer-Aided Design of Integrated Circuits and Systems,* 16(8), 882–893.

Tu, K.N. (2003). *Solder Joint Technology: Materials, Properties, and Reliability.* Berlin, Germany: Springer.

Tu, K.N. (2003), Recent advances on electromigration in very-large-scale-integration of interconnects, *J. Appl. Phys.,* 94, 5451–5473.

Tu, K.N. (1994). Irreversible processes of spontaneous whisker growth in bimetallic Cu-Sn thin-film reactions, *Phys. Rev. B,* 49(3), 2030–2034.

Tunga, K., Pyland, J., Pucha, R.V., and Sitaraman, S.K. (2003). Field-use conditions vs. thermal cycles: A physics-based mapping study, *Proc. 53rd Electronic Components and Technology Conference,* May 27–30, 2003, pp. 182–188.

Wiese, S., Schubert, A., Walter, H., Dudek, R., Feustel, F., Meusel E., and Michel, B. (2001). Constitutive behaviour of lead-free solders vs. lead-containing solders–Experiments on bulk specimens and flip-chip joints, *Proc. 51st Electronic Components and Technology Conference,* pp. 890–902.

Zahn, B.A. (2002). Finite element based solder joint fatigue life predictions for a same die size-stacked-chip scale-ball grid array package, *SEMICON West, International Electronics Manufacturing Technology (IEMT) Symposium,* pp. 274–284.

Zahn, B.A. (2003). Solder joint fatigue life model methodology for 63Sn37Pb and 95.5Sn4Ag0.5Cu materials, *Proc. 53rd Electronic Components and Technology Conference,* May 27–30, 2003, pp. 83–94.

6

Lead-Free Electronic Reliability: Higher Temperature

> Crack growth has been invariably seen to occur within the bulk of the solder joint. No separation between solder and intermetallic layers has been observed. Crack initiation has been observed at the device corners for both capacitors and resistors, as well as at the toe of the joint for the resistors. This has been in agreement with the results of finite element modeling (FEM) simulations, which has indicated maximum creep strain accumulation at these locations.

The development of lead-free (Pb-free) solders faces several challenges because they are not just drop-in substitutes for traditionally used leaded solders. These challenges may be related to the solder melt temperature, processing temperature, wettability, mechanical and thermomechanical fatigue behaviors, etc. The knowledge base on leaded solders gained by experience over a long period of time is not directly applicable to Pb-free solders. As a result, a database for modeling reliability predictions of Pb-free solders is not currently available. Most Pb-free solder developments for electronic applications are aimed at arriving at suitable alloy compositions. The tin–silver (Sn–Ag) alloy system, with or without small alloy additions such as copper (Cu), is believed to have significant potential. The binary Sn–Ag eutectic temperature is 221°C, and the ternary Sn–Ag–Cu (SAC) eutectic temperature is 217°C, both being reasonably higher than the Sn–Ag binary eutectic temperature of 182°C. Although the processing parameters must be modified to accommodate this increase in eutectic temperature, such solders provide higher service temperature capability to the solder joints. Results of several studies on such solders are recently reported in the published literature.

6.1 Computer Coupling of Phase Diagrams and Thermochemistry and Differential Thermal Analysis

Tin–silver alloys were developed as an alternative to Pb-containing solders. Typical commercial Sn–Ag alloys contain between 3 and 5% Ag.

Tin–silver solders are used for high-temperature, high-reliability inter-connect applications. Solder joints using Sn–Ag alloys maintain better high-temperature strength than Sn–Pb solders. The design of industrial processes requires reliable thermodynamic data. CALPHAD (Computer Coupling of Phase Diagrams and Thermochemistry) aims to promote computational thermodynamics through the development of models to represent thermodynamic properties for various phases that permit prediction of properties of multicomponent systems from those of binary and ternary subsystems, critical assessment of data and their incorporation into self-consistent databases, development of software to optimize and derive thermodynamic parameters, and the development and use of databanks for calculations to improve our understanding of various industrial and technological processes.

Tin-rich alloys in the Sn–Ag–Cu (SAC) system are being studied for their potential as Pb-free solders. Thus, the location of the ternary eutectic involving L, (Sn), Ag3Sn, and Cu6Sn5 phases is of critical interest. Phase diagram data in the Sn-rich corner of the SAC system have been measured. The ternary eutectic was confirmed to be at a composition of 3.5 wt% Ag, 0.9 wt% Cu at a temperature of 217.2 ± 0.2°C (2σ). A thermodynamic calculation of the Sn-rich part of the diagram from the three constituent binary systems and the available ternary data using the CALPHAD method was conducted. The best fit with the experimental data was 3.66 wt% Ag and 0.91 wt% Cu at a temperature of 216.3°C. Using the thermodynamic description to obtain the enthalpy-temperature relation, the DTA signal is simulated and used to explain the difficulty of liquidus measurements in these alloys. Here, differential thermal analysis (DTA) is a "fingerprinting" technique that provides information on the chemical reactions, phase transformations, and structural changes that occur in a sample during a heat-up or a cool-down cycle. DTA measures the differences in energies released or absorbed, and also the changes in the heat capacity of materials as a function of temperature. All materials behave in certain predictable ways when exposed to certain temperatures, so the resulting DTA curve is an indication of the materials and phases present in the sample. For example, DTA is used to indicate the relative magnitude of reactions and phase transitions of ceramic materials or batches that can be destructive so that safe drying and firing schedules can be determined. DTA identifies the temperature regions and the magnitude of critical events during a drying or firing process such as drying, binder burnout, carbon oxidation, sulfur oxidation, structural clay collapse, alpha- to beta-quartz transition, carbonate decompositions, recrystallizations, melting and cristobalite transitions, melting, solidification or solidus temperature, glass transition temperatures (T_g), Curie point, energy of reaction, heat capacity, and others.

Although current approaches for high-temperature Pb-free solders tend to evaluate the alloy systems with a melting temperature of about 220°C, a significant increase in service temperature capability may not result as a consequence of the thermomechanical behavior of such systems. Many industrial projects deal with Pb-free solders for applications with a service temperature of 150°C. For example, in automotive under-the-hood applications, there is significant interest from designers to mount the electronic circuit boards on the engine manifold. This will significantly decrease the amount of wiring and minimize several complications in the electrical circuitry. Similar conditions also exist in aerospace and defense applications.

Automotive under-the-hood electronics often have to withstand temperatures up to 175°C in combination with harsh environmental conditions. Several LBGA (low-profile ball grid array) configurations, defined in terms of both materials systems and structural parameters, were submitted to thermal cycling testing in automotive under-the-hood environmental conditions and event detection to assess the second-level assembly (i.e., package-to-board interconnect) reliability. Package types tested included body size (23 mm × 23 mm, 15 mm × 15 mm, 10 mm × 12 mm, and 12 mm × 16 mm), ball pitch (1.0 mm) and diameter (0.76 mm and 0.5 mm), and other (die size, substrate thickness, etc.) parameter selections of materials systems or structural parameters, selected to enhance the robustness of such packages in this severe environment, and, consequently, the likelihood of meeting customer requirements in this market segment. These LBGA configurations are intended to house dedicated ICs as well as memory devices. Results indicate that not only are some configurations very robust, but there are also clues to package structural variables that could lead to significant enhancement in solder joint reliability performance.

A more severe environment experienced by the solder joint is in high-current/high-temperature applications such as in an automotive alternator or rectifier. At present there is no suitable and economical substitute for the high-Pb solders used for such applications. Solders based on Sn–Au alloys are cost prohibitive in large-scale automotive-type manufacturing situations. This will be an area that will be the focus of a significant number of investigations in the near future. Because regular electronic solders, such as in electronic components or computers, do not experience such severe environments, this aspect of Pb-free solders has not been addressed to date.

6.2 Solder Joint Integrity in Accelerated Thermal Cycling 0°C to 90°C

Incorporation of dispersoids to improve the mechanical and thermomechanical behavior of solders, even in leaded solder systems, has been an avenue that has been pursued to improve the service temperature capability without significantly altering the processing parameters. Research has been performed to address the role of dispersoids, their suitability, methods to incorporate them in the solder, etc. These research efforts are related to *solder joint integrity in accelerated thermal cycling* (ATC) 0°C to 90°C. For example, the Bi-43%Sn eutectic solder alloy is a candidate for Pb-free replacement of the widely used Pb–Sn solders. The alloy exhibits a microstructural instability at elevated service temperatures, causing extensive coarsening and nonuniformity in microstructure and severe creep deformation. The addition of insoluble dispersoid particles using a novel magnetic distribution technique has been found to significantly reduce the coarsening and onset of tertiary creep. With improved microstructural stability, the useful service range of Bi–Sn eutectic solder can be raised to a higher homologous temperature.

The thermal profile shown in Figure 6.1 was simulated for the analysis of solder joint reliability when the flip chip was subjected to 0°C to 90°C ATC. The cycle starts at room temperature (25°CC), has a dwell of 12 minutes at the extreme temperatures of 0°C and 90°C, and the ramp time between the extreme temperatures is 3 minutes; therefore the total time period for the cycle is 30 minutes.

Simulations for the eighteen different configurations were run for this profile and the simulation results are listed in Table 6.1. Because the Kester underfill material has a T_g (glass transition temperature) in the range of 70°C to 75°C, it behaves very differently as compared to the other underfill materials whose T_g is above 90°C. The normalized inelastic strain energy density (ISED) listed in Table 6.1 has been evaluated considering the T_g of Kester above 90°C so that the T_g does not affect the relative comparison of the reliability of various configurations. The effect of T_g of the underfill material on solder joint reliability was analyzed separately.

Higher temperatures are being studied as well. For example, the solder joint reliability of ceramic chip resistors assembled to laminate substrates

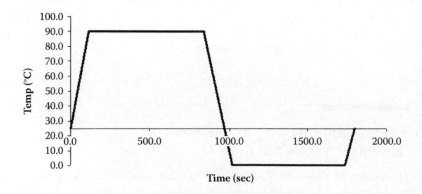

FIGURE 6.1
0°C to 90°C accelerated thermal cycle profile.

TABLE 6.1

Simulation Results for 0°C to 90°C Accelerated Thermal Cycle

Test Case	Solder Composition	Underfill	Bump Gap Height	Bump Size (mils)	Normalized Inelastic Strain Energy Density per Cycle
1	Eutectic	Kester	Low	4	3.354
2	Eutectic	FP 4526	High	4	0.931
3	Eutectic	Kester	High	4	2.678
4	Eutectic	Kester	Mid	4	2.780
5	Eutectic	FP 4526	Low	4	1.093
6	Eutectic	3M 3667	Mid	4	2.710
7	Eutectic	FP 4549	High	4	1.489
8	Eutectic	FP 4550	Mid	3	1.548
9	Eutectic	3M 3667	Mid	3	2.732
10	Pb-Free	FP 4549	High	4	1.653
11	Pb-Free	FP 4549	Mid	3	1.720
12	Pb-Free	Kester	High	4	2.834
13	Pb-Free	Kester	Mid	3	3.204
14	Pb-Free	3M 3667	High	4	2.346
15	Pb-Free	3M 3667	Mid	3	2.940
16	Pb-Free	FP 4526	Low	4	1.115
17	Pb-Free	FP 4526	Mid	4	1.030
18	Pb-Free	FP 4526	High	4	1.000

has been a long-time concern for systems exposed to harsh environments such as those found in automotive and aerospace applications. This is due to a combination of the extreme temperature excursions experienced by the assemblies, along with the large coefficient of thermal expansion (CTE) mismatches between the alumina bodies of the chip resistors and the glass-epoxy composites of the printed circuit boards (PCBs). These reliability challenges are exacerbated for components with larger physical size (distance to neutral point) such as the 2512 resistors used in situations where higher voltages or currents lead to power dissipations up to 1 Watt.

Here, the thermal cycling reliability of several 2512 chip resistor Pb-free solder joint configurations has been investigated. In an initial study, a comparison was made between the solder joint reliabilities obtained with components fabricated with both Sn–Pb and pure Sn solder terminations. In the main portion of the reliability testing, two temperature ranges (−40°C to 125°C and −40 to 150°C) and five different solder alloys were examined. The investigated solders included the normal eutectic Sn–Ag–Cu (SAC) alloy recommended by earlier studies (95.5Sn-3.8Ag-0.7Cu), and three variations of the Pb-free ternary SAC alloy that include small quaternary additions of bismuth (Bi) and indium (In) to enhance fatigue resistance. For each configuration, thermal cycling failure data were gathered and analyzed using two-parameter Weibull models to rank the relative material performances. The obtained Pb-free results were compared to data for standard 63Sn-37Pb joints. In addition, a second set of thermally cycled samples was used for microscopy studies to examine crack propagation, changes in the microstructure of the solders, and intermetallic growth at the solder-to-PCB pad interfaces.

6.2.1 Effect of Solder Joint Material Composition

Solders, in general, operate at high homologous temperature ranges. During turning on and off operations of the electrical circuitry when heat-up or cool-down occurs, they also experience low cycle thermomechanical fatigue due to stresses that develop as a consequence of CTE mismatches among the solder, substrate, and components. Mechanical vibration of other entities to which the electronic components are mechanically attached, such as automotive engines, can create higher frequency vibrational fatigue conditions. An automobile/tank hitting a pothole/major obstruction, or an airplane making a landing can impose impact loading on the solder joint.

Although a fine-grained microstructure may be beneficial for mechanical fatigue considerations, it may not be ideal for creep resistance because creep deformation at the service temperature (high homologous temperature for the solder alloys) will be by grain boundary sliding. In addition to these opposing requirements, the highly inhomogeneous as-joined solder joint microstructure coarsens during service. This aging process causes growth of the solder/substrate interface intermetallic layer and coarsening of microstructural constituents within the solder joint. Such evolving microstructure continuously alters the mechanical properties of the solder joint, resulting in significant hurdles in reliability prediction modeling. The presence of fine, stable, compatible dispersoids at the grain boundaries can retard coarsening, enhance mechanical fatigue behavior, and decrease the creep rate by decreasing grain boundary sliding tendency by keying the grain boundaries.

It is observed from Figure 6.2 that the Pb-free solder bumps accumulate a higher ISED per cycle as compared to the eutectic solder bumps for all the configurations. Thus, there is a reversal in the trend for the 0°C to 90°C ATC as compared to the trend for the 0°C to 70°C ATC. This, however, does not necessarily mean higher reliability for the Pb-free solder bumps, as the reliability prediction depends on the damage relationship, which is different for both solder compositions. This means that despite the higher ISED value, the flip chip with Pb-free bumps might exhibit higher reliability. Although one would prefer a strong solder joint for creep resistance purposes, it may not be ideal in electronic applications. If the solder in the joint is not able to dissipate the stresses that develop, the failure of the electronic components will result. One would prefer a reasonably strong and pliable solder joint. Although these two requirements appear to be mutually exclusive, both of

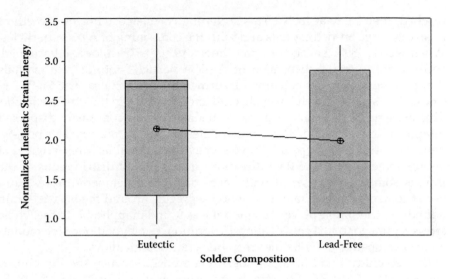

FIGURE 6.2
Normalized inelastic strain energy density versus solder composition.

them can be satisfied by appropriate microstructural engineering of solders with suitable dispersoids.

Eutectic Au–Sn solder is increasingly used in high-reliability and/or high-temperature applications where conventional Sn–Pb and Pb-free solders exhibit insufficient strength, creep resistance, and other characteristics. These applications include hybrid microelectronics (particularly flip chips), MEMS (micro-electro-mechanical systems), optical switches, light-emitting diodes (LEDs), laser diodes, RF devices, and hermetic packaging for commercial, industrial, military, and telecommunications applications. For most of these applications, Au–Sn solder provides the additional benefit of not requiring flux during reflow, thus significantly reducing the potential for contamination and pad corrosion. However, the materials and processing considerations are substantially different from those for conventional solders. Many companies struggle with issues such as poor solder flow, excessive void formation, variable reflow temperature (arising from off-eutectic compositions), heterogeneous phase distribution, and others, all contributing to development delays, process yield loss, and field reliability issues.

6.2.2 Effect of Underfill Composition

Aging effects and creep behavior along with microstructure changes in eutectic Pb–Sn and the Pb-free solder alloy Sn-3.9Ag-0.6Cu were studied. The room-temperature aged Sn-3.9Ag-0.6Cu alloy continually age-softened due to the growth of relatively large Sn-rich crystals. The 180°C aged Sn-3.9Ag-0.6Cu alloy initially age-softened, and the minimum flow strength was reached after 1 day at 180°C. This softening correlated with the growth of relatively large Sn-rich crystals and with the coarsening of Ag-3Sn particles. When aged at 180°C beyond one day, the Sn-3.9Ag-0.6 Cu alloy age-hardened, corresponding to the dispersion of Ag-3Sn particles into Sn-rich crystals that previously had not contained intermetallic precipitates. The Sn-3.9Ag-0.6Cu alloy showed much lower absolute creep rates than the Pb–Sn eutectic. This extreme increase in creep resistance may result from finely dispersed intermetallic compound (IMC) precipitates in the Sn matrix. The beta-Sn dendrites after creep appear to have some orientational features (aligned at approximately 45° to the flow direction). The size and distributions of the IMC is somewhat coarsened with increasing creep temperature. A number of coarsened precipitates of Cu6Sn5 segregate around the beta-Sn grain boundary. The simple power-law model was found inapplicable to the whole stress regime for both types of alloy. In addition, two hyperbolic-sine models were developed to describe the creep behavior of both alloys.

The reliability trend for the underfill composition when the flip chip is subjected to 0°C to 90°C ATC remains the same as in case of 0°C to 70°C ATC. Figure 6.3 illustrates the trend of increased value of normalized ISED from capillary to reflow encapsulant for various flip-chip configurations.

Lead–free solder has seen increasing use in interconnection systems for electronic packages due to environmental and business concerns. Recommended by NEMI (National Electronics Manufacturing Initiative), the Sn-3.9Ag-0.6Cu alloy, and a number of close variations, have already seen commercial use. From previous studies, it was found that Sn–Ag–Cu solder alloys exhibit intermetallic structures within the Sn matrix. These intermetallic structures

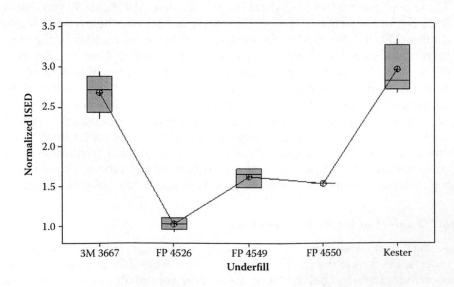

FIGURE 6.3
Normalized inelastic strain energy density versus underfill composition.

are composed of Ag3Sn and Cu6Sn5. The intermetallic microstructures do not take on a constant form; rather, their morphology varies according to the thermal process history. Aging studies of Sn–Ag–Cu solder alloys have been performed in order to evaluate microstructural changes, especially focusing on the growth kinetics of the intermetallic layers that formed between the different solder alloys and substrates. In these studies, phase coarsening during thermomechanical cycling was observed. There are a number of research papers published on Sn–Ag binary alloy creep behavior, and a good part of the data was measured with bulk samples. For example, research has shown that steady-state creep rates are controlled by the dislocation-pipe diffusion in the Sn matrix for the Sn-3.5Ag alloy. In addition, aging affects the secondary creep rates. However, the creep data available on Sn–Ag–Cu ternary alloys are very scarce. This does not allow for reliable predictions by finite element analysis (FEA) in electronics packages.

6.2.3 Effect of Bump Gap Height

The methodology employed to enhance the service performance of the solder joint should not affect the well-laid-out current metallurgical processes in electronics manufacturing. It is essential that the incorporation of dispersoids does not significantly alter the solderability, solder/substrate wetting characteristics, melting temperature, etc. Solder development for high-current/high-temperature service environments is not currently being pursued vigorously. Electromigration, which can cause a significant decrease in solder reliability, could probably be controlled by fine dispersoids in the grain boundaries. Fine dispersion of Cu atoms in aluminum causes a significant decrease in electromigration in computer circuitry. However, such aspects will not be addressed in this chapter due to the lack of available information in the literature.

It is observed from Figure 6.4 that the normalized ISED is much more sensitive to bump height in the case of Pb-free solder bumps as compared to eutectic solder bumps. However, the effect of bump height on the solder joint reliability for both eutectic and Pb-free solder in the case of the 0°C to 90°C ATC remains the same as for the 0°C to 70°C ATC. The ISED value decreases with an increase in the bump gap height. Microstructures of the solders before thermal fatigue tests can be classified into a single-crystal like and a fine-grain type. However, this classification, which affects the amount of thermal strain by the anisotropic nature of beta-Sn, cannot accurately describe the observed thermal fatigue lives. On the other hand, the Vickers microhardness of the solders, which resulted from fine dispersoids, showed a good relationship with observed thermal fatigue endurance.

Researchers have investigated composite solders obtained by adding Sn-3.0Ag-0.5Cu (SAC) nanoparticles to conventional eutectic Sn-58Bi solder paste. The microstructure analysis and measurement of the Vickers microhardness have been carried out. Utilizing the self-developed consumable-electrode

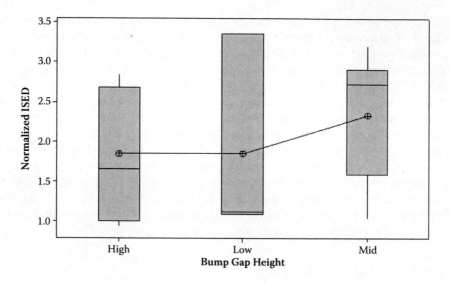

FIGURE 6.4
Normalized inelastic strain energy density versus the effect of bump gap height.

direct current arc (CDCA) technique, Sn-3.0Ag-0.5Cu nanoparticles with an average particle size between 20 and 80 nm are prepared. The reinforced Pb-free Sn–Bi solder was prepared by thoroughly blending the nanometer-sized SAC particles into the eutectic Sn–Bi solder paste. The SAC-reinforced Sn–Bi composite solder paste was printed onto ENIG (electroless nickel immersion gold)/Cu metallized substrate and reflowed in a conventional reflow oven. After reflow, the morphology of the as-solidified reinforced composite solder was observed by means of SEM (scanning electron microscope) and TEM (transmission electron microscope). The Vickers microhardness measurements indicated that the addition of SAC nanoparticles enhanced the overall strength of the eutectic solder, and the results agreed well with the theory of dispersion strengthening.

6.2.4 Effect of Bump Size

The effect of bump size has been widely investigated for Pf-free electronics. A consortium of eleven industrial corporations in the United States carried out the "lead-free solder project," led by the National Center for Manufacturing Sciences (NCMS), in order to evaluate alternatives to eutectic Sn–Pb solder. Five alloys emerged from the down-selection as viable candidates to replace 80% of the currently used near-eutectic Sn–Pb solder. It was found during the NCMS project that Sn-58Bi eutectic, Sn-3.4Ag-4.8Bi, and Sn-3.5Ag eutectic solders performed substantially better than the eutectic Sn–Pb solder in certain surface-mount applications. The project also showed that Sn-58Bi and Sn-3.4Ag-4.8Bi had fatigue lives comparable to or better than eutectic Sn–Pb, possessing similar bulk properties at room temperature and very high tensile and yield strengths combined with moderate to high elongation. However, the NCMS project concluded that a "drop-in" replacement for eutectic Sn–Pb solder had still not been identified by the end of the "1992–1996" four-year project.

As shown by Figure 6.5, the 4-mil bump exhibits higher reliability than the 3-mil bump, which was also true in the case of the 0°C-70°C ATC. Also, the bump of larger size will have larger crack length so it will take longer for the crack to propagate throughout the solder joint before the failure occurs. Thus, both effects will be coupled and will add up to the solder joint reliability. The UK Government–Department of Trade & Industry (DTI)-funded "lead-free soldering" project focused on the suitability for electronic assembly and supply potential during their Pb-free alloy selections. Five solder alloys, namely, Bi-42Sn, Sn-9Zn, Sn-5Sb, Sn-3.5Ag, and Sn-0.7Cu, were tested and the Bi-42Sn, Sn-9Zn, and Sn-5Sb solders were rejected for various reasons, such as poor mechanical properties, corrosion of the Zn phase, or the melting temperature being too high. The Sn-3.5Ag and Sn-0.7Cu solder alloys showed better performance among the five selected alloys.

The JIPE project was undertaken by Senju Metals, Alpha Metals Japan, Nihon Handa, Ishikawa Metals, etc. in Japan. This Pb-free soldering project focused on the Sn–Ag–Bi (Bi, 7–25%) alloy. It was found that any solder alloy containing more than 7% Bi was very brittle, and fillet lifting became a serious concern even though the melting temperature of the solder alloy was

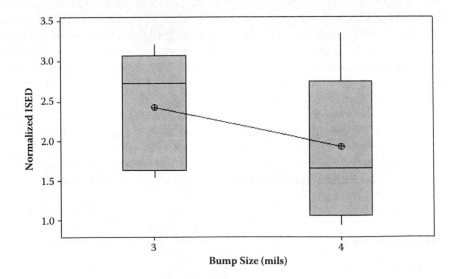

FIGURE 6.5
Normalized inelastic strain energy density versus bump size.

relatively low. The Sn–Ag–Cu alloys with or without a few percent bismuth proved useful. It was also discovered that Sn–Zn and Sn–Cu alloys had great potential for use as solder alloys. Studies dealing with these alloys are currently in progress.

6.2.5 Summary of Simulation Results

For solder joint integrity in accelerated thermal cycling, simulation results for 0°C to 70°C and 0°C to 90°C accelerated thermal cycle are shown by Table 6.2.

- For the 0°C to 70°C ATC, the eutectic solders had higher ISEDs as compared to the Pb-free solder.
- In the case of the 0°C to 90°C ATC, a reversal in the previous trend was observed.
- The Pb-free solder had a higher ISED as compared to the eutectic solder.
- A decrease in gap height increased the ISED, and this had greater impact on the Pb-free solders for both eutectic and Pb-free solders.
- An increase in the bump size reduced the ISED.

Figure 6.6 shows the normalized ISED versus thermal cycle. FEM utilizes finite element analysis to calculate the ISED accumulated per cycle during thermal or power cycling. The strain energy density is then used with crack growth data to calculate the number of cycles to initiate a crack, and the number of cycles to propagate cracks through a joint. The measured crack growth data are correlated with the calculated ISED per cycle in the solder.

The composite approach was developed mainly to improve the service performance, including service temperature capability. In other words, the basic purpose of this methodology was to engineer and stabilize a fine-grained microstructure, and homogenize solder joint deformation, so as to improve the mechanical properties of the solder joint, especially creep and thermomechanical fatigue resistance. Also, the added reinforcements do not change the melting point of the solder matrix, but may effectively increase the service temperature of the base solder materials by improving the creep or thermomechanical fatigue properties of the solder matrix.

6.3 Field Profiles

Several efforts have been made to improve the comprehensive properties of Pb-bearing solders using a composite approach. Microstructural analysis as

TABLE 6.2

Simulation Results for 0°C to 70°C and 0°C to 90°C Accelerated Thermal Cycles

Test Case	Temperature (°C)	Solder Composition	Underfill	Bump Gap Height	Bump Size (mils)	Normalized ISED
1	0–70	Eutectic	Kester	Low	4	4.415
2	0–70	Eutectic	FP 4526	High	4	1.208
3	0–70	Eutectic	Kester	High	4	3.543
4	0–70	Eutectic	Kester	Mid	4	3.716
5	0–70	Eutectic	FP 4526	Low	4	1.423
6	0–70	Eutectic	3M 3667	Mid	4	3.487
7	0–70	Eutectic	FP 4549	High	4	1.964
8	0–70	Eutectic	FP 4549	Mid	3	2.024
9	0–70	Eutectic	3M 3667	Mid	3	3.55
10	0–70	Pb-Free	FP 4549	High	4	1.795
11	0–70	Pb-Free	FP 4549	Mid	3	1.89
12	0–70	Pb-Free	Kester	High	4	3.444
13	0–70	Pb-Free	Kester	Mid	3	3.89
14	0–70	Pb-Free	3M 3667	High	4	3.047
15	0–70	Pb-Free	3M 3667	Mid	3	3.483
16	0–70	Pb-Free	FP 4526	Low	4	1.045
17	0–70	Pb-Free	FP 4526	Mid	4	1.012
18	0–70	Pb-Free	FP 4526	High	4	1
19	0–90	Eutectic	Kester	Low	4	3.354
20	0–90	Eutectic	FP 4526	High	4	0.931
21	0–90	Eutectic	Kester	High	4	2.678
22	0–90	Eutectic	Kester	Mid	4	2.78
23	0–90	Eutectic	FP 4526	Low	4	1.093
24	0–90	Eutectic	3M 3667	Mid	4	2.71

(Continued)

TABLE 6.2 (CONTINUED)

Simulation Results for 0°C to 70°C and 0°C to 90°C Accelerated Thermal Cycles

Test Case	Temperature (°C)	Solder Composition	Underfill	Bump Gap Height	Bump Size (mils)	Normalized ISED
25	0–90	Eutectic	FP 4549	High	4	1.489
26	0–90	Eutectic	FP 4550	Mid	3	1.548
27	0–90	Eutectic	3M 3667	Mid	3	2.732
28	0–90	Lead-Free	FP 4549	High	4	1.653
29	0–90	Lead-Free	FP 4549	Mid	3	1.72
30	0–90	Lead-Free	Kester	High	4	2.834
31	0–90	Lead-Free	Kester	Mid	3	3.204
32	0–90	Lead-Free	3M 3667	High	4	2.346
33	0–90	Lead-Free	3M 3667	Mid	3	2.94
34	0–90	Lead-Free	FP 4526	Low	4	1.115
35	0–90	Lead-Free	FP 4526	Mid	4	1.03
36	0–90	Lead-Free	FP 4526	High	4	1

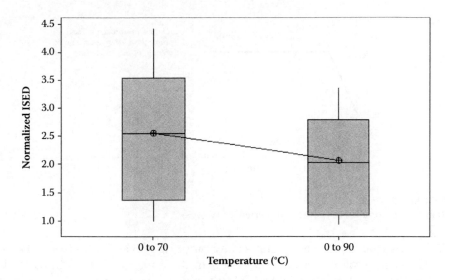

FIGURE 6.6
Normalized inelastic strain energy density versus thermal cycle.

well as mechanical testing of such composite solders have been reported. Certain composite solders did show improved mechanical properties sought by electronic/automobile industries. Here, three different field profiles are provided for the simulation and analysis of solder joint reliability. Simulations were run with all three profiles. Again, the ISED was used as the relative index of damage.

6.3.1 Field Profile-1

Research studies have investigated the micro-characterization of composite solders. These composite solders were primarily prepared by mixing Cu6Sn5 (10, 20, 30 wt%), Cu3Sn (10, 20, 30 wt%), Cu (7.6 wt%), Ag (4 wt%), or Ni (4 wt%) particles with the eutectic Sn-37Pb solder paste. The microstructural features of these bulk composite solder specimens showed Cu–Sn, Ag–Sn, and Ni–Sn intermetallics developed in the composite solders around Cu, Ag, and Ni particles, respectively. A Cu6Sn5 layer formed around Cu3Sn particles in the Cu3Sn-reinforced composite solder, while no more new intermetallic formed in the Cu6Sn5 particle-reinforced composite solder. The microstructural analysis showed good bonding of the particulate reinforcements to the solder matrix, suggesting that the resulting composite solders might exhibit enhanced strength.

Field Profile-1 is an averaged thermal duty cycle for 4 hours of write and 30 minutes of read. The temperature profile shown in Figure 6.7 was simulated for the analysis. This thermal cycle has a 4-hour dwell time at the high temperature of 87°C, two quick ramp-ups of 1 second from 63°C to 87°C and

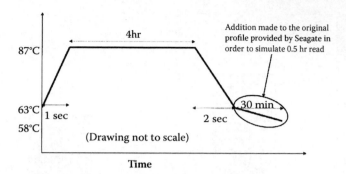

FIGURE 6.7
Temperature profile of the averaged thermal duty cycle 1.

2 seconds from 87°C to 63°C, and a slow ramp-up of 30 minutes from 63°C to 58°C. Because the Kester material has T_g in the range of 70 to 75°C, which is below the operating range of the field profile, the configurations with Kester underfill material test cases were run with the actual material model, that is, T_g in the range of 70 to 75°C and the ideal case with T_g above the operating range.

Intermetallic formation at the solder/copper interface was studied for the above composite solder samples aged at 140°C for 0 to 16 days. Intermetallic formation near the Cu substrate was greatly affected by these particle additions. Ag and Au retarded and Ni suppressed the formation of Cu3Sn and enhanced the growth of Cu6Sn5, as compared to pure solder with the Cu substrate. Addition of Cu-containing particles to the solder resulted in a decrease in both the Cu6Sn5 and Cu3Sn interface intermetallic thickness relative to the pure solder. This effect was believed to be due to the particles acting as Sn sinks. Similar studies were carried out with aging temperatures of 110°C to 160°C for 0 to 64 days. The Cu-containing reinforcements resulted in increased activation energies for Cu6Sn5 formation and decreased activation energies for Cu3Sn formation as compared to pure solder. The activation energy for Cu3Sn formation decreased relative to the eutectic solder for the Ag and Au composite solders even though less Cu3Sn was formed at the substrate interface. Both Ni and Pd drastically reduced the Cu3Sn thickness and increased the Cu6Sn5 thickness. Two mechanisms were proposed for the effects of Cu-containing particles and Ag particles on the kinetics of intermetallic formation. First, the particles act as Sn sinks that remove Sn from the solder and decrease the amount of Sn available for reaction at the interface. Second, the particles reduce the solder cross-sectional area available for Sn diffusion, which also reduces the amount of Sn available at the interface for reaction.

6.3.2 Field Profile-2

Dispersion-strengthened in-situ composite solders of Sn–Pb–Ni and Sn–Pb–Cu alloys containing 0.1- to 1.0-μm dispersoids/reinforcements were produced by induction melting and inert gas atomization. It was found that, upon reflow of the solder specimens, the fine spherical dispersoids in rapidly solidified Sn–Pb–Cu alloys coarsen to >1- μm platelets; however, the dispersoids in Sn–Pb–Ni alloys remain spherical and also stable with a size <1 μm. The difference in stability among dispersoids in Cu- and Ni-containing solders was explained on the basis of the difference in solubilities and diffusivities of Cu and Ni in the Sn–Pb matrix. These composite solders showed an increase of 25% to 180% in yield stress and 20% to 80% in the modulus values compared to eutectic Sn-37Pb solder.

The temperature profile of the averaged thermal duty cycle 2, Field Profile-2, is shown in Figure 6.8.

Another type of dispersion-strengthened composite solder was formulated by adding 2.2 wt% Ni3Sn4 intermetallic particles into the Sn-40Pb solder matrix. Mechanical alloying, a solid-state high-energy milling process developed for superalloy manufacture, provided the means to process such dispersion-strengthened solders. The presence of Ni3Sn4 dispersoids resulted in a smaller grain size in the as-cast microstructure and after aging at 100°C for 29 hours. The subsequent study of Cu9NiSn3 intermetallic particle-reinforced Sn-40Pb composite solder showed an increase in the strain to failure in shear of 40%, while the ultimate shear strength remained essentially unchanged. This was claimed to be an indication of improved fatigue resistance because it was believed that fine, uniformly dispersed phases would stabilize microstructures by pinning grain boundary dislocations and by restricting grain boundary motion.

6.3.3 Field Profile-3

Other composite solders were prepared by mixing 3 vol% of 10-nm-sized Al_2O_3 powders or 3 vol% of 5-nm-sized TiO_2 powders with 35-nm-sized eutectic Sn-37Pb solder powder. Nanosized, nonreacting, noncoarsening oxide particles formed uniform coatings of solder after repeated plastic deformation for rearrangement of the particles. A three-order decrease in

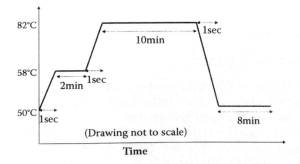

FIGURE 6.8
Temperature profile of the averaged thermal duty cycle 2.

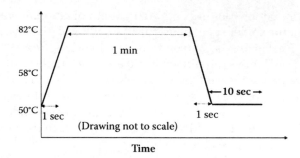

FIGURE 6.9
Temperature profile of the averaged thermal duty cycle 3.

the steady-state creep rate was achieved using this approach. Such composite solder was found to be much more creep resistant than the control sample, the eutectic Sn-80Au solder. This has great significance in replacing the conventional high-melting-point (278°C) Sn-80Au solder for its creep-resistant applications such as in optical or opto-electronic packaging.

The temperature profile of the averaged thermal duty cycle 3, Field Profile-3, is shown in Figure 6.9.

Researchers have reported that, with properly controlled porosity, Cu6Sn5 particle-reinforced eutectic Sn-37Pb solders exhibited twice the yield strength without significant ductility loss. It was also shown in that study that the creep rate of the composite solder was nearly an order of magnitude less than that of unreinforced solders. The boundary-layer fracture behavior was studied using single shear lap specimens using the same composite solder. The specimens failed, as shear fracture ran in from opposite edges about 10 μm inside the interfaces. These boundary-layer fractures were characterized, and a fracture model was developed. Composite strengthening was shown to significantly improve the ductility, creep life, and properties associated with improved reliability and creep–fatigue life.

6.4 Relative Damage Index

The development of a solder-joint-stress-based relative damage index has been investigated to establish a method for damage equivalency. Damage proxies for failure mechanisms at the Cu-to-solder, solder-to-PCB, and Cu-to-package substrate have been developed. A relative damage index based on transient strain history has been developed to show the damage progression with respect to number of drops to failure, as investigated using damage superposition (Miner's rule). The relative damage index gives an indication of the comparative damage inflicted per cycle on the solder joint by the three different field profiles. Figure 6.10 shows the plot of the relative damage index for the three field profiles for different flip-chip configurations. It is observed that for most of the configurations, the damage done by Field Profile-1 is the highest, and the damage done by Field Profile-3 is lowest. Table 6.3 shows simulations results for evaluation of relative damage index of the three field profiles. The reason behind this lies in the difference in

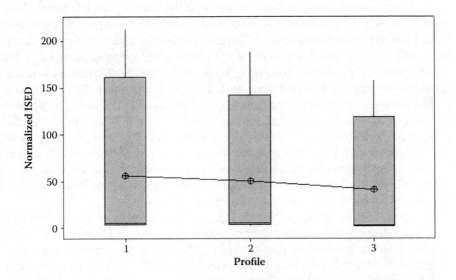

FIGURE 6.10
Normalized inelastic strain energy density versus profile.

TABLE 6.3

Simulations Results for Evaluation of Relative Damage Index of the Three Field Profiles

Underfill	Solder	Bump Size (mils)	Gap	Profile	Normalized ISED
FP 4549	Pb-free	4	High	1	2.7
3M 3667	Pb-free	4	High	1	4.61
Kester	Pb-free	4	High	1	5.13
Kester	Pb-free	4	High	1	213.36
FP 4549	Pb-free	4	High	2	2.61
3M 3667	Pb-free	4	High	2	4.73
Kester	Pb-free	4	High	2	5.23
Kester	Pb-free	4	High	2	188.13
FP 4549	Pb-free	4	High	3	1
3M 3667	Pb-free	4	High	3	2.24
Kester	Pb-free	4	High	3	2.32
Kester	Pb-free	4	High	3	157.31

dwell period among the profiles. Field Profile-1 has the maximum dwell period as compared to the other two profiles. Because the solder joint creeps during the dwell period, which then results in the accumulation of the non-linear plastic work, the ISED accumulated by the solder joint due to Field Profile-1 is maximum. This means that Field Profile-1 will have the highest acceleration factor in terms of life in cycles; however, this does not mean that Field Profile-1 will have the highest acceleration factor in terms of the time.

The life prediction in Chapter 5 shows that the solder joint life in terms of time is lowest for Field Profile-3. In spite of the lowest damage index value of Field Profile-3, it accumulates the highest damage in terms of time because the ISED accumulates itself in the solder joint after every successive cycle; and because the cycle time for Field Profile-3 (1 minute, 12 seconds) is minimum when compared to Field Profile-1 (4 hours, 30 minutes, 3 seconds) and Field Profile-2 (20 minutes, 3 seconds), the component is subjected to a much higher number of cycles in the case of Field Profile-3 as compared to the other two profiles. This results in the accumulation of higher amounts of ISED in the case of Field Profile-3 for the same period of time, which means that the components would take a greater number of cycles to fail when subjected to Field Profile-3, but still in terms of time the Field Profile-3 will cause failures much earlier than for the other two profiles.

As a result of recent research activities, a large number of Pb-free solders have been developed and a number of patent applications have been filed for various alloy compositions. Although not all these alloys are commercially available, there is still a wide range from which to choose. The most convenient way to separate the available Pb-free alloys is to consider their melting temperatures. Some typical examples of Pb-free solders under research and development are listed in Table 6.4, along with their melting temperature ranges.

In summary, there are several Pb-free alloys available; however, there is no universal drop-in replacement for leaded solders identified thus far. At the

TABLE 6.4

Examples of Pb-Free Solder Alloys under Research and Development

Category	Alloy System	Composition (wt%)	Melting Range (°C)
Low melting temp. (<180°C)	Sn–Bi	Sn-58Bi	138
	Sn–In	Sn-52In	118
Melting temp. (183–200°C) equivalent to eutectic Sn–Pb solder	Sn–Zn	Sn-9Zn	198.5
	Sn–Bi–Zn	Sn-8Zn-3Bi	189–199
	Sn–Bi–In	Sn-20Bi-10In	143–193
Mid-range melting temp. (200–230°C)	Sn–Ag	Sn-3.5Ag	221
	Sn–Cu	Sn-0.7Cu	227
	Sn–Ag–Cu	Sn-3.8Ag-0.7Cu	217
High melting temp. (230–350°C)	Sn–Sb	Sn-5Sb	232–240
	Sn–Au	Sn-80Au	280

moment, the most promising alloys for general electronic soldering appear to be those based on Sn-3.5Ag and Sn–Ag–Cu. Other alloys with potential are Sn-0.7Cu and Sn–Ag–Bi. The need for higher process temperatures is another important technological issue that must be addressed when companies change over to Pb-free soldering because higher process temperatures will impact existing soldering technology in such key areas as materials stability/reliability, equipment reliability, and higher energy cost issues. No applicable Pb-free solder alloys can be used at high service temperatures except the expensive Sn-80Au solder, which will not be practical for large-scale soldering operations. How to increase the service temperature of existing Pb-free solders for such applications or arrive at new solder compositions is one of the serious problems that Pb-free solder researchers have yet to address.

Bibliography

Amagai, M., Watanabe, M., Omiya, M., Kishimoto, K., and Shibuya, T. (2002). Mechanical characterization of Sn–Ag based lead-free solders, *Micoelectronics Reliability*, 42, 951–966.

Antolovich, S.D., and Antolovich, B.F. (1996). *An Introduction to Fracture Mechanics in ASM Handbook. 19 Fatigue and Fracture,* Materials Park, OH: ASM International®, 1996.

Arora, N.D., Raol, K.V., Schumann, R., and Richardson, L.M. (1996). Modeling and extraction of interconnect capacitances for multilayer VLSI circuits, *IEEE Trans. Computer Aided Design of Integrated Circuits and Systems,* 15(1), 58–66.

Bilotti, A.A. (1974). Static temperature distribution in IC chips with isothermal heat sources, *IEEE Trans. Electron Devices,* ED-21(March), pp. 217–226.

Black, J.R. (1969). Electromigration failure models in aluminium metallization for semiconductor devices, *Proc. IEEE,* 57(9), 1587–1594.

Blech, I.A., and Herring, C. (1976). Stress generation by electromigration, *Appl. Phys. Lett.,* 29, 131–133.

Chen, C., and Liang, S.W. (2007). Electromigration issues in lead-free solder joints, *J. Mater. Sci.,* 18, 259–268.

Darveaux, R. (2000). Effect of simulation methodology on solder joint crack growth correlation, *Proc. 50th ECTC,* May 2000, pp. 1048–1058.

Darveaux, R. (1996). How to use finite element analysis to predict solder joint fatigue life. *Proc. VIII Int. Congress on Experimental Mechanics,* Nashville, TN, June 10–13, 1996, pp. 41–42.

Drake, J. L. (2007). "Thermo-Mechanical Reliability Models for Life Prediction of Ball Grid Arrays on Cu-Core PCBs in Extreme Environments," Auburn University Thesis (Master of Science), Auburn, AL, August.

Dreezen, G., Deckx, E., and Luyckx, G. (2003). Solder alternative: Electrically conductive adhesives with stable contact resistance in combination with non-noble metallization. *CARTS Europe 2003,* pp. 223–227.

Gale, W.F., and Totemeier, T.C. (2004). *Smithells Metals Reference Book, (8th edition)*. Maryland Heights, MO: Elsevier.

Galyon, G.T. (2003). *Annotated Tin Whisker Bibliography*, Herndon, VA: NEMI publication, July.

Hunter, W.R. (1997). Self-consistent solutions for allowed interconnect current density. I. Implication for technology evolution, *IEEE Trans. Electron Devices*, 44(2), 304–309.

Hunter, W.R. (1997). Self-consistent solutions for allowed interconnect current density. II. Application to design guidelines, *IEEE Trans. Electron Devices*, 44(2), 310–316.

Lall, P., Hariharan, G., Tian, G., Suhling, J., Strickland, M., and Blanche, J. (2006). "Risk Management models for flip-chip electronics in extreme environments." *ASME International Mechanical Engineering Congress and Exposition*. Chicago, IL, November, pp. 1–13.

Lall, P., Singh, N, Suhling, J., Strickland, M., and Blanche, J. (2005). "Thermo-mechanical reliability tradeoffs for deployment of area array packages in harsh environments," *IEEE Transactions on Components and Packaging Technologies*, 28, 3, September, pp. 457–466.

Lall, P., Islam, N., Shete, T., Evans, J., Suhling, J., and Gale, S. (2004). "Damage mechanics of electronics on metal-backed substrates in harsh environments," *Proceedings of 54th Electronic Components & Technology Conference*, IEEE, Las Vegas, NV, June 1–4, pp. 704–711.

Lall, P., Islam, N., Suhling, J., and Darveaux, R. (2003). "Model for BGA and CSP reliability in automotive underhood applications," in *Proc. 53rd Electronic Components Technology Conf.*, May 27–30, 2003, pp. 189–196.

Lall P., Pecht, M., and Hakim, E. (1997). *Influence of Temperature on Microelectronic and System Reliability*. Boca Raton, FL: CRC Press.

O'Connor, P. and Kleyner, A. (2012). *Practical Reliability Engineering, (5th edition)*. Hoboken, NJ: Wiley.

Pang, J.H.L., and Chong, D.Y.R. (2001). Flip chip on board solder joint reliability analysis using 2-D and 3-D FEA models, *IEEE Trans. Advanced Packaging*, 24(4), 499–506.

Pang, J.H.L., Xiong, B.S., and Low, H. (2004). Creep and fatigue characterization of lead free 95.5Sn-3.8Ag-0.7Cu solder, *Proc. 54th ECTC*, June 2004, pp. 1333–1337.

Shi, X.Q., Pang, H.L.J., Zhou, W., and Wang, Z.P. (1999). A modified energy-based low cycle fatigue model for eutectic solder alloy, *Scripts Material*, 41(3), 289–296.

Strauss, R. (1998). *SMT Soldering Handbook, (Second edition)*. Maryland Heights, MO: Elsevier/Newnes.

Suo, Z. (2004). A continuum theory that couples creep and self-diffusion, *J. Appl. Mechanics*, 71, 646–651.

Syed, A.R. (2004). Accumulated creep strain and energy density based thermal fatigue life prediction models for SnAgCu solder joints, *Proc. 54th ECTC*, June 2004, pp. 737–746.

Syed, A.R. (1995). Creep crack growth prediction of solder joints during temperature cycling: An engineering approach, *Trans. ASME*, 117 (June), pp. 116–122.

Teng, C.C., Cheng, Y.K., Rosenbaum, E., and Kang, S.M. (1997). iTEM: A temperature-dependent electromigration reliability diagnosis tool, *IEEE Trans. Computer-Aided Design of Integrated Circuits and Systems*, 16(8), 882–893.

Tu, K.N. (2003). *Solder Joint Technology: Materials, Properties, and Reliability*. Berlin, Germany: Springer.

Tu, K.N. (2003), Recent advances on electromigration in very-large-scale-integration of interconnects, *J. Appl. Phys.*, 94, 5451–5473.

Tu, K.N. (1994). Irreversible processes of spontaneous whisker growth in bimetallic Cu–Sn thin-film reactions, *Phys. Rev. B*, 49(3), 2030–2034.

Tunga, K., Pyland, J., Pucha, R.V., and Sitaraman, S.K. (2003). Field-use conditions vs. thermal cycles: A physics-based mapping study, *Proc. 53rd Electronic Components and Technology Conference*, May 27–30, 2003, pp. 182–188.

Wiese, S., Schubert, A., Walter, H., Dudek, R., Feustel, F., Meusel E., and Michel, B. (2001). Constitutive behaviour of lead-free solders vs. lead-containing solders–Experiments on bulk specimens and flip-chip joints, *Proc. 51st Electronic Components and Technology Conference*, pp. 890–902.

Zahn, B.A. (2002). Finite element based solder joint fatigue life predictions for a same die size-stacked-chip scale-ball grid array package, *SEMICON West, International Electronics Manufacturing Technology (IEMT) Symposium*, pp. 274–284.

Zahn, B.A. (2003). Solder joint fatigue life model methodology for 63Sn37Pb and 95.5Sn4Ag0.5Cu materials, *Proc. 53rd Electronic Components and Technology Conference*, pp. 83–94, May 27–30, 2003.

7

Fatigue Design of Lead-Free Electronics and Weibull Distribution

The new breed of environmentally friendly solder alloys contain high proportions of tin, typically over 90%, and differ markedly from the traditional Sn–Pb eutectic solder alloy containing 37% lead. The key microstructural feature in tin-based alloys is the presence of small intermetallic phase-particles, most notably those that form between the alloying components in the popular Sn–Ag, Sn–Cu, and Sn–Ag–Cu systems.

The analysis of cracks within structures is an important application if the damage tolerance and durability of structures and components are to be predicted. As part of the engineering design process, engineers have to assess not only how well the design satisfies the performance requirements, but also how durable the product will be over its life cycle. Often, cracks cannot be avoided in structures; however, the fatigue life of the structure depends on the location and size of these cracks. To predict the fatigue life for any component, a fatigue life and crack growth analysis must be performed. The solder joint failure data of a given component type, component finish, and solder alloy were evaluated using Weibull analysis. Lifetime analysis of the second-level fatigue failures was performed using two-parameter Weibull distributions with median rank regression. The characteristic lifetime, or Eta, is the number of cycles at which 63.2% of the sample set has failed.

7.1 Fatigue Design of Lead-Free Electronics

As part of the engineering design process, engineers have to assess not only how well the design satisfies the performance requirements, but also how durable the product will be over its life cycle. A major cause of failure is the growth of cracks that grow due to fatigue loadings to the point where the product fails. This chapter describes how to predict crack growth, which combines the best features of boundary element and finite element technology. The crack and the crack growth are simulated using the boundary element model (BEM), and the finite element model (FEM) is used to represent the remaining part of the structure. Examples are presented showing how this can be applied in the FEM environment to a complex three-dimensional fitting. The chapter also demonstrates how technology can be used to provide higher resolution stress data near small details essential for fatigue calculations.

The two-parameter Weibull failure plots indicated that the eutectic SAC alloy (95.5Sn-3.8Ag-O.7Cu) recommended by earlier studies has similar reliability to standard 63Sn-37Pb for testing from −40°C to 125°C. However, 63Sn-37Pb joints dramatically outperformed the lead free Sn–Ag–Cu alloy joints for the more extreme −40°C to 150°C testing. This result agrees with earlier observations for CBGA (ceramic ball grid array) and non-underfilled flip chip on laminate assemblies (configurations with high stiffness components, large CTE [coefficient of thermal expansion] mismatches, and large temperature excursions). Solder joints of electronic assemblies are complex elements that cannot be studied using the traditional techniques of structural analysis or fatigue of engineering metals. The acknowledged complexity of the mechanics of solder joints arises from the following:

- The problem is three-dimensional (3-D), with solder joints subjected to a system of distributed, multi-axial forces and moments exerted by the interconnected parts. Even when taking advantage of symmetries, all joints are not equal because of varying distances to the neutral axis of an assembly and variability in joint geometry and metallurgy.

- A solder joint is a multilayered, nonhomogeneous structure. Reflowed solder is sandwiched between thin layers of intermetallic compounds. In the case of Sn–Pb, the solder itself is made up of Pb- and Sn-rich phases with variations in composition near the intermetallic layers (e.g., Sn-depleted regions on the board side due to the formation of Cu–Sn intermetallic compounds during reflow). Moreover, the Sn–Pb microstructure evolves in service. The microstructure coarsens due to thermally activated grain growth, a phenomenon that takes place under stress or at constant temperature.

- The mechanical behavior of solder is highly nonlinear and temperature dependent. Solder creeps readily at ambient temperature (and below), and creep rates increase with temperature.

- Failure of Sn–Pb solder joints is a complex sequence of events involving microstructural coarsening, matrix creep, grain boundary sliding, micro-void formation and linking, crack initiation, and crack growth. In the case of SAC and Sn–Ag solder joints, the damage accumulation process leads to much less coarsening of the microstructure, if any.

- Most often, electrical opens resulting from solder joint failures are intermittent and may be difficult to detect accurately. Electrical continuity may still be maintained when a solder joint is fully cracked because of contacts between asperities on the opposite surfaces of the crack. This may result in hard-to-detect failures and "No Trouble Found" (NTF) diagnoses during troubleshooting.

Despite all this, significant progress has been made in the understanding of Sn–Pb solder joint mechanics, fatigue, and failure. Although the assessment of Sn–Pb assembly reliability is a semi-empirical science, a vast body of knowledge, simplified engineering models, test data, and experimental findings has accumulated that provides useful insight into the mechanical behavior of solder joints of real assemblies.

The fatigue life prediction model, also known as the damage relationship, correlates the life of the solder joint to the strain energy density or the strain (plastic, creep, or total strain) accumulated during each accelerated thermal cycle (ATC) that the package is subjected to. Several strain energy and strain-based life prediction models for the eutectic solder have been used by researchers in the past and published in the literature. The life prediction models developed for the Pb-free solder are very few.

7.2 Weibull Distribution for Life Testing Data Analysis

Reliability and failure data, both from life testing and from in-service records, are often modeled by the Weibull or Lognormal distributions so as to be able to interpolate and extrapolate results. As the Weibull and Lognormal distributions are not from the same mathematical family, we are unable to derive mathematical relationships between their shape and scale parameters directly.

Research tries to use a simulation method to determine their approximate relationship. The simulation was carried out using Weibull analysis software, which uses the Monte Carlo method to generate samples from the Weibull (or the Lognormal) distribution and can then fit the Lognormal (or the Weibull) line to the sample points. Researchers have found approximate equations of shape parameters (or scale parameters) for complete sample sizes vary from 3 to 99 and from 100 to 1,000 between both distributions. For $\eta = 1$; $\beta = 0.5, 1, 3, 5$; and $n = 10, 25, 50, 100$, the residuals between true ρ (shape parameter of the Lognormal) and estimated ρ are less than ±0.003. Again under the same conditions as above, the residuals between true θ (scale parameter of the Lognormal) and estimated θ are less than ±0.0005. The approximate equations are simple, direct, and their accuracy is acceptable. They are, therefore, recommended for use.

In statistics, there are many distributions that can be used in various areas. Such distributions include the Normal, Chi-squared, Exponential, Rayleigh, Weibull, Erlang, Gamma, Extreme-Value, Lognormal, and others. The relationships between the various distributions are shown in Figure 7.1 where the

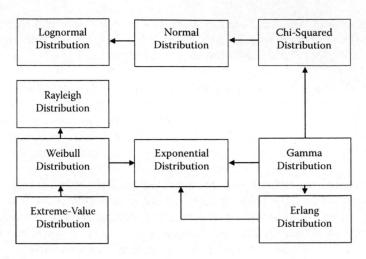

FIGURE 7.1
The relationship between various distributions.

direction of each arrow represents going from the general to a special case. Here we focus only on the Weibull and Lognormal distributions; both distributions can, interestingly, have skewed frequency curves (the Weibull can be positively or negatively skewed but the Lognormal can only be positively skewed). These two distributions are not in the same mathematical family, yet each can be made to fit life data with acceptable accuracy. Several writers in the field mention that the Lognormal is a distribution that competes against the Weibull distribution in the area of reliability.

7.2.1 Mathematical Model

The Weibull probability distribution has three parameters: η, β, and t_0. It can be used to represent the failure probability density function (PDF) with time, so that

$$f_w(t) = \frac{\beta}{\eta}\left(\frac{t - t_0}{\eta}\right)^{\beta - 1} e^{\left[-\left(\frac{t - t_0}{\eta}\right)^{\beta}\right]} \tag{7.1}$$

for $\eta > 0$, $\beta > 0$, $t > 0$, $-\infty < t_0 < t$, where β is the shape parameter (determining what the Weibull PDF looks like) and is positive, and η is a scale parameter (representing the characteristic life at which 63.2% of the population can be expected to have failed), which is also positive; t_0 is a location (or shift or threshold) parameter (sometimes called a guarantee time, failure-free time, or minimum life). t_0 can be any real number. If $t_0 = 0$, then the Weibull distribution is said to be two-parameter.

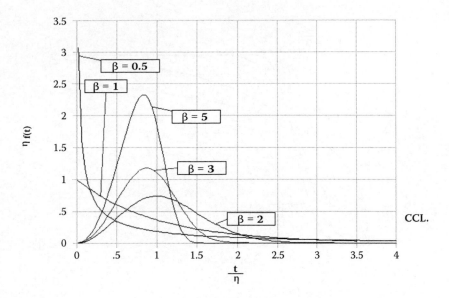

FIGURE 7.2
The Weibull PDF.

Figure 7.2 shows the diverse shape of the Weibull PDF with $t_0 = 0$ and various values of η and β (= 0.5, 1, 2, 3, 5). Note that figures are all based on the assumption that $t_0 = 0$.

The cumulative distribution function (CDF), denoted by $F(t)$, is

$$F_w(t) = 1 - e^{\left[-\left(\frac{t-t_0}{\eta}\right)^{\beta}\right]}$$ (7.2)

Figure 7.3 shows the Weibull CDF with $t_0 = 0$ and various values of η and β (=0.5, 1, 2, 3, 5). All curves intersect at the point of (1, 0.632), the characteristic point for the Weibull CDF.

The PDF for three-parameter Lognormal distribution is

$$f_L(t) = \frac{\rho}{\sqrt{2\pi}(t-t_0)} e^{\left\{-\frac{\left[\ln\left(\frac{t-t_0}{\theta}\right)^{\rho}\right]^2}{2}\right\}}$$ (7.3)

for $\theta > 0$, $\rho > 0$, $t > 0$, $-\infty < t_0 < t$, where ρ is the shape parameter, θ is the scale parameter, and t_0 is the location parameter. The units of ρ, θ, and t_0 are the same as in the Weibull case. The Lognormal is said to be a two-parameter distribution when $t_0 = 0$. Figure 7.4 shows the diverse shape of the Lognormal PDF with $t_0 = 0$ and various values of θ and ρ (= 0.4, 0.8, 1.6, 2.5, 4).

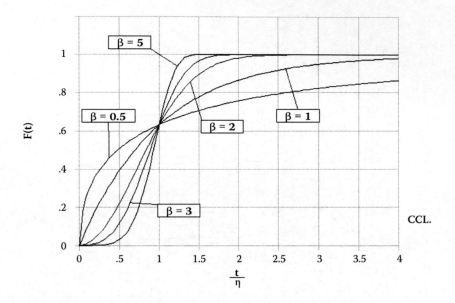

FIGURE 7.3
The Weibull CDF.

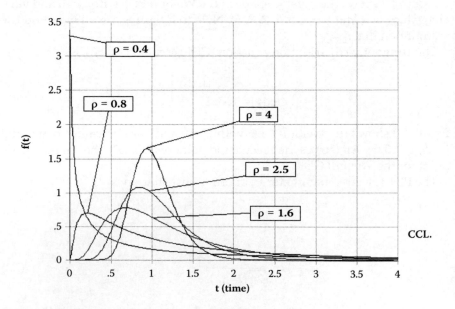

FIGURE 7.4
The Lognormal PDF.

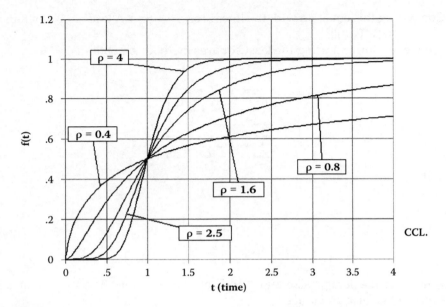

FIGURE 7.5
The Lognormal CDF.

The corresponding Lognormal CDF is the integral of the PDF from 0 to time-to-failure t. It can be written in terms of the standard Normal CDF as

$$F_L(t) = \Phi\left[\ln\left(\frac{t-t_0}{\theta}\right)^\rho\right] \tag{7.4}$$

where $\Phi(\bullet)$ is the CDF of the standard Normal distribution defined as

$$\Phi(z) = \int_{-\infty}^{z} \frac{1}{\sqrt{2\pi}} e^{\left(-\frac{\mu^2}{2}\right)} d\mu \tag{7.5}$$

where $\mu = \ln\theta$. $\Phi(\bullet)$ is tabulated in O'Conner (2012). Figure 7.5 shows the Lognormal CDF with $t_0 = 0$ and various values of θ and ρ ($= 0.4, 0.8, 1.6, 2.5, 4$). It is clear that all curves intersect at the point (1, 0.5), the characteristic point of the Lognormal CDF.

7.2.2 Data Fitting Method

The linear form of the resulting Weibull CDF can be represented by a rearranged version of Equation (7.2), that is,

$$\ln t = \frac{1}{\beta}\ln\ln\left(\frac{1}{1-F_W(t)}\right) + \ln\eta \tag{7.6}$$

Comparing this equation with the linear form $y = Bx + A$ leads to $y = \ln t$ and $x = \ln \ln\{1/[1-F_W(t)]\}$.

If we minimize it using the least squares method, we obtain

$$\hat{\beta} = \frac{n\sum\limits_{i=1}^{n} x_i^2 - \left(\sum\limits_{i=1}^{n} x_i\right)^2}{n\sum\limits_{i=1}^{n} x_i y_i - \sum\limits_{i=1}^{n} x_i \sum\limits_{i=1}^{n} y_i}, \tag{7.7}$$

and

$$\hat{\eta} = e^{\left(\frac{\sum\limits_{i=1}^{n} y_i}{n} - \frac{\sum\limits_{i=1}^{n} x_i}{n\beta}\right)} \tag{7.8}$$

where n is the sample size and $\wedge$ indicates an estimate. The mathematical expressions for x_i and y_i are

$$x_i = \ln \ln \left[\frac{1}{1 - F_W\left(t_i\right)}\right] \tag{7.9}$$

and

$$y_i = \ln t_i \tag{7.10}$$

$F(t_i)$ can be estimated using Bernard's formula, which is a good approximation of the median rank estimator. Researchers have used Bernard's median rank because it shows the best performance and it is the most widely used to estimate $F(t_i)$. The procedure for ranking complete data is the following:

1. List the time-to-failure data from small to large.
2. Use Bernard's formula to assign median ranks to each failure.
3. Estimate the β and η by Equations (7.7) and (7.8), respectively.

The Lognormal CDF, when plotting against appropriate probability axes, appears linear and, therefore, can be represented by a rearranged version of Equation (7.3) as

$$\ln t = \frac{z_p}{\rho} + \ln \theta \tag{7.11}$$

Comparing this with the linear form $y = Bx + A$ leads to $y = \ln t$ and $x = z_p$.

The same least squares as used for the Weibull distribution yields

$$\hat{\rho} = \frac{n\sum\limits_{i=1}^{n} x_i^2 - \left(\sum\limits_{i=1}^{n} x_i\right)^2}{n\sum\limits_{i=1}^{n} x_i y_i - \sum\limits_{i=1}^{n} x_i \sum\limits_{i=1}^{n} y_i}, \tag{7.12}$$

and

$$\hat{\theta} = e^{\left(\frac{\sum\limits_{i=1}^{n} y_i}{n} - \frac{\sum\limits_{i=1}^{n} x_i}{n\hat{\rho}}\right)} \tag{7.13}$$

where $x_i = z_p$ and $y_i = \ln t_i$, and $z_p = \Phi^{-1}(z_p)$ is the percentile of the standard Normal CDF that is widely tabulated. Again, $F(t_i) = \Phi(z_p)$ can be estimated using Bernard's formula. The same ranking procedure as was used for the Weibull distribution was also used for the Lognormal distribution. The only difference is that Step 3 in the ranking procedures above should be replaced by "estimate the ρ and θ by Equations (7.12) and (7.13)."

7.2.2.1 Relationship between Shape Parameters of Weibull and Lognormal Distributions

Because the Weibull and Lognormal distributions are not from the same mathematical family, we are unable to derive mathematical relationships between their shape and scale parameters directly. However, we can use a simulation method to determine their approximate relationship. The simulation was carried out using Weibull analysis software, which uses the Monte Carlo method to generate samples from the Weibull (or the Lognormal) distribution and can then fit the Lognormal (or the Weibull) line to the sample points. For example, twenty-five random samples were generated from W(1, 1) and the resulting data fitted to both the Weibull and Lognormal lines using MRR (median rank regression) as shown in Figures 7.6 and 7.7, respectively. As we can see from these two graphs, W(1, 1) corresponds to L(0.577, 0.822), where the 0.822 is the value of ρ, the reciprocal of σ, and muAL = θ and sdF = e^σ = 3.377. Therefore, the approximate relationships between the shape parameters and scale parameters of the Weibull and Lognormal distributions can be determined.

Following this procedure, we can find other sample sizes, from 3(1) 100(10) 1000, the corresponding values of β, ρ and η, θ. The relationship between the shape parameters of the two distributions is studied first. Figure 7.8 shows the relationship between ρ and n for different values of β using MRR. It is clear that for all β, the relationship between ρ and n is nonlinear.

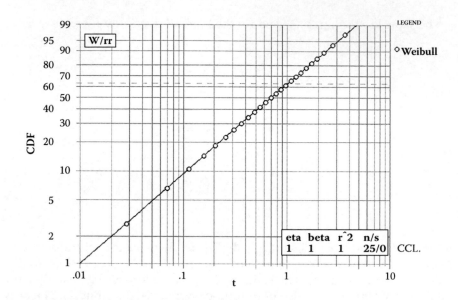

FIGURE 7.6
Samples from the Weibull (1, 1) and fitted to the Weibull line.

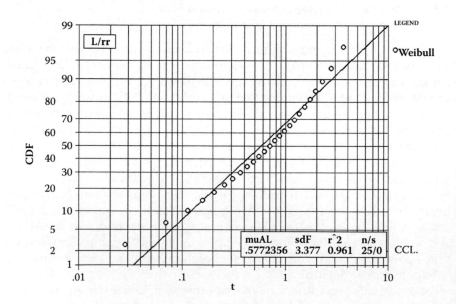

FIGURE 7.7
Samples from the Weibull (1, 1) and fitted to the Lognormal line.

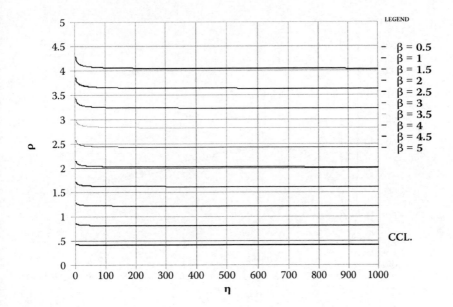

FIGURE 7.8
The relationship between ρ and n for different values of β using MRR.

With the Weibull analysis software (using the MRR method) and by trial and error, general equations for the relationships between ρ and β for different sample sizes can be obtained.

For complete sample sizes, n, from 3 to 99, the relationship is represented by

$$\rho = \frac{\beta}{2.235939 - e^{\left(\frac{0.546524}{n+5.23201}\right)}} \tag{7.14}$$

For complete sample sizes, n, from 100 to 1,000, the relationship is represented by

$$\rho = \frac{\beta}{2.241233 - e^{\left(\frac{1.522754}{n+58.32657}\right)}} \tag{7.15}$$

Equations (7.14) and (7.15) show that for a given sample size n, the approximate relationship between ρ and β is a straight line passing through the origin, both equations being of the form $\rho = c\,\beta + 0$, where c is a constant.

The residuals for estimating ρ using Equations (7.14) and (7.15) are shown in Figures 7.9 and 7.10, respectively.

Particular values of the residuals of ρ for different sample sizes and shape parameters used for the Monte Carlo simulation are shown in Table 7.1. From Figures 7.9 and 7.10, and Table 7.1 it is clear that the maximum

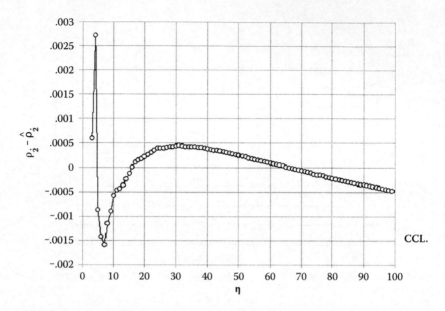

FIGURE 7.9
The residuals for estimating ρ using Equation (7.14).

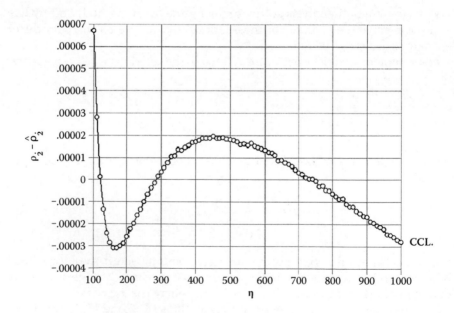

FIGURE 7.10
The residuals for estimating ρ using Equation (7.15).

TABLE 7.1

The Residuals between True ρ and Estimated ρ (using MRR)

n	η	β	ρ	Estimated ρ	Residual
10	1	0.5	0.4166	0.4169	−0.0003
25	1	0.5	0.4108	0.4106	0.0002
50	1	0.5	0.4080	0.4078	0.0001
100	1	0.5	0.4060	0.4060	0.0000
10	1	1	0.8332	0.8337	−0.0006
25	1	1	0.8216	0.8212	0.0004
50	1	1	0.8159	0.8157	0.0003
100	1	1	0.8120	0.8120	0.0001
10	1	3	2.4995	2.5012	−0.0017
25	1	3	2.4648	2.4637	0.0012
50	1	3	2.4477	2.4470	0.0008
100	1	3	2.4361	2.4359	0.0002
10	1	5	4.1658	4.1687	−0.0029
25	1	5	4.1081	4.1061	0.0020
50	1	5	4.0796	4.0783	0.0013
100	1	5	4.0602	4.0599	0.0003

residual for estimating ρ is less than ±0.003. Table 7.1 gives corresponding values of β and ρ, together with their residuals, for sample sizes of $n = 10$, 25, 50, and 100.

However, Equations (7.14) and (7.15) were derived from an "ideal" situation. That is, the life data came from the Weibull distribution and fitted the Weibull line exactly. Such "ideal" data were then fitted to the Lognormal line to produce the corresponding Lognormal parameters. In practice, this "ideal" situation is unlikely to occur. So, it is essential to check how accurate these equations are when used in real situations. Here, a small Monte Carlo simulation was run to check the accuracy of Equations (7.14) and (7.15).

For four different sample sizes (without any suspensions or censoring), points that had been randomly generated from a Weibull distribution were fitted to both a Weibull line and a Lognormal line. For the Weibull line, values of β and η were computed; for the Lognormal line, values of ρ and θ were also computed using the software. Then plotting the values of β on the x-axis and ρ on the y-axis, their relationship was a straight line when the sample size was constant.

The curve-fitting method MRR was used. For each of the sample sizes, the MRR of fitting 100 replications were made. The data of β versus ρ are linear so that the least squares method could be used to find the best fit line. Table 7.2 gives the least squares equations of β versus ρ for 100 replication data. Table 7.3 gives the SE values of the LS equation and of Equation (7.14) or (7.15) for 100 replication data.

TABLE 7.2

The Least Squares Equations of β versus ρ for 100 Replication Data

β	n	LS Equations
0.5	10	$\rho = 0.02179 + 0.79064\beta$
0.5	25	$\rho = 0.02751 + 0.76500\beta$
0.5	50	$\rho = 0.03200 + 0.75102\beta$
0.5	100	$\rho = 0.04783 + 0.71534\beta$
1	10	$\rho = 0.02619 + 0.80429\beta$
1	25	$\rho = 0.03924 + 0.78114\beta$
1	50	$\rho = 0.08183 + 0.73181\beta$
1	100	$\rho = 0.06359 + 0.74814\beta$
3	10	$\rho = 0.09696 + 0.79720\beta$
3	25	$\rho = 0.10504 + 0.75796\beta$
3	50	$\rho = 0.22223 + 0.74056\beta$
3	100	$\rho = 0.19458 + 0.74524\beta$
5	10	$\rho = 0.25115 + 0.78014\beta$
5	25	$\rho = 0.24746 + 0.77018\beta$
5	50	$\rho = 0.47449 + 0.72013\beta$
5	100	$\rho = 0.43847 + 0.72605\beta$

TABLE 7.3

The SE Values of ρ Using the LS Equation and Equation (7.14) or (7.15) for 100 Replication Data

β	n	SE_1 of LS Eq.	SE_2 of Eq. (7.14) or (7.15)	SE_1-SE_2
0.5	10	0.014	0.017	−0.003
0.5	25	0.008	0.010	−0.002
0.5	50	0.007	0.008	−0.001
0.5	100	0.005	0.007	−0.002
1	10	0.026	0.028	−0.002
1	25	0.019	0.021	−0.002
1	50	0.013	0.017	−0.004
1	100	0.011	0.013	−0.002
3	10	0.060	0.072	−0.012
3	25	0.058	0.071	−0.013
3	50	0.044	0.054	−0.010
3	100	0.030	0.036	−0.006
5	10	0.126	0.185	−0.059
5	25	0.095	0.110	−0.015
5	50	0.069	0.098	−0.029
5	100	0.047	0.064	−0.017

As we can see from Table 7.3, the LS equations outperform the approximate equations in all cases. However, as shown in Table 7.2, the LS equations change slightly for different sample sizes and values of β. So, we have not found a general LS equation that applies to all practical cases. The maximum SE of ρ for the LS equation is 0.126, whereas for the approximate equation it is 0.185 when $\beta = 5$, $n = 10$. The difference between SE values is always less than 0.06. The approximate equations are simple and their accuracy is acceptable. They are, therefore, recommended for use.

7.2.2.2 Relationship between Scale Parameters of Weibull and Lognormal Distributions

The procedures for determining the relationship between the scale parameters θ and η of the two distributions are similar to the procedures for the shape parameters in the previous section. However, the relationship between θ and η depends on sample size n and the shape parameter β. Figure 7.11 shows the relationship between θ, n, η, and β.

With Weibull distribution analysis and by trial and error, general equations for the relationships between θ, n, η, and β for different sample sizes can be determined.

For complete sample sizes, n, from 3 to 99, the relationship is represented by

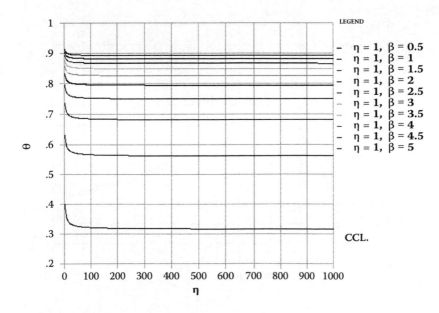

FIGURE 7.11
The relationship between θ, n, η, and β.

$$\theta \cong \left\{ \left[e^{\left(\frac{0.36843}{n+2.653} \right)} - 0.43652 \right]^{\frac{1}{\beta}} \times \eta \right\} \tag{7.16}$$

For complete sample sizes, n, from 100 to 1,000, the relationship is represented by

$$\theta \cong \left\{ \left[e^{\left(\frac{0.64311}{n+26.9355} \right)} - 0.4384 \right]^{\frac{1}{\beta}} \times \eta \right\} \tag{7.17}$$

Equations (7.16) and (7.17) show that for a given sample size n and shape parameter β, the approximate relationship between θ and η is a straight line passing through the origin, both equations having the form $\theta = c *\eta + 0$. The residuals for estimating θ using Equations (7.16) and (7.17) are shown in Figures 7.12 and 7.13, respectively.

Figure 7.14 shows Weibull plots of hours, comparing Darveaux model with CAVE constants and Morrow model for Field Profile-1. Particular values of the residuals of θ for different sample sizes and parameters to be used for the Monte Carlo simulation are shown in Table 7.4. From Figure 7.12 it is clear that the maximum residual for estimating θ is less than ±0.0014. From Table 7.4 it also can be seen that the corresponding values for η and θ depend on n and β. (With constant η and β: the larger the sample size, the

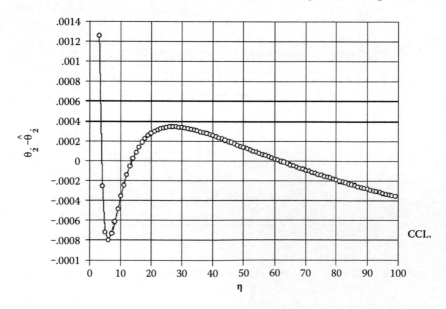

FIGURE 7.12
The residuals for estimating θ using Equation (7.16).

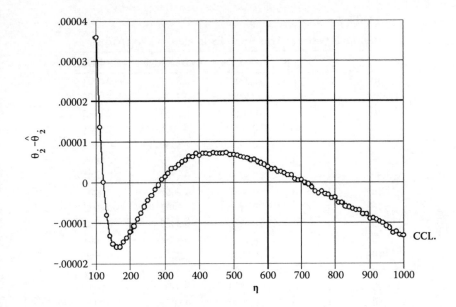

FIGURE 7.13
The residuals for estimating θ using Equation (7.17).

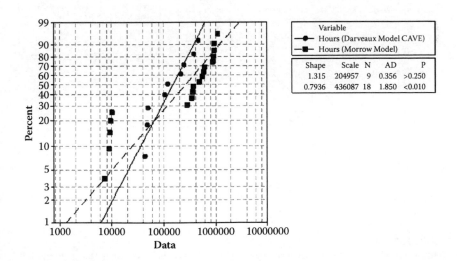

FIGURE 7.14
Weibull plots of hours (Darveaux model with CAVE constants and hours (Morrow model) for Field Profile-1.

TABLE 7.4

Residuals between True θ and Estimated θ (using MRR)

n	η	β	θ	Estimated θ	Residual
10	1	0.5	0.3513	0.3517	−0.0004
25	1	0.5	0.3332	0.3328	0.0004
50	1	0.5	0.3256	0.3255	0.0002
100	1	0.5	0.3212	0.3211	0.0001
10	1	1	0.5927	0.5930	−0.0003
25	1	1	0.5772	0.5769	0.0003
50	1	1	0.5706	0.5705	0.0001
100	1	1	0.5667	0.5667	0.0000
10	1	3	0.8400	0.8402	−0.0002
25	1	3	0.8326	0.8325	0.0001
50	1	3	0.8295	0.8294	0.0001
100	1	3	0.8275	0.8275	0.0000
10	1	5	0.9007	0.9008	−0.0001
25	1	5	0.8959	0.8958	0.0001
50	1	5	0.8939	0.8938	0.0000
100	1	5	0.8926	0.8926	0.0000

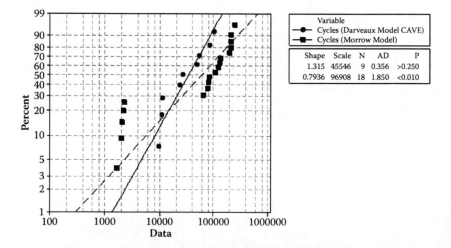

FIGURE 7.15
Weibull plot of cycles (Darveaux model with CAVE constants) and cycles (Morrow model) for Field Profile-1.

smaller the scale parameter θ. With constant n and η: the larger the shape parameter β, the greater the scale parameter θ.)

Figure 7.15 shows Weibull plots of cycles, comparing Darveaux model with CAVE constants and Morrow model for Field Profile-1. Equations (7.16) and (7.17), for estimating θ, were derived from "ideal" conditions. In practice, conditions may not be ideal, so a Monte Carlo simulation (as described in the previous section) was used to assess its accuracy.

Figure 7.16 shows Weibull plots of hours, comparing Darveaux model with CAVE constants and Morrow model for Field Profile-2. Figure 7.17 shows Weibull plots of cycles, comparing Darveaux model with CAVE constants and Morrow model for Field Profile-2. The data of η versus θ is linear so the

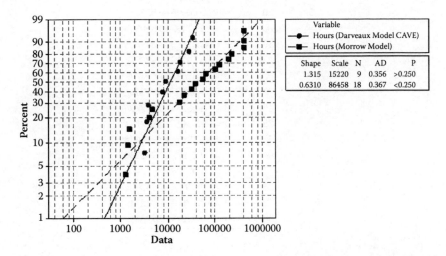

FIGURE 7.16
Weibull plot of hours (Darveaux model with CAVE constants) and hours (Morrow model) for Field Profile-2.

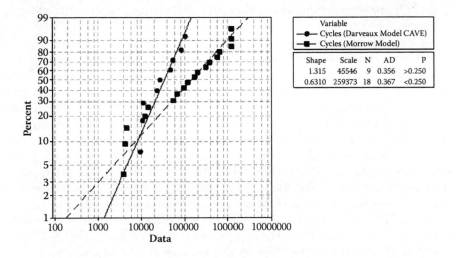

FIGURE 7.17
Weibull plot of cycles (Darveaux model with CAVE constants) and cycles (Morrow model) for Field Profile-2.

TABLE 7.5

Least Squares Equations of η versus θ for 100 Replication Data

β	n	LS Equations
0.5	10	$\theta = 0.01230 + 0.34887\eta$
0.5	25	$\theta = -0.06776 + 0.40481\eta$
0.5	50	$\theta = -0.00995 + 0.33505\eta$
0.5	100	$\theta = -0.01497 + 0.33035\eta$
1	10	$\theta = -0.05680 + 0.64560\eta$
1	25	$\theta = -0.04155 + 0.62384\eta$
1	50	$\theta = -0.03632 + 0.60417\eta$
1	100	$\theta = -0.02558 + 0.59346\eta$
3	10	$\theta = -0.08183 + 0.91275\eta$
3	25	$\theta = -0.02446 + 0.85374\eta$
3	50	$\theta = -0.12891 + 0.95694\eta$
3	100	$\theta = -0.07577 + 0.90124\eta$
5	10	$\theta = -0.07408 + 0.97003\eta$
5	25	$\theta = -0.04632 + 0.94002\eta$
5	50	$\theta = -0.15173 + 1.04274\eta$
5	100	$\theta = -0.03295 + 0.92334\eta$

LS method can be used to find the best fit line. Table 7.5 gives the LS equations of η versus θ for 100 replication data.

Table 7.6 shows the SE of θ using the LS equations and Equations (7.16) and (7.17) for 100 replication data. As we can see from Table 7.6, Equations (7.16) and (7.17) are very competitive, although the LS equations are still better in all cases.

Figure 7.18 shows Weibull plot of hours, comparing Darveaux model with CAVE constants and Morrow model for Field Profile-3. As shown in Table 7.6, the maximum values of SE of θ for the LS equation is 0.1527, and for the approximate equation is 0.1529, when $\beta = 0.5$ and $n = 10$. The maximum difference between the LS equation and Equation (7.16) is −0.0020. As a consequence, predicting values of θ using Equations (7.16) or (7.17) gives us results that are very close to the values from the LS equation.

Figure 7.19 shows Weibull plots of cycles, comparing Darveaux model with CAVE constants and Morrow model for Field Profile-3. Again, an approximate equation such as Equations (7.16) and (7.17) is simple and the standard error is acceptable. It is, therefore, recommended for use.

Figure 7.20 shows boxplot of maximum Von Mises stress from the model with fillet and without fillet. For sample sizes $n = 3(1)99$ and $100(10)1000$, the approximate relationships between the Weibull β and Lognormal ρ were derived by a trial-and-error method. These nonlinear relationships, Equations (7.14) and (7.15), are functions of the sample size n. To see if the equations fit the data well, a small Monte Carlo simulation (100 replications) was made. The

TABLE 7.6

SEs of θ using the LS Equation and Equation (7.16) or (7.17) for 100 Replication Data

β	n	SE_1 of LS Eq.	SE_2 of Eq. (7.16) or (7.17)	SE_1-SE_2
0.5	10	0.1526	0.1529	−0.0003
0.5	25	0.0729	0.0729	−0.0000
0.5	50	0.0508	0.0509	−0.0001
0.5	100	0.0366	0.0372	−0.0006
1	10	0.1038	0.1052	−0.0014
1	25	0.0683	0.0697	−0.0014
1	50	0.0415	0.0419	−0.0004
1	100	0.0334	0.0335	−0.0001
3	10	0.0492	0.0512	−0.0020
3	25	0.0323	0.0325	−0.0002
3	50	0.0195	0.0205	−0.0010
3	100	0.0151	0.0155	−0.0004
5	10	0.0317	0.0326	−0.0009
5	25	0.0198	0.0201	−0.0003
5	50	0.0146	0.0154	−0.0008
5	100	0.0097	0.0100	−0.0003

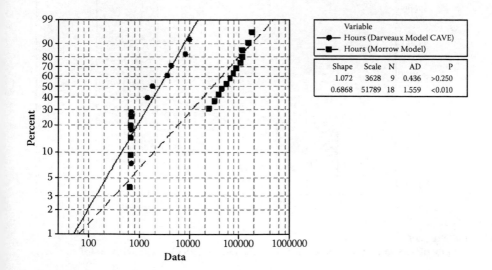

FIGURE 7.18
Weibull plot of hours (Darveaux model with CAVE constants) and hours (Morrow model) for Field Profile-3.

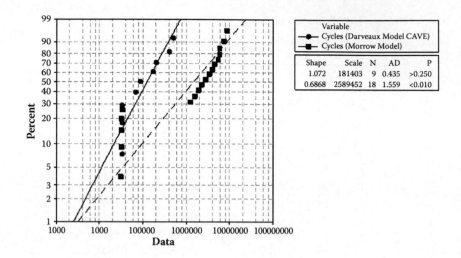

FIGURE 7.19
Weibull plot of cycles (Darveaux model with CAVE constants) and cycles (Morrow model) for Field Profile-3.

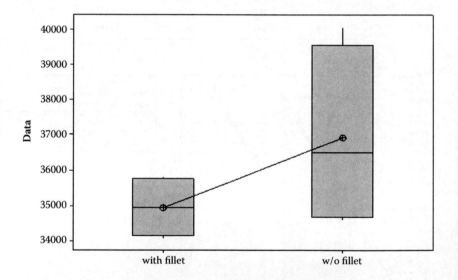

FIGURE 7.20
Boxplot of maximum Von Mises stress from the model with fillet and without fillet.

result shows that the approximate equations fit the data very well if the MRR method is used. The approximate expressions using MRR can readily determine the other distribution's shape parameter if the sample size is known.

Similarly, the relationship between θ and η of the two distributions is also derived by a trial-and-error method. For sample sizes $n = 3(1)99$ and

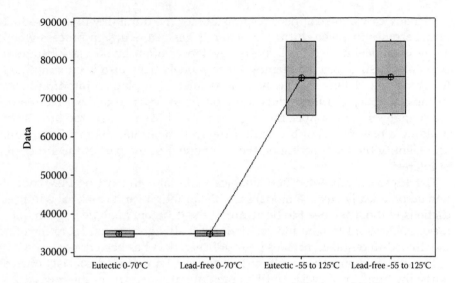

FIGURE 7.21
Boxplot of maximum Von Mises stress based on simulation results for two different ATCs.

100(10)1000 using MRR, approximate relationships were obtained. The non-linear approximate expressions vary according to sample size and shape parameter. The general expressions that represent their relationship are found in Equations (7.16) and (7.17). A small Monte Carlo simulation (100 replications) was made previously. Figure 7.21 shows boxplot of maximum Von Mises stress based on simulation results for two different ATCs.

These four approximate equations were then compared with LS equations. The standard errors for comparison between the approximate equations and LS equations are quite small. As a result, these approximate equations are recommended for use. Henceforth, the approximate expressions derived from the Weibull side. Parameters of the Weibull were used to obtain expressions for parameters of the Lognormal. For further research study, approximate expressions worked in the other direction should be obtained.

7.3 Fatigue Life Prediction Based on Field Profile

The boundary element method is an ideal solution for performing crack growth analysis due to the high accuracy of the stress results computed on the surface of the structure and its ability to represent the stress field singularities near the crack front. In addition, because only the boundary of the body needs to be discretized, the complexity in meshing small details and features to obtain high fidelity data can be significantly reduced.

In many cases, the engineer responsible for the durability of the product only becomes involved during the later stages of the design process when analysis models have already been developed, often involving substantial parts of the structure. Frequently these models have also been simplified to remove the details such as holes and fillets that play an important role in the durability of the product. Previously, in order to make a fatigue or crack analysis, for example, the engineer would have to start the process of building a new model of the area of interest, including the missing detail, and identify the loads acting on the component or the part of the structure of interest.

The approach presented here enables us to take an existing FEM model and automatically create a model suitable for fatigue and crack growth predictions without the need to be aware of the different analytical techniques used. Solder joint fatigue life prediction for all eighteen configurations for the three different field profiles per published data has been done.

The life of a component subject to a fatigue type of load consists of two parts: the number of cycles until a crack initiates and the number of cycles required for the crack to grow to a size where it become unstable. The first so-called life to initiation can normally be predicted based on the stress history, which can be obtained from a suitable stress analysis. However, the prediction of the life once the crack has initiated requires a model that can simulate the crack path and the fracture properties.

7.3.1 Field Profile-1

The solder joint fatigue life prediction for the test cases with eutectic bumps has been done using both Darveaux's model modified with CAVE constants as well as Morrow's model; but because the constants for Darveaux's model are not available for the Pb-free solder bumps, the life prediction was performed using Morrow's model only. Table 7.7 lists the simulation results for all eighteen test cases subjected to Field Profile-1.

In practical applications, a simple cyclic loading is not able to represent the conditions the component or structure will experience during its working life. Therefore, software has been linked to a comprehensive multi-axial loading module that enables real-life loading data to be applied to the model. Another important element is the crack growth model that is used to predict the crack growth rate (da/dn). The analysis code allows a range of fatigue growth laws to be represented (e.g., Paris, NASGRO) and the code is linked to the NASGRO database of fatigue crack growth data for fatigue analysis. This also allows the use of retardation models for crack growth.

In the analysis of a crack, at each iteration a series of new elements are added to the crack front. These elements are formed from the positions of the old crack front and the predicted positions of the new crack front. Where the crack intersects the surface of the component or structures, the surface mesh must be modified to represent the new geometry as the crack grows. In some

TABLE 7.7

Predicted Life for All Eighteen Test Cases for Field Profile-1

Test Case	Scaled Fatigue Life Based on Darveaux's Model with CAVE Constants		Scaled Fatigue Life Based on Morrow's Model	
	(Cycles)	(Hours)	(Cycles)	(Hours)
1	9,700	43,648	1,620	7,288
2	103,923	467,651	254,340	1,144,528
3	11,439	51,476	2,073	9,327
4	10,954	49,295	1,976	8,893
5	85,899	386,546	212,992	958,462
6	26,928	121,177	81,922	368,651
7	54,527	245,371	138,221	621,993
8	48,907	220,080	133,425	600,415
9	23,738	106,819	80,533	362,397
10			124,250	559,125
11			109,929	494,680
12			2,273	10,230
13			2,160	9,721
14			78,006	351,027
15			64,103	288,464
16			203,231	914,539
17			210,716	948,223
18			211,694	952,622

models, the crack re-meshing can be very complex. Consequently, when performing the meshing "manually," time constraints and model complexity can prohibit crack growth beyond a few iterations. The aim of automatic re-meshing in this software is to remove this manual work from the user to allow more detailed assessment of crack growth models to be performed automatically.

7.3.2 Field Profile-2

Table 7.8 summarizes predicted life for all 18 test cases for Field Profile-2. One important extension of this technique is that it is possible to use the same algorithm to add a crack into a model that has simply been defined to perform a stress analysis. This allows the generation of the model without having the task of modeling the crack. The crack can then be added anywhere on the model and a new data file automatically generated containing the crack.

As stated previously, finite element models are developed for a number of purposes, and consequently the engineer responsible for durability may be presented with the task of rebuilding or substantially modifying an existing FEM model in order to obtain the information needed. This is particularly

TABLE 7.8

Predicted Life for All Eighteen Test Cases for Field Profile-2

Test Case	Scaled Fatigue Life Based on Darveaux's Model with CAVE Constants		Scaled Fatigue Life Based on Morrow's Model	
	(Cycles)	(Hours)	(Cycles)	(Hours)
1	9700	3241	3,916	1,305
2	103923	34727	197,111	65,704
3	11439	3823	4,694	1,565
4	10954	3661	4,242	1,414
5	85899	28705	168,581	56,194
6	26928	8998	66,960	22,320
7	54527	18221	116,764	38,921
8	48907	16343	95,725	31,908
9	23738	7932	53,544	17,848
10			672,521	22,4174
11			591,457	197,152
12			14,149	4,716
13			12,029	4,010
14			375,020	125,007
15			310,930	103,643
16			1,231,714	41,0571
17			1,238,548	412,849
18			1,245,464	415,155

onerous in the case of cracks, as they are generally a very small feature compared with the size of the structure or component and typically require special modeling treatment.

The approach presented here overcomes many of these difficulties by combining the technologies in FEM and BEM. Consider the case where the engineer has an FEM model and wishes to predict if a crack will grow if it develops at a particularly highly stressed area. Depending on the application, the user has a number of choices on how the FEM data can be used to simulate the behavior of the crack. For example,

- The complete model, including all the loading and restraints, can be automatically transferred.
- A representative part of the model can be selected, including the loads and restraints.
- In the case where there are complex nonlinear stresses (e.g., residual stress fields), the stress field can be transformed to a part or the whole of the model.

In some applications, the original FEM model may not contain sufficient detail in the area where the crack is located (i.e., some of the geometrical details missing, poorly refined mesh, etc.). In this case, the user can build a new geometric model of a sub-section and transfer the loads and restraints.

7.3.3 Field Profile-3

Table 7.9 summarizes predicted life for all 18 test cases for Field Profile-3. A general approach to modeling the durability of components and structures has been developed that combines BEM and FEM. This approach removes the requirement of rebuilding FEM models in order to capture the important stress raising features that significantly affect fatigue life predictions. This approach enables simple and accurate predictions of stress intensity factors and the automatic simulation of single and multiple crack growth.

This method is ideally suited for predicting data for fatigue life calculations in components and structures. Example applications have been presented demonstrating some of the capabilities of the method. Life distribution is a theoretical population model used to describe the lifetime of a solder joint and is defined as the CDF for the population. Thus, the objective of reliability tests is to obtain failures (the more the better) and to best fit the failure data to determine the parameters of the CDF of a chosen probability

TABLE 7.9

Predicted Life for All Eighteen Test Cases for Field Profile-3

Test Case	Scaled Fatigue Life Based on Darveaux's Model with CAVE Constants		Scaled Fatigue Life Based on Morrow's Model	
	(Cycles)	(Hours)	(Cycles)	(Hours)
1	34226	685	33,266	665
2	504542	10091	8,821,451	176,429
3	35062	701	33,716	674
4	34779	696	33,426	669
5	422478	8450	7,678,171	153,563
6	90517	1810	2,243,723	44,874
7	215185	4304	4,509,119	90182
8	177162	3543	3,953,683	79,074
9	71780	1436	1,923,949	38,479
10			3,349,661	66,993
11			2,825,215	56,504
12			34,945	699
13			32,334	647
14			1,584,396	31,688
15			124,8814	24,976
16			5,471,989	109,440
17			5,965,388	119,308
18			5,965,388	119,308

distribution (e.g., exponential, lognormal, and Weibull). The number of items (sample size) to be tested should be such that the final data are statistically significant.

The life test data is best fitted to the Weibull distribution. Also, the sample mean, population mean, sample characteristic life, true characteristic life, sample Weibull slope, and true Weibull slope for some of the high-density packages are provided and discussed. Furthermore, the relationship between the reliability and the confidence for a life distribution is established. Finally, the confidences for comparing the quality (mean life) of Pb-free solder joints of high-density packages are determined.

Here, fatigue life prediction of solder joints in electronics packages with FEM describes the method in great detail starting from a theoretical basis. Specific steps in the analysis method are discussed through examples. Fatigue life prediction of solder joints in electronic packages with FEM allows engineers to conduct fatigue reliability analysis of solder joints in electronics packages.

7.4 Copper Trace Integrity

Copper trace failure mode is another critical failure mode in flip-chip devices and other electronics packages. Because the copper trace has a negligible effect on solder joint reliability, it was ignored in the solder joint integrity analysis.

A copper trace of 0.7-mil thick and 2-mil wide has been added to the 3-D diagonal slice model. Four models with low gap height bump and 4-mil pad have been generated for all four different underfill materials for the copper trace integrity analysis. For the simplicity of the model, the underfill fillet was initially not considered in the model. Later, an underfill fillet was added for the accuracy of the result.

The Von Mises stress in the case of the model without underfill fillet shows the stress concentration in the copper trace in the region outside the underfill away from the die edge. The actual copper trace failure, however, occurs at the die edge just below the underfill fillet. Because the FEA without the underfill fillet was not able to simulate the actual failure, the model was recreated with an underfill fillet for better results.

Simulation with the underfill fillet showed the high stress region below the underfill fillet, which is also the location of failure in field and experimental testing. This generates confidence in the model. The values of maximum Von Mises stress from the simulation run without fillet is higher by approximately 10% than the values of the model with fillet (Table 7.10).

For the models with fillet for 0°C to 70°C ATC, the high stress region occurs just below the underfill fillet in all cases. Von Mises stress has been used for

TABLE 7.10

Maximum Von Mises Stress from the Model with
Fillet and without Fillet

	Max Stress (psi) Von Mises	
Underfill Material	With Fillet	Without Fillet
FP 4526	35,821	40,057
FP 4549	35,577	38,073
3M UF3667	34,092	34,604
Kester 9110S	34,344	34,945

TABLE 7.11

Simulation Results for Two Different ATCs

	Max Stress (psi) Von Mises 0°C–70°C		Max Stress (psi) Von Mises −55°C–125°C	
Underfill Material	Eutectic	Pb-free	Eutectic	Pb-free
FP 4526	35,821	35,802	85,718	85,626
FP 4549	35,577	35,575	83,444	83,372
3M UF3667	34,092	34,166	67,691	68,195
Kester 9110S	34,344	34,396	65,031	65,594

the plots and the comparison of stresses because Von Mises stress is the failure criterion for the failure of ductile materials under fatigue.

In all, sixteen simulations were run for two different ATC profiles for all four underfill materials with eutectic as well as Pb-free solder bumps. The results are listed in Table 7.11. The results indicate that there is hardly any difference in the value of the maximum stress of the copper trace between the eutectic and the corresponding Pb-free cases, the stresses for lead-free being slightly higher than the eutectic cases.

The ultimate tensile strength of oxygen-free electronic copper is in the range of 32,054 to 65,993 psi, and the tensile yield strength is in the range of 10,008 to 52,940 psi. Thus, the copper trace failure is predicted for all test cases in the case of −55°C to 125°C ATC but it may not fail when the flip chip is subjected to 0°C to 70°C ATC.

7.5 Fatigue Validation of Lead-Free Circuit Card Assembly

This section gives a briefing on the methodology of using life prediction models and life prediction calculations for solder joint reliability. Recommendations on the selection of the accelerated test methodologies are also provided at the end of the section. Solder joint life acceleration factors depend on various parameters:

- The ATC (accelerated thermal cycle) selected for the field life prediction
- The field profile for which the life is to be predicted
- Material, geometry, and assembly parameters
- Duty cycle in terms of average number of field cycles that the component is subjected to during its application on a per-day basis

These factors must be kept in mind while performing any life prediction calculations. One must be extremely careful and wise while using the modeling and simulation tools, as if not wisely used it may give the user a set of completely irrelevant results that may prove fatal in the design and selection of components, instead of helping the designer.

7.5.1 Circuit Card Assembly and Drive-Level Tests

7.5.1.1 Life Prediction Calculation Using Darveaux's Energy-Based Model

For simplicity of understanding the various steps involved in solder joint life prediction, calculations for a sample test case were demonstrated for the Test Case-1 with Field Profile-3. The acceleration factor was also calculated for the 0°C to 90°C ATC based on Field Profile-3.

Because the Kester material has a *Tg* below the maximum temperature (82°C) of Field Profile-3, the weighted average of the ISED was used to calculate the scaled ISED and also used for the life calculation. A weight factor (WF) of 2 and a scaling factor (SF) of 10 for Field Profile-3 were used in the calculation.

$$\text{SISED } (\Delta W) = (WF)*(SF)*(ISED)$$

$$= 2*0.1*4.71 \tag{7.18}$$

$$= 0.942 \text{ psi}$$

Now from Darveaux's CAVE-modified model,

$$N_0 = K_1(\Delta W)^{K_2}$$

$$= 25768 \text{ cycles} \tag{7.19}$$

$$\frac{da}{dN} = K_3(\Delta W)^{K_4}$$

$$= 4.73\text{E-}7$$

$$N_e = N_0 + \frac{a}{da/dN} \tag{7.20}$$

$$= 34{,}226 \text{ cycles}$$

If we assume that the flip chip encounters twenty averaged thermal duty cycles of Field Profile-3 per day, then the field life of the component would be 1,711 days or 4 years, 8.3 months.

Now using the same calculations with appropriate values of the constants and ISED for the 0°C to 70°C ATC, the characteristic life (N) turns out to be 8,854 cycles. This means that 8,854 cycles of 0°C to 70°C test cycles correspond to 4 years, 8.3 months of drive-level field life. Because the cycle time for the 0°C to 70°C ATC is 30 minutes, therefore the characteristic life is 4,427 hours, or 184 days.

Thus, the acceleration factor (AF) for the 0°C to 70°C circuit card assembly (CCA)-level test cycle w.r.t. drive-level Field Profile-3 is 9.3X (AF = 1711/184) in terms of time. Similarly, the calculations can be done for any drive-level profile and CCA-level accelerated test using any of the life prediction models for which the results are presented in Section 5.4.

7.5.2 CCA Component-Level Accelerated Test

The following recommendations can be made for the accelerated test methodologies at the CCA component level to predict solder joint reliability based on FEA.

- Before selecting the accelerated test cycle for thermomechanical testing of the component, it is extremely critical to make sure that the failure mode and the mechanism in the ATC testing remain identical to the failure mode and mechanism encountered in the field application.
- It is observed that for the Kester 9110S underfill material, the ISED accumulated in the solder joint increases drastically for thermal cycles having maximum temperature above the Tg (70°C to 75°C).
- It is extremely important to check the maximum temperature encountered by the solder joint in the field application and select an underfill material that has a Tg value above the maximum field temperature.
- The ATC should also be selected so that the maximum temperature in the test cycle is not above the Tg of the underfill material.
- The 0°C to 70°C ATC is recommended for comprehensive accelerated testing of the flip chip with all the underfill materials. The solder joint failure for the 0°C to 70°C test cycle remains independent of the Tg of the underfill material for all capillary flow and re-flow

underfills. However, the 0°C to 90°C ATC can be used for the testing of flip chips with capillary flow underfills.

Bibliography

Amagai, M., Watanabe, M., Omiya, M., Kishimoto, K., and Shibuya, T. (2002). Mechanical characterization of Sn–Ag based lead-free solders, *Micoelectronics Reliability*, 42, 951–966.

Antolovich, S.D., and Antolovich, B.F. (1996). *An Introduction to Fracture Mechanics in ASM Handbook. 19 Fatigue and Fracture*, Materials Park, OH: ASM International®, 1996.

Arora, N.D., Raol, K.V., Schumann, R., and Richardson, L.M. (1996), Modeling and extraction of interconnect capacitances for multilayer VLSI circuits, *IEEE Trans. Computer Aided Design of Integrated Circuits and Systems*, 15(1), 58–66.

Bilotti, A.A. (1974). Static temperature distribution in IC chips with isothermal heat sources, *IEEE Trans. Electron Devices*, ED-21 (March), pp. 217–226.

Black, J.R. (1969). Electromigration failure models in aluminium metallization for semiconductor devices, *Proc. IEEE*, 57(9), 1587–1594.

Blech, I.A., and Herring, C. (1976). Stress generation by electromigration, *Appl. Phys. Lett.*, 29, 131–133.

Chen, C., and Liang, S.W. (2007). Electromigration issues in lead-free solder joints, *J. Mater. Sci.*, 18, 259–268.

Darveaux, R. (2000). Effect of simulation methodology on solder joint crack growth correlation, *Proc. 50th ECTC*, May 2000, pp. 1048–1058.

Darveaux, R. (1996). How to use finite element analysis to predict solder joint fatigue life. *Proc. VIII Int. Congress on Experimental Mechanics*, Nashville, TN, June 10–13, 1996, pp. 41–42.

Dreezen G., Deckx E., and Luyckx G. (2003). Solder alternative: Electrically conductive adhesives with stable contact resistance in combination with non-noble metallization. *CARTS Europe 2003*, pp. 223–227.

Gale, W.F., and Totemeier, T.C. (2004). *Smithells Metals Reference Book, (8th edition)*. Maryland Heights, MO: Elsevier.

Galyon, G.T. (2003). Annotated Tin Whisker Bibliography, a NEMI Publication, July.

Hunter, W.R. (1997). Self-consistent solutions for allowed interconnect current density. I. Implication for technology evolution, *IEEE Trans.Electron Devices*, 44(2), 304–309.

Hunter, W.R. (1997). Self-consistent solutions for allowed interconnect current density. II. Application to design guidelines, *IEEE Trans. Electron Devices*, 44(2), 310–316.

Lall, P., Hariharan, G., Tian, G., Suhling, J., Strickland, M., and Blanche, J. (2006). "Risk Management models for flip-chip electronics in extreme environments." *ASME International Mechanical Engineering Congress and Exposition*. Chicago, IL, November, pp. 1–13.

Lall, P., Islam, N., Shete, T., Evans, J., Suhling, J., and Gale, S. (2004). "Damage mechanics of electronics on metal-backed substrates in harsh environments," *Proceedings of 54th Electronic Components & Technology Conference*, IEEE, Las Vegas, NV, June 1–4, pp. 704–711.

Lall, P., Islam, N., Suhling, J., and Darveaux, R. (2003). "Model for BGA and CSP reliability in automotive underhood applications," in *Proc. 53rd Electronic Components and Technology Conference*, May 27–30, 2003, pp. 189–196.

Lall P., Pecht, M., and Hakim, E. (1997). *Influence of Temperature on Microelectronic and System Reliability*. Boca Raton, FL: CRC Press.

Lall, P., Singh, N, Suhling, J., Strickland, M., and Blanche, J. (2005). "Thermo-mechanical reliability tradeoffs for deployment of area array packages in harsh environments," *IEEE Transactions on Components and Packaging Technologies*, 28, 3, September, pp. 457–466.

O'Connor, P. and Kleyner, A. (2012). *Practical Reliability Engineering, (5th edition)*. Hoboken, NJ: Wiley.

Pang, J.H.L., and Chong, D.Y.R. (2001). Flip chip on board solder joint reliability analysis using 2-D and 3-D FEA models, *IEEE Trans. Advanced Packaging*, 24(4), 499–506.

Pang, J.H.L., Xiong, B.S., and Low, H. (2004). Creep and fatigue characterization of lead free 95.5Sn-3.8Ag-0.7Cu solder, *Proc. 54th ECTC*, June 2004, pp. 1333–1337.

Shi, X.Q., Pang, H.L.J., Zhou, W., and Wang, Z.P. (1999). A modified energy-based low cycle fatigue model for eutectic solder alloy, *Scripts Material*, 41(3), 289–296.

Shirgaokar, A. (2009). "Principal Component Regression Models for Thermomechanical Reliability of Plastic Ball Grid Arrays on Cu-Core and No Cu-Core PCB Assemblies in Harsh Environments," Auburn University Thesis (Master of Science), Auburn, AL, August.

Strauss, R. (1998). *SMT Soldering Handbook, (Second edition)*. Maryland Heights, MO: Elsevier/Newnes.

Suo, Z. (2004). A continuum theory that couples creep and self-diffusion, *J. Appl. Mechanics*, 71, 646–651.

Syed, A.R. (2004). Accumulated creep strain and energy density based thermal fatigue life prediction models for SnAgCu solder joints, *Proc. 54th ECTC*, June 2004, pp. 737–746.

Syed, A.R. (1995). Creep crack growth prediction of solder joints during temperature cycling: An engineering approach, *Trans. ASME*, 117 (June), pp. 116–122.

Teng, C.C., Cheng, Y.K., Rosenbaum, E., and Kang, S.M. (1997). iTEM: A temperature-dependent electromigration reliability diagnosis tool, *IEEE Trans. Computer-Aided Design of Integrated Circuits and Systems*, 16(8), 882–893.

Tu, K.N. (2003). *Solder Joint Technology: Materials, Properties, and Reliability*. Berlin, Germany: Springer.

Tu, K.N. (2003). Recent advances on electromigration in very-large-scale integration of interconnects, *J. Appl. Phys.*, 94, 5451–5473.

Tu, K.N. (1994). Irreversible processes of spontaneous whisker growth in bimetallic Cu–Sn thin-film reactions, *Phys. Rev. B*, 49(3), 2030–2034.

Tunga, K., Pyland, J., Pucha, R.V., and Sitaraman, S.K. (2003). Field-use conditions vs. thermal cycles: A physics-based mapping study, *Proc. 53rd Electronic Components and Technology Conference*, May 27–30, 2003, pp. 182–188.

Wiese, S., Schubert, A., Walter, H., Dudek, R., Feustel, F., Meusel E., and Michel, B. (2001). Constitutive behaviour of lead-free solders vs. lead-containing solders–Experiments on bulk specimens and flip-chip joints, *Proc. 51st Electronic Components and Technology Conference,* pp. 890–902.

Zahn, B.A. (2002). Finite element based solder joint fatigue life predictions for a same die size-stacked-chip scale-ball grid array package, *SEMICON West, International Electronics Manufacturing Technology (IEMT) Symposium,* pp. 274–284.

Zahn, B.A. (2003). Solder joint fatigue life model methodology for 63Sn37Pb and 95.5Sn4Ag0.5Cu materials, *Proc. 53rd Electronic Components and Technology Conference,* pp. 83–94, May 27–30, 2003.

8

Enhancing Reliability of Ball Grid Array

In modern electronics packaging, especially surface-mount technology (SMT), thermal strain is usually induced between components during processing, and in service, by a mismatch in the thermal expansion coefficients. Because solder has a low melting temperature and is softer than other components in electronics packaging, most of the cyclic stresses and strains take place in the solder. Fatigue crack initiation and fatigue crack propagation are likely to occur in the solder even when the cyclic stress is below the yield stress.

Ball grid array (BGA) is a package with one face covered with balls in a grid pattern. These balls, or solder spheres, conduct electrical signals from the integrated circuit (IC) to the printed circuit board (PCB) on which it is placed. These solder spheres can be placed manually or with automated equipment. These solder spheres are held in place with a tacky flux until soldering occurs. The device is placed on a PCB with copper pads in a pattern that matches the solder balls. The assembly is then heated, either in a reflow oven or by an infrared heater, causing the solder balls to melt. Surface tension causes the molten solder to hold the package in alignment with the circuit board, at the correct separation distance, while the solder cools and solidifies. A BGA is a type of surface-mount packaging used for ICs.

As the eager demand increases for high power-handling capability in smaller packages, thermally enhanced BGA (TEBGA) packages provide a very attractive solution for improving the poor thermal performance problems of conventional over molded plastic BGA packages. In this chapter, with solder joint reliability a concern during the initial package design stage, an engineering empirical approach using a finite-volume-weighted averaging technique is applied for characterizing the strain concentration field around the corners of solder joints due to a dramatic geometry/material change. Furthermore, a parametric finite element analysis (FEA) is performed over a number of geometry/material design parameters to investigate the dependence on the fatigue lives of the thermally loaded solder joint in a typical TEBGA assembly. Through the parametric design, together with a rational characterization of the fatigue lives of the solder joints, the reliability characteristics of the TEBGA package can then be effectively identified.

8.1 Thermally Enhanced BGA

Due to progressive demand of high-performance electronic applications, in recent years researchers in the electronics community have been devoting tremendous efforts in developing an innovative solution to build cost-effective, high-density, high-I/O devices with better high-power handling capability. Under these driving factors, area array surface-mount technologies (SMT), such as BGA packages or flip-chip (FP) interconnection techniques, are quickly becoming a considerably noticeable and attractive alternative and, more importantly, are gradually overriding, in some aspects, conventional peripheral packaging techniques, such as the plastic quad flat package (PQFP).

Among these emerging area array surface mount packaging technologies, the BGA is a solution to the problem of producing a miniature package for an IC with many hundreds of pins. Pin grid arrays and dual-in-line surface mount (SOIC) packages were being produced with more and more pins, and with decreasing spacing between the pins, but this was causing difficulties for the soldering process. As package pins got closer together, the danger of accidentally bridging adjacent pins with solder grew. BGAs do not have this problem when the solder is factory applied to the package. Thermally enhanced BGA (TEBGA) packages are being developed due to the following considerations:

- BGA technologies, specifically including plastic BGA (PBGA), ceramic BGA (CBGA), and tape BGA (TBGA), have been extensively applied in industry in the past few years.

- In contrast, PBGA implies several advantages over CBGA, such as lower cost of substrate, smaller global thermal expansion mismatch from PCB than ceramic, and lower profile.

- However, conventional over molded plastic BGA packages have drawbacks in comparison with CBGA, such as excessive moisture sensitivity of substrate as well as poor thermal performance of over molded compound.

- Driven by more and more requirements of high power-handling applications, of great and eager demand is an improved, cost-effective PBGA that can enhance thermal performance.

Thermally enhanced BGA packages (TEBGA, or the so-called Super BGA) (Figure 8.1) provide a very engaging solution that can improve the poor thermal performance problems of conventional over molded plastic BGA packages and, in addition, are relatively cost effective. As opposed to the conventional PBGA face-up design of chips, TEBGA packages adopt a chip face-down configuration, in which a copper plate is attached to the chip and

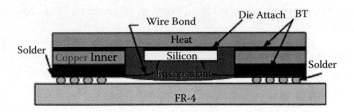

FIGURE 8.1
Thermally enhanced BGA packages (TEBGA or the so-called Super BGA).

the organic substrate to increase the cooling rate or capability. For TEBGA, package key features often include:

- Liquid encapsulation protection
- Unique and optimized Cu paste process/material
- Fully grounded chip can guarantee maximum electrical performance
- Direct thermal dissipation path to system board is possible

Similar to a conventional face-up PBGA assembly, the solder joints in the TEBGA assembly function not only as a thermal passage system to conduct the heat generated from the chip, but also as a mechanical mechanism to withstand the thermal-deformation mismatch between the package and the PCB. It is observed that the solder joints in the packages will likely suffer from a crucial inelastic strain under thermal cycling loading; consequently, potential solder joint reliability problems may still occur in the TEBGA package. With solder joint reliability a concern, many studies on PBGA reliability problems have been reported in the literature; however, only few have investigated the reliability of cavity-down, thermally enhanced PBGA packages. Furthermore, among these few reports, most of them simply focus on the prediction of the fatigue life of solder joints under thermal cyclic loading using nonlinear finite element analysis (FEA). However, it is important to note that the reliability of the solder joints is basically a function of many variables, for example, material constants and geometry parameters. With appropriate combinations of these parameters and constants, it is expected that the reliability of the packages can be accordingly enhanced. In general, there are two main techniques that can be applied for this purpose:

1. Design of experiments (DOE)
2. Parametric FEA

The DOE approach requires the utilization of a conventional, very time-consuming trial-and-error procedure in pursuit of an optimal design; hence, a parametric FEA is consequently presented here. The other implied advantage of a parametric FEA lies in its capability of reflecting the effect of each individual design parameter on the problem of concern. The design parameters include

- Thickness of the PCB
- Silicon chip
- Die attach
- Die size
- Young's modulus
- CTE of PCB, etc.

In addition, during a nonlinear FEA, a high, mesh-sensitive stress–strain concentration field may be generated around the corners of solder joints because of a significant geometry/material change; and more importantly, the fatigue life of the solder ball as well as overall package reliability are extensively determined by the maximum effective inelastic strain range. It is, therefore, essential to effectively characterize the strain concentration. Theoretically, the problem has been resolved or eased using either an explicit geometry representation, fillet, or a material-nonlinear modeling.

A volume-weighted averaging technique has been introduced to characterize the stress–strain response at the geometry/material singular point. The fatigue lives of thermally loaded solder joints are predicted using the finite element method. An appropriate constitutive relationship to model the time-dependent inelastic deformation of the near-eutectic solder is implemented in a commercial finite element code, and the stress–strain responses of different electronic assemblies under the applied temperature cycles are calculated. The FEA results are coupled with a newly developed approach for fatigue life predictions using a volume-weighted averaging technique instead of an approach based on the maximum stress and strain locations in the solder joint. Volume-weighted average stress and strain results of three electronic assemblies are related to the corresponding experimental fatigue data through least-squares curve-fitting analyses for determination of the empirical coefficients of two fatigue life prediction criteria. The coefficients thus determined predict the mean cycles-to-failure value of the solder joints. Among the two prediction criteria, the strain range criterion uses the inelastic shear strain range and the total strain energy criterion uses the total inelastic strain energy calculated over a stabilized loading cycle. The obtained coefficients of the two fatigue criteria are applied to the FEA results of two additional cases obtained from the literature. Good predictions are achieved using the total strain energy criterion; however, the strain range criterion underestimated the fatigue life. It is concluded that strain information alone is not sufficient to model fatigue behavior; rather, a combination of stress and strain information is required, as in the case of total inelastic strain energy. However, employment of this approach into this work may unfortunately underestimate the stress–strain response due to the fact that it is volume-weighted and averaged within the entire domain and, as a result, leads to a very nonconservative fatigue life prediction.

In this chapter, a finite-volume-weighted averaging technique is proposed. Instead of averaging the response within the entire domain, the improved approach averages the structural response in a finite zone. The dimension of the finite zone can be substantially characterized using an engineering empirical criterion. Based on the average plastic strain, the fatigue life of the solder joint can then be rationally predicted and applied in the parametric FEA. Through the above design parametric analysis in combination with a rational characterization of the strain response of the solder joints, the reliability characteristics of the TEBGA package can then be effectively identified.

8.2 Typical TEBGA Package and Finite Element Modeling

The package considered here is a 256-pin, cavity-down TEBGA that is mounted on a multilayer PCB. It contains four rows of perimeter BGA eutectic solder joints (i.e., 63Sn-37Pb) with a pitch of 1.27 mm. A center cross-section of the package is shown in Figure 8.1. Main dimensions are summarized as follows:

- The geometry of the baseline package is square, with a package size of 27×27 mm.
- The chip size of the package is 6×6 mm, with the thickness of 0.25 mm, and is directly attached to the 0.28-mm thick copper plate by the silver-filled die attach adhesive with a thickness of 0.047 mm.
- The substrate is composed of two layers of BT (bismaleimide triazine) material, each with a thickness of 0.047 mm and 0.235 mm, respectively, and one layer of copper inner ring with a thickness of 0.274 mm.
- The dimension of the cavity that is embedded with a silicon chip is 15.328 mm square and filled with epoxy encapsulant to shield the chip as well as the wire bond.
- In addition, the package is laminated with 0.047-mm-thick solder mask.
- The size of the PCB that is made of FR-4 material is 32.4×32.4 mm with a thickness of 1.0 mm, which is slightly larger than the package.

Because one of the major issues in this chapter is to discuss the reliability of the solder joints, it is essential to include PCBs in the analysis. Accordingly, the elastic material properties of these components are listed in Table 8.1. It should be noted that all the materials in the package are assumed

TABLE 8.1

Material Properties of a TEBGA Package

Materials	Young's Modulus(GPa)	Poisson Ratio	CTE(ppm/°C)
Copper heat spreader/Inner ring	121.0	0.38	16.3
BT substrate	186	0.36	14.2~16.7 (X,Y)
			57.0 (Z)
Silicon chip	162.4	0.23	3.3
Epoxy solder mask	6.87	0.35	19.0
Eutectic solder ball	29.8	0.43	25.1
Silver-filled epoxy die attach	2.62	0.42	90.0 (< 0°C)
			160.0 (> 0°C)
Epoxy encapsulant	8.96	0.35	19.0
FR-4 PCB	17.2	0.28	16.0 (X,Y)
			72.0 (Z)

to be linearly elastic except for eutectic solder joints. The eutectic solder is considered viscoplastic and temperature dependent because creep may occur even at room temperature.

In addition, it is reported that the temperature- and time-dependent inelastic strain is the major factor that causes the failure of solder joints. Hence, a nonlinear finite element model that accounts for both the temperature-dependent plasticity and the time-dependent creep is constructed to characterize the inelastic stress/strain response of the solder joint.

Analysis of the strain-stress data with both power law creep and Garofalo's hyperbolic sine relation shows the transition to a low-stress exponent creep regime with decreasing stress and/or increasing testing temperature. In this chapter, the Garofalo's hyperbolic sine law is applied to model the creep behavior, in which it is given as follows:

$$\dot{\varepsilon}^{cp} = A(\sinh B\sigma)^n \exp\left(\frac{-\Delta H}{RT}\right) \tag{8.1}$$

where
 $\dot{\varepsilon}^{cp}$ denotes the uniaxial equivalent creep strain rate
 A is equal to 147.9 (sec^{-1})
 B is 0.0805 (MPa^{-1})
 n is a stress exponent equal to 3
 σ is the equivalent stress
 ΔH is the activation energy equal to 52961 (J.mol^{-1})
 R is the gas constant that is 8.31 (J.mol^{-1}.k^{-1})
 T is the absolute temperature

In addition, the Prandtl–Reuss formulation is employed to simulate the rate-independent plastic deformation. Consider that a typical eutectic solder joint holds a tensile stress–strain relationship, as shown in Figure 8.2.

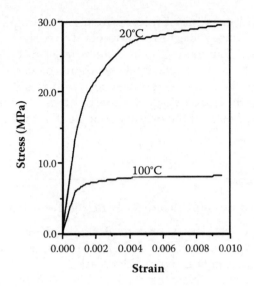

FIGURE 8.2
The stress–strain relationship of eutectic solder.

8.2.1 Fatigue Life Prediction

The fatigue life of the eutectic solder joint can be predicted using an empirical Coffin–Manson relationship:

$$N_f = \frac{1}{2} \left(\frac{\varepsilon^{IE}}{2\varepsilon_f} \right)^{(1/c)} \tag{8.2}$$

where
  N_f is the mean cycle to failure
  ε^{IE} is the inelastic strain range in one cycle of thermal loading
  ε_f is the fatigue ductility coefficient that is 0.325
  c is the fatigue ductility exponent. The exponent is generally in between −0.5 and −0.7, and is assumed to be −0.5 in this chapter

Once the cyclic shear strain range of the solder is derived, the corresponding fatigue life can then be estimated.

As mentioned previously, the main goal is to explore the impact on TEBGA reliability in terms of a number of specified design parameters. Therefore, for simplicity and preference we wanted to model the problem using a 2-D plane-strain finite element model as long as the corresponding failure mechanisms would not be altered. Essentially, the 2-D finite element model consists of all major segments aforementioned, including one copper plate for heat dissipation, chip, BT substrate, copper inner ring, solder joints, die attach, solder mask, and PCB, in which the geometry profile of the solder joint is determined using Surface Evolver. Surface Evolver is a computer program that minimizes the energy of a surface subject to constraints. The surface is represented as a simplicial complex. The energy can include surface tension, gravity, and other forms. Constraints can be geometrical constraints on vertex positions or constraints on integrated quantities such as body volumes. The minimization is done by evolving the surface down the energy gradient. Due to the symmetry of the package, only half of the package is applied in the FEA.

8.3 Finite-Volume-Weighted Averaging Technique

Before continuing with the parametric design of TEBGA assemblies over a number of design parameters, the stress–strain information near the interfaces in between the solder joint and the solder mask as well as the PCB must be first effectively characterized. Note that the interfaces of these components usually involve an abrupt geometry change that inevitably forms a sharp corner and, more importantly, induces a singular (concentrated) stress–strain response. Because the material nonlinearity is considered such that the stress concentration can be eventually eased, techniques that can be applied to the characterization of the strain response around the singular point are in critical demand.

In this chapter, an improved volume-weighted averaging technique is presented to effectively characterize the strain response in the corner of the solder joint. Instead of averaging the state variables in the entire material domain, a specific zone is introduced and applied to perform the averaging technique as follows:

$$\tilde{\varepsilon} = \sum_{e=1}^{n} \int_{\Omega_e} \varepsilon_e d\Omega \Bigg/ \sum_{e=1}^{n} \int_{\Omega_e} d\Omega \qquad (8.3)$$

where
  n is the total number of elements in the zone
  ε_e is the strain of the e-th element
  Ω_e is the volume (area) of the e-th element
  $\tilde{\varepsilon}$ is the volume-weighted averaging strain in a specific zone

The dimension of the finite zone is determined in an empirical manner such that it should be small enough to capture the maximal strain field and, on the other hand, large enough to obtain a converging solution as the mesh density increases. To explore the dimension of a particular finite zone, four different finite element models, each with a different mesh density in the fan-shaped circular sector (i.e., radius is equal to 0.12 mm, as shown in Figure 8.3) at the corners of each ball, are applied for modeling the solder joint. The total number of elements in one fan-shaped circular sector are 18, 66, 192, and 504,

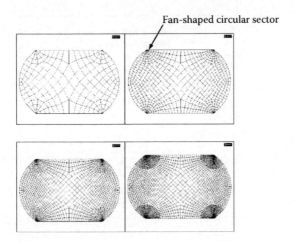

Fan-shaped circular sector

FIGURE 8.3
Four different finite element models.

with respect to these four finite element models. In addition, four different radii, starting from the center of the fan-shaped circular sector, are defined; they are 0.01, 0.02, 0.04, and 0.06 mm respectively. For simplicity, the creep behavior is not considered in the analysis; and in addition, an isothermal net temperature swing of 100°C is used as the thermal loading.

Parametric reliability analysis has been applied to the no-underfill flip-chip package. Finite element modeling (FEM)-based simulation enables a parametric analysis for a flip-chip package with a constraint-layer structure. Previous research has shown that flip-chip type packages with organic substrates require underfill for achieving adequate reliability life. Although underfill encapsulant is needed to improve the reliability of flip-chip solder joint interconnects, it will also increase the difficulty of reworkability, increase the packaging cost, and decrease the manufacturing throughput. The FEM is based on the fact that if the thermal mismatch between the silicon die and the organic substrate could be minimized, then the reliability of the solder joint could be enhanced accordingly. This research presents a structure using a ceramic-like material with a coefficient of thermal expansion (CTE) close to silicon, mounted on the backside of the substrate to constrain the thermal expansion of the organic substrate. The ceramic-like material could reduce the thermal mismatch between the silicon die and substrate, thereby enhancing the reliability life of the solder joint. Furthermore, in order to achieve better reliability design of this flip-chip package, a parametric analysis using FEA is performed for package design. The design parameters of the flip-chip package include die size, substrate size and material, constraint-layer size and material, etc. The results show that this constraint-layer structure could make the solder joints of the package achieve the same range of reliability as the conventional underfill material. More importantly, the flip-chip package without underfill material

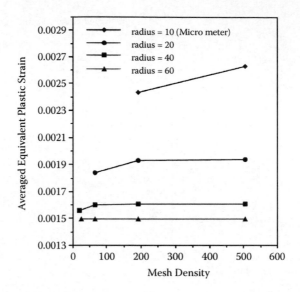

FIGURE 8.4
The averaged equivalent plastic strain versus mesh density within four different finite zones.

could easily solve the reworkability problem, enhance the thermal dissipation capability, and also improve the manufacturing throughput.

Figure 8.4 presents the volume-weighted averaging equivalent plastic strains associated with these four different zones. It can be easily seen from Figure 8.4 that the 20-mm zone seems to empirically provide considerably more agreement with the currently proposed criterion in the selection of the dimension of the finite zone than the others. In addition, the curve of the 20-mm zone shows that the average equivalent plastic strain is well converged as the mesh density is up to about 200.

This implies that the strain response in the geometry singularity area can be substantially characterized and, more importantly, would no longer be mesh sensitive while the mesh is adequately dense. Thus, the characterized 20-mm zone with the mesh density model of 192 (i.e., Model C) as shown in Figure 8.4 will be applied. However, this numerically empirical result should be extensively substantiated using experimental studies in order to accurately characterize the "exact" fatigue life of the solder joints.

8.4 Parametric Design of TEBGA Reliability

Flip-chip and wafer-level packaging technologies are currently attracting increasing attention from the electronics packaging industry owing to their good thermal performance, smaller size, lower profile, lighter weight, higher

I/O density, etc. Conventionally, flip-chip packages used an underfill encapsulant to reduce the strain of the solder joints between the IC chip and its substrate. This encapsulant not only provides dramatic reliability enhancement with minimal impact on the flip-chip manufacturing process flow, but also extends to a variety of organic substrate materials, resulting in a 10-to 100-fold improvement in reliability compared to a flip-chip package without underfilling. However, the major shortcomings posed by the underfilling process remain unresolved, that is, significantly increased difficulty of reworkability and the decrease in manufacturing throughput.

TEBGA bridges the thermal performance gap between PBGA and cavity-down thermal BGA or equivalent flip-chip BGA. As designs migrate to 0.13 μm to provide higher speed and integration, the power dissipation requirements increase. Parametric design of TBEGA assemblies using FEA can effectively provide designers with not only a clear, comprehensive picture of the reliability–parameter relationship, but also a reliability-enhanced design rule. In general, the stress–strain response and distribution depend extensively on the configuration of structures; thus, it is important to investigate the dependence of the package configuration on the corresponding fatigue life. Moreover, considerable differences in the epoxy-based materials' properties, such as BT and FR-4, may be perceived among various vendors. Assessment of the effect of these variations on package reliability is also one of the most important issues in the initial stage of package design. Reports have indicated that CTE and Young's modulus are two of the most pronounced material properties governing the reliability of the packages.

Conclusively, the design parameters undergoing this parametric analysis can be categorized into two main categories: the geometry parameters and the material properties of components. The geometry design parameters comprise the thickness of die/die attach adhesive/PCB/BT/heat spreader, and the size of die; on the other hand, the material design parameters consist of the CTE and Young's modulus of FR-4 and/or BT. For the parametric analysis, the inelastic strain range that is assumed to be stabilized within two thermal cycles is employed for evaluating the fatigue life of the solder joint based on the empirical Coffin–Manson relationship. The numerically accelerated test condition is composed of a 5 minute linear temperature loading/unloading ramp and 20 minute low/high temperature dwell periods, and the testing temperature range is from −40 to 125°C. The stress-free condition is assumed to be at room temperature (25°C). Using nonlinear FEA, the results of the parametric reliability design of TEBGA assemblies are shown in the following.

8.4.1 Effect of Die Thickness

There are a total of nine different packages, each with a different die thickness, employed for investigating the effect of die thickness on solder joint reliability. They are from 0.15 to 0.55 mm thick, respectively; on the other hand, all the other material property constants and geometry parameters

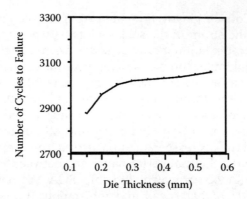

FIGURE 8.5
Effect of die thickness on fatigue life.

of the package are kept unchanged. The results are shown in Figure 8.5, in which there exists a nonlinear relationship in between the fatigue life of the solder joint and the die thickness. In addition, from Figure 8.5, it is evident that the fatigue life of the solder joint increases as the chip thickness gets thicker even though the maximum discrepancy is only about 6.5%.

8.4.2 Effect of Die Size

To examine the dependence of the die size on the solder joint's fatigue life, nine different die sizes (see Figure 8.6) are conducted in the parametric analysis. The results are shown in Figure 8.6, with a significantly distinct result presented in contrast to that of the die thickness. It reveals that the fatigue life of the solder joint is first enhanced as the die size is enlarged. As soon as the die size approaches beyond 7.0 mm, the fatigue life decreases while the die size increases. In other words, for this particular case, the 7-mm die size will provide an engaging, optimal reliability design of the assembly.

However, the nonlinear effect shown by Figure 8.6 contradicts most of the reliability studies of the face-up PBGA assembly. For a face-up PBGA assembly, reliability studies have focused on the solder joint located right beneath the chip because the largest CTE mismatch of the whole assembly occurs at that location. As the die size or die thickness is larger, the fatigue life of the solder joint becomes shorter. For the cavity-down TEBGA assembly, the x-directional expansion of the PCB is greater than that of the package because of a larger CTE in that direction. In other words, the farther the distance from the center of the package, the larger the deformation. As a result, the endmost solder joint will likely undergo the largest deformation.

However, according to Figure 8.7, the above postulates can only be partially true. It presents the relationship of the maximum average equivalent inelastic strain range versus the solder joints for different die sizes (i.e., 2.0, 6.0, and 10.0 mm), in which the value "1" on the abscissa represents the joint nearest

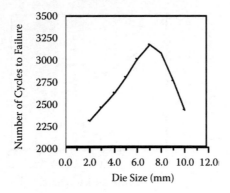

FIGURE 8.6
Effect of die size on fatigue life.

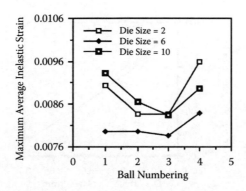

FIGURE 8.7
Maximum average equivalent inelastic strain range in solder joints.

the die and "4" denotes the farthest one. As can be seen from Figure 8.7, the maximum inelastic response for the 2.0-mm die size case occurs at Joint "4." This result well satisfies the preceding statements that the farthest joint will suffer from the highest inelastic strain response. On the other hand, for the 10.0-mm die size case in Figure 8.7, Joint "1" becomes the most critical one. This implies that there exist more than one thermal-mechanical deformation mechanisms in the problem. Essentially, the dominant deformation mechanism for the 2.0-mm die size is due to the global CTE mismatch between the PCB and the package; however, that for the 10.0-mm die size is attributed to the local CTE mismatch between the die and the heat spreader.

8.4.3 Effect of PCB Thickness

A total of nine finite element models, each with a different PCB thickness ranging from 0.6mm to 1.4mm, are applied in the parametric analysis.

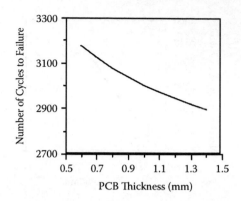

FIGURE 8.8
Effect of PCB thickness on fatigue life.

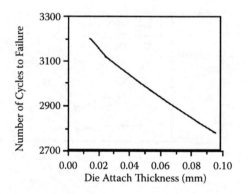

FIGURE 8.9
Effect of die attach thickness on fatigue life.

It should be noted that both the CTE and Young's modulus are kept constant while the thickness of the PCB increases. As shown in Figure 8.8, the relationship between them turns out to be relatively linear, and more importantly, the fatigue life of the solder joint decreases as the thickness of the PCB increases. The maximum decrement is only about 9.0% and is due to the fact that the increase in the PCB thickness will inevitably enlarge the expansion mismatch associated with the package.

8.4.4 Effect of Die Attach Adhesive Thickness

As shown in Figure 8.9, a linear relationship is presented in between the fatigue life of the solder joint and the thickness of the die attach adhesive. There is about a 13.0% decrease in solder joint fatigue life as the thickness of the die attaches increases from 0.015 to 0.095 mm.

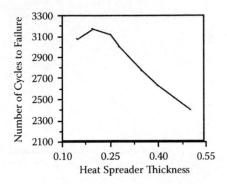

FIGURE 8.10
Effect of heat spreader thickness on fatigue life.

8.4.5 Effect of Heat Spreader Thickness

The dependence of the thickness of the heat spreader on solder joint reliability is presented in Figure 8.10. It can be easily observed that this implies a complicated deformation mechanism governing the solder joint response as the heat spreader thickness is relatively thin. The complicated deformation mechanism may stem from both the local CTE mismatch between the chip and the heat spreader, as well as the global CTE mismatch between the package and the PCB. However, as the thickness becomes thicker, the effect of the global CTE mismatch between the package and the PCB becomes dominant. It can be seen that the optimal thickness of the heat spreader would be 0.2 mm, such that the most reliable solder joint can be achieved.

8.4.6 Effect of BT Substrate Thickness

Figure 8.11 demonstrates a nonlinear dependence of the BT substrate thickness with respect to solder joint reliability. It is found that the fatigue life of the solder joint is, to some extent, insensitive to the variation in BT substrate thickness; moreover, the maximum difference between these packages is about 6.0%. Initially, the fatigue life of the solder joint is promoted as the thickness of the substrate increases. As soon as the thickness of the substrate reaches a value of 0.35 mm, the fatigue life of the solder joint gradually decreases as the thickness of the BT substrate continually expands. A further analysis of the complicated deformation mechanisms will be conducted.

8.4.7 Effect of BT/FR-4 CTE

To explore the effect of the BT and FR-4 CTE on solder joint fatigue life, six different packages, each with a different BT and FR-4 CTE as shown in Figure 8.12 and Figure 8.13, are employed. The results are presented in both

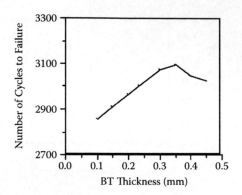

FIGURE 8.11
Effect of BT substrate thickness on fatigue life.

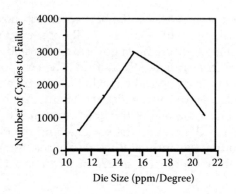

FIGURE 8.12
Effect of BT CTE on fatigue life.

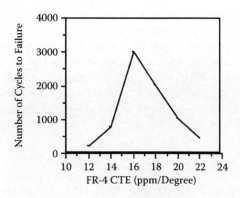

FIGURE 8.13
Effect of FR-4 CTE on fatigue life.

Figure 8.12 and Figure 8.13, respectively; both results exhibit a nonlinear relationship in between the solder joint fatigue life and the materials' CTE (i.e., BT and FR-4).

More importantly, the results also demonstrate that the present design of the TEBGA assembly provides the best reliability performance of the solder joint, in which the maximum fatigue life cycle is about 3,000 cycles in both cases. Furthermore, considerable discrepancies in the predicted fatigue life are detected between these models. By comparing Figure 8.12 with Figure 8.13, the sensitivity of the solder joint fatigue life with respect to the FR-4 CTE turns out to be much more significant than that of the BT CTE; on the other hand, in contrast to the other aforementioned geometry parameters, the effect of these material parameters on the solder joint fatigue life is exceedingly pronounced.

8.4.8 Effect of FR-4 Young's Modulus

It is reported that the Young's modulus of the organic materials in the assembly is one of the most important factors that may extensively affect solder joint fatigue life. In this work, the influence of the Young's modulus of the FR-4 material (for PCB) is investigated in order to comprehend its dependence on solder joint reliability. As shown in Figure 8.14, there are six TEBGA models applied for this particular purpose, each consisting of a different value of Young's modulus. As well, Figure 8.14 indicates that the relationship of solder joint fatigue life and the FR-4 Young's modulus is relatively linear. As the FR-4 Young's modulus increases, there is a decrease in the number of cycles to failure of the solder joint; in addition, there is a maximal 10.0% discrepancy between these models.

A finite-volume-weighted averaging technique is proposed to characterize the strain/stress response of the solder joint at the material/geometry singularity point. The dimension of the finite zone is determined using

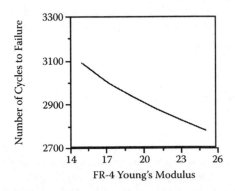

FIGURE 8.14
Effect of FR-4 Young's modulus on fatigue life.

an effectively empirical approximation method. It is found that the strain response near the geometry/material singularity point can be substantially characterized and, more importantly, would no longer be mesh sensitive while the mesh is adequately dense. In addition, a parametric analysis of the reliability of the TEBGA assembly in terms of these material property constants and geometry parameters is performed using a nonlinear FEA. It is found that the reliability impact of these design parameters on solder joint reliability has a wide range, in which the effect of the organic materials' CTE significantly dominates the other factors and the thickness of the BT substrate seems to provide a minimal influence. In conclusion, based on the solutions of the parametric analysis, the design guideline for obtaining the optimal reliability performance of a typical TEBGA assembly can be educed as follows: a larger die and BT substrate thickness, a thinner PCB/die attach adhesive/heat spreader, a smaller FR-4 (PCB) Young's modulus, and more importantly, a 7.0-mm die size together with a currently employed BT/FR-4 CTE. However, a design optimization technique should be employed for a more accurate conclusion. In addition, the current analysis lacks investigation of the effect of other failure mechanisms, such as die cracking, on the solder joint reliability in terms of these or other design parameters. A further analysis of these effects, together with substantial validation by existing experimental data, should be performed in order to truly pursue the optimal reliability design of a TEBGA assembly.

Finite element analysis has been generally accepted by the electronics packaging research community. Many structural/material behaviors of electronic packages are material/geometrically nonlinear, thermally dependent, strain rate dependent, time dependent (creep/relaxation), and even size, layout, and process dependent. Therefore it is unlikely to set all the nonlinear material properties as inputs for finite element analysis. Thus, one needs to carefully examine the analysis methodologies, procedures, mechanics/ numerical algorithm selection, geometry configuration, boundary conditions and loading condition, etc. Usually, parametric analysis with 2D/3D finite element nonlinear analysis is quite feasible for electronic package analysis and design, and it has been widely adopted in the engineering design of lead-free electronic packages.

Bibliography

Akay, H.U., Paydar, N.H., and Bilgic, A. 1997. Fatigue life predictions for thermally loaded solder joints using a volume-weighted averaging technique, *ASME Trans., J. Electronic Packaging,* 119 (December) 228–234.

Amagai, M., Watanabe, M., Omiya, M., Kishimoto, K., and Shibuya, T. (2002). Mechanical characterization of Sn-Ag based lead-free solders, *Micoelectronics Reliability,* 42, 951–966.

Antolovich, S.D., and Antolovich, B.F. (1996). *An Introduction to Fracture Mechanics in ASM Handbook. 19 Fatigue and Fracture,* ASM International®, 1996.

Arora, N.D., Raol, K.V., Schumann, R., and Richardson, L.M. (1996). Modeling and extraction of interconnect capacitances for multilayer VLSI circuits, *IEEE Trans. Computer Aided Design of Integrated Circuits and Systems,* 15(1), 58–66.

Bilotti, A.A. (1974). Static temperature distribution in IC chips with isothermal heat sources, *IEEE Trans. Electron Devices,* ED-21 (March), 217–226.

Black, J.R. (1969). Electromigration failure models in aluminium metallization for semiconductor devices, *Proc. IEEE,* 57(9), 1587–1594.

Blech, I.A., and Herring, C. (1976). Stress generation by electromigration, *Appl. Phys. Lett.,* 29, 131–133.

Box, G.E.P., Hunter, W.G., and Hunter, J.S. (1978). *Statistics for Experimenters—An Introduction to Design, Data Analysis, and Model Building.* New York: John Wiley & Sons, Inc.

Brakke, K.A. (1994). *Surface Evolver Manual.* University of Minnesota, Geometry Center.

Chen, C., and Liang, S.W. (2007). Electromigration issues in lead-free solder joints, *J. Mater. Sci.,* 18, 259–268.

Coffin, L.F., Jr. (1954). A study of the effects of cyclic thermal stresses on a ductile metal, *ASME Transactions,* 76, 931–950.

Darveaux, R. (1996). How to use finite element analysis to predict solder joint fatigue life, *Proc. VIII Int. Congress on Experimental Mechanics,* Nashville, TN, June 10–13, 1996, pp. 41–42.

Darveaux, R. (2000). Effect of simulation methodology on solder joint crack growth correlation, *Proc. 50th ECTC,* Las Vegas, NV, May 2000, pp. 1048–1058.

Dreezen G., Deckx, E., and Luyckx G. (2003). Solder alternative: Electrically conductive adhesives with stable contact resistance in combination with non-noble metallization, *CARTS Europe 2003,* pp. 223–227.

Gale, W.F., and Totemeier, T.C. (2004). *Smithells Metals Reference Book, (8th edition).* Maryland Heights, MO: Elsevier.

Galyon, G.T. (2003). *Annotated Tin Whisker Bibliography,* Herndon, VA: a NEMI Publication, July.

Guo, Q., Cuttiongco, E.C., Keer, L.M., and Fine, M.E. (1992). Thermomechanical fatigue life prediction of 63Sn/37Pb solder, *ASME Trans., J. Electronic Packaging,* 114, 145–150.

Hong, B.Z. (1997). Finite element modeling of thermal fatigue and damage of solder joints in a ceramic ball grid array package, *J. Electronic Materials,* 26(7), 814–820.

Hunter, W.R. (1997). Self-consistent solutions for allowed interconnect current density. I. Implication for technology evolution, *IEEE Trans. Electron Devices,* 44(2), 304–309.

Hunter, W.R., Self-consistent solutions for allowed interconnect current density. II. Application to design guidelines, *IEEE Trans. Electron Devices,* 44(2), 310–316.

Ju, T.H., Chan, Y.W., Hareb, S.A., and Lee, Y.C. (1995). An integrated model for ball grid array solder joint reliability, Structural analysis in microelectronic and fiber optic systems, *ASME,* EEP-12, 83–89.

Jung, W., Lau, J.H., and Pao, Y.-H. (1997). Nonlinear analysis of full-matrix and perimeter plastic ball grid array solder joints, *ASME Trans., J. Electronic Packaging,* 119 (September), 163–170.

Lall, P., Islam, N., Suhling, J., and Darveaux, R. (2003). Model for BGA and CSP reliability in automotive underhood applications, *Proc. 53rd Electronic Components and Technology Conference, 2003,* May 27–30, pp. 189–196.

Lall, P., Pecht, M., and Hakim, E. (1997). *Influence of Temperature on Microelectronic and System Reliability*. Boca Raton, FL: CRC Press.

Lau, J.H., and Pao, Y.H. (1997). *Solder Joint Reliability of BGA, CSP, Flip Chip and Fine Pitch SMT Assemblies*. New York: McGraw-Hill.

Lee, S.-W.R. and Lau, J.H. 1998. Solder joint reliability of cavity-down plastic ball grid array assemblies, *Soldering & Surface Mount Technology*, 10(1), 26–31.

Manson, S.S. (1965). A complex subject—some simple approximations, *Experimental Mechanics*, 5(7), 193–226.

Nagaraj, B., and Mahalingam, M. (1993). Package-to-board attachment reliability methodology and case study on OMPAC package, *ASME Advances in Electronic Packaging*, 4-1, 537–543.

Pan, T.-Y. (1994). Critical Accumulated Strain Energy (CASE) failure criterion for thermal cycling fatigue of solder joints, *ASME Trans., J. Electronic Packaging*, 116 (September), 163–170.

Pang, J.H.L., and Chong, D.Y.R. (2001). Flip Chip on Board Solder Joint Reliability Analysis Using 2-D and 3-D FEA Models, *IEEE Trans. Advanced Packaging*, 24(4), 499–506.

Pang, J.H.L., Xiong, B.S., and Low, H. (2004). Creep and fatigue characterization of lead free 95.5Sn-3.8Ag-0.7Cu solder, *Proc. 54th ECTC*, June 2004, pp. 1333–1337.

Pao, Y.-H., Jih, E., Adams, R., and Song, X. (1998). BGAs in automotive applications, *SMT*, January, pp. 50–54.

Paydar, N., Tong, Y., and Akay, H.U. (1994). A finite element study of factors affecting fatigue life of solder joints, *ASME Trans., J. Electronic Packaging*, 116 (December), 265–273.

Racz, L.M., and Szekely, J. (1993). An analysis of the applicability of wetting balance measurements of components with dissimilar surfaces, *Advances in Electronic Packaging*, ASME, EEP-4-2, 1103–1111.

Shi, X.Q., Pang, H.L.J., Zhou, W., and Wang, Z.P. (1999). A modified energy-based low cycle fatigue model for eutectic solder alloy, *Scripts Material*, 41(3), 289–296.

Strauss, R. (1998). *SMT Soldering Handbook, (Second edition)*. Maryland Heights, MO: Elsevier/Newnes.

Suo, Z. (2004). A continuum theory that couples creep and self-diffusion, *J. Appl. Mechanics*, 71, 646–651.

Syed, A.R. (2004). Accumulated creep strain and energy density based thermal fatigue life prediction models for SnAgCu solder joints, *Proc. 54th ECTC*, June 2004, pp. 737–746.

Syed, A.R. (1995). Creep crack growth prediction of solder joints during temperature cycling: An engineering approach, *Trans. ASME*, 117 (June), 116–122.

Teng, C.C., Cheng, Y.K., Rosenbaum, E., and Kang, S.M. (1997). iTEM: A temperature-dependent electromigration reliability diagnosis tool, *IEEE Trans. Computer-Aided Design of Integrated Circuits and Systems*, 16(8), 882–893.

Tu, K.N. (2003). *Solder Joint Technology: Materials, Properties, and Reliability*. Berlin, Germany: Springer.

Tu, K.N. (2003). Recent advances on electromigration in very-large-scale integration of interconnects, *J. Appl. Phys.*, 94, 5451–5473.

Tu, K.N. (1994). Irreversible processes of spontaneous whisker growth in bimetallic Cu-Sn thin-film reactions, *Phys. Rev. B*, 49(3), 2030–2034.

Tunga, K., Pyland, J., Pucha, R.V., and Sitaraman, S.K. (2003). Field-use conditions vs. thermal cycles: A physics-based mapping study, *Proc. 53rd Electronic Components and Technology Conference*, May 27–30, 2003, pp. 182–188.

Wiese, S., Schubert, A., Walter, H., Dudek, R., Feustel, F., Meusel, E., and Michel, B. (2001). Constitutive behaviour of lead-free solders vs. lead-containing solders-experiments on bulk specimens and flip-chip joints, *Proc. 51st Electronic Components and Technology Conference*, pp. 890–902.

Winter, P.R., and Wallach E.R. (1997). Microstructural modeling and electronic interconnect reliability, *Int. J. Microcircuits and Electronic Packaging*, 20(2), 124–129.

Zahn, B.A. (2002). Finite element based solder joint fatigue life predictions for a same die size-stacked-chip scale-ball grid array package, *SEMICON West, International Electronics Manufacturing Technology (IEMT) Symposium*, pp. 274–284.

Zahn, B.A. (2003). Solder joint fatigue life model methodology for 63Sn37Pb and 95.5Sn4Ag0.5Cu materials, *Proc. 53rd Electronic Components and Technology Conference*, May 27–30, 2003, pp. 83–94.

Yeh, C.-P., Zhou, W.X., and Wyatt, K. (1996). Parametric finite element analysis of flip chip reliability, *Int. J. Microcircuits and Electronic Packaging*, 19(2), 120–127.

9

Finite Element Modeling under High-Vibration and High-Temperature Environments

Solder joint fatigue failure is, at its essence, a fatigue crack growth problem. It is therefore natural that a nonempirical understanding of this problem can only result from adopting a fracture mechanics framework. While linear elastic fracture mechanics (LEFM) does provide approaches such as the Paris law that deals with fatigue crack growth, the assumptions made in these approaches are almost always not valid for studying crack growth in solder interconnections.

This chapter targets the reliability of electronics components under high random vibration conditions. Examples include electronics in vehicles, construction equipment, and aircraft. A fatigue life estimation procedure is presented, and each step of procedure is explained. A finite element model of the test vehicle is built in ANSYS. The model is first validated by correlating the natural frequencies, mode shapes, and transmissibility functions from simulation with experimentally measured ones. The model is then used to simulate a random vibration response spectrum at critical solder interconnects. The obtained results are then transformed into time domain, and the number of cycles is counted to determine the effective number of cycles that can then be used to predict fatigue life with Miner's damage rule. In this chapter we take a look at some of the major issues with electronics packaging under high-vibration environments, namely the application of ball grid arrays (BGAs) on rack-style printed wiring boards (PWBs). There are many factors that affect the BGA components on the circuit board, including

- Placement on the PWB
- Vibration level
- Temperature level
- Solder type

In addition to all these variables, every design is slightly different, with different resonances and requirements. Rack card assemblies do not have the ability to control the PWB resonances around the BGAs like other PWB designs without significant design improvements; the goal is to keep the design simple. Take

ACTEL 484 BGA as an example; BGAs are used for programmable logic devices as well as processors for high amounts of input and output. It is critical to make sure that these programmable logic devices and processors operate under all conditions. Finally, BGAs are unreliable in higher-vibration environments, so a proper understanding of how each of the above variables affects the component is critical to having a good design the first time, and to not run development testing under common environments that are shown in this chapter.

9.1 Lead versus Lead-Free Solder

In modern automotive control modules, mechanical failures of surface-mounted electronic components such as microprocessors, crystals, capacitors, transformers, inductors, and ball grid array packages, etc., are major roadblocks to design cycle time and product reliability. This chapter presents a general methodology of failure analysis and fatigue prediction of these electronic components under automotive vibration environments. The mechanical performance of these packages is studied through a finite element modeling (FEM) approach for given vibration environments in automotive applications. Vibration simulation provides system characteristics such as modal shapes and transfer functions, and dynamic responses including displacements, accelerations, and stresses. The system-level model is correlated through vibration experiments. Using the results of vibration simulation, fatigue life is predicted based on cumulative damage analysis and material durability information. A detailed model of solder/lead joints is built to correlate the system-level model and obtain solder stresses. The predicted failure mechanism of the leads agrees with experimental observations. On the test vehicle with multiple components, one of the 160-pin gull-wing lead plastic quad flat packages was chosen as an example to illustrate the approach of failure analysis and fatigue life prediction.

The primary reason for looking at lead-free solder is to transform automotive electronics to green, lead-free solder. Each of the packages in the ACTEL 484 BGA is built as lead-free. This is because these components are not only for aerospace applications, but also for cell phones and computers. The automotive electronics field is looking to go to lead-free components for many reasons, including green compliance, ROHS (restriction of hazardous substances) compliance, and to avoid the obsolescence of key parts, mostly due to the fact that aerospace products' lifespan is up to 20 years of service. The difficulty is that high-volume products whose lifespan is around 1 to 2 years, which means high risk or part obsolescence, drive the electronics industry. In addition to different mechanical and thermal properties, lead-free solder has concerns with tin whisker growth. The last reason that certain industries are behind in making the change to lead-free solder is due to the different techniques and procedures to assemble these parts to the PWBs. Lead-based solder will not be around forever, hence the importance of studying the differences of lead versus lead-free solder and their mechanical responses in BGAs.

This chapter develops an assessment methodology based on vibration tests and finite element analysis (FEA) to predict the fatigue life of electronic components under random vibration loading. A specially designed printed circuit board (PCB) with BGA packages attached was mounted to an electrodynamic shaker and subjected to different vibration excitations at the supports. An event detector monitored the resistance of the daisy-chained circuits and recorded the failure time of the electronic components. In addition, accelerometers and dynamic signal analyzers were utilized to record the time-history data of both the shaker input and the PCB's response. The finite element-based fatigue life prediction approach consists of two steps:

1. The first step aims at characterizing the fatigue properties of the Pb-free solder joint (SAC305/SAC405) by generating the S-N (stress-life) curve. A sinusoidal vibration over a limited frequency band centered at the test vehicle's first natural frequency was applied and the time to failure was recorded. The resulting stress was obtained from the FEM through harmonic analysis in ANSYS.

2. Spectrum analysis specified for random vibration, as the second step, was performed numerically in ANSYS to obtain the response power spectral density (PSD) of the critical solder joint.

The volume-averaged Von Mises stress PSD was calculated from the FEA results and then transformed into time-history data through an inverse Fourier transform. Rainflow cycle counting was used to estimate cumulative damages of the critical solder joint. The calculated fatigue life based on the rainflow cycle counting results, the S-N curve, and the modified Miner's rule agreed with actual testing results. There are five different types of lead-free solder bump interconnections for flip-chip electronic packaging applications. Lead-free solder bumps are fabricated from

1. Pure tin (Sn)
2. Tin–bismuth (Sn–Bi)
3. Eutectic tin–copper (Sn–Cu)
4. Eutectic tin–silver (Sn–Ag)
5. Ternary tin–silver–copper (Sn–Ag–Cu) alloys

In this chapter, both FEM and experimental techniques are employed to assess the reliability of electronics components under random vibration loading conditions. A fatigue life estimation procedure helps analysts make relatively accurate predictions of an electronic component's fatigue life. For example, for fatigue life prediction, the Finite Element Modeling (FEM) of the test vehicle can be developed using ANSYS software. The model is first validated by correlating the natural frequencies, mode shapes, and transmissibility functions from simulation with experimentally measured ones. The model is then used to simulate both sinusoidal and random vibration tests to obtain the stress and the response spectrum at critical solder interconnects, respectively. Results show that the corner solder ball experiences the highest stress level under both tests; hence, averaged stress at this solder ball was used to construct the S-N curve of the solder material and later to calculate the fatigue life using the Steinberg method. A comparison between simulation and experimental results is conducted at the end and a good correlation is obtained.

9.2 Analytical Model: PCB Normal Modes and Displacement

Vibration loading has become very important in the reliability assessment of modern electronic systems. The objective of this analysis is to develop a rapid assessment methodology that can determine the solder joint fatigue life of BGAs and chip-scale packages (CSPs) under vibration loading. The current challenge is how to execute the vibration fatigue life analysis rapidly and accurately. The approach in this chapter involves global (entire PWB) and local (particular component of interest) modeling approaches. In the global model approach, the vibration response of the PWB is determined. This global model gives us the response of the PWB at specific component locations of interest. This response is then fed into a local stress analysis for accurate assessment of the critical stresses in the solder joints of interest. The stresses are then fed into a fatigue damage model to predict the life. The goal is to maintain as much accuracy and physical insight as possible while retaining computation efficiency.

The analysis is performed based on a 6×9-inch PCB. Of course there are many sizes and configurations that can be used to optimize the board design, but boards of this size are common in the lower fuselage electronics bays of aircraft. Placements will include the center of the board, upper-right corner, lower-left corner and left center, as shown in Figure 9.1. Lead-free assembly with fine-pitch BGA components can be prone to latent defects at the solder joint level. These defects could arise due to design, variations in the quality of incoming components or PCB, assembly process, shipping, or handling. Latent solder joint defects are often not detected in end-of-line inspections and electrical functional testing, and they pose a serious reliability risk of early life failures in the end-use application.

Each placement on the board will have different results based on the curvature of the board and the board displacement at that point. This means that every single solder ball will have a unique max stress based on the curvature and placement. This is one key item to find in the analysis. The first thing about doing a Finite Element Model, analytical calculations are invaluable

FIGURE 9.1
PWB BGA placement.

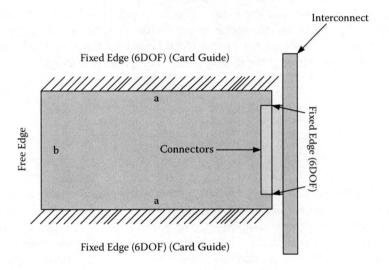

FIGURE 9.2
PWB and interconnect diagram.

to determining if the solution is correct. Equation (9.1) is used to solve for the first mode natural frequency of a PWB that is fixed on three edges and has one free edge (fixed means that the edge is controlled in all six degrees of freedom). The FEM boundary conditions are modeled to be exactly like a circuit card in a chassis. Figure 9.2 shows a PWB and interconnect diagram. The fixed positions are from the connectors at one end of the PWB that connect to the interconnect, and also the two card guides on the long edge of the PWB. BGA packages are a relatively new package type and have rapidly become the package style of choice. Many high-density, high-I/O count semiconductor devices are now only offered in this package style. Designers are naturally concerned about the robustness of BGA packages in a vibration environment when their experience base is with products using more traditional, compliant gull or J leaded surface-mount packages. Because designers simply do not have the experience, tools are needed to assess the vibration fatigue life of BGA packages during early design stages so as not to wait for product qualification testing, or field returns, to determine if a problem exists.

$$f_n = \frac{\pi}{3} * \frac{D}{\rho} * \sqrt{\left[\frac{.75}{a^4} + \frac{2}{a^2 b^2} + \frac{12}{b^4}\right]}$$ (9.1)

$$D = \frac{E * h^3}{12 * (1 - \upsilon^2)}$$ (9.2)

Equation (9.1) shows that the first mode natural frequency is calculated based on the plate stiffness D, the density (ρ), the length (a), and width (b). Equation (9.2) shows that the plate stiffness is calculated based on Young's Modulus (E), thickness (h), and Poisson's ratio (υ). The second equation is to solve for the displacement of the PWB based on the first mode natural frequency at the center of the board, Lucic (2010) and Steinberg (2000):

$$Z_{RMS} = \frac{9.81 * G}{f_n^2}$$ (9.3)

$$G = \sqrt{\frac{\pi}{2} * P * Q * f_n}$$ (9.4)

The displacement is really a dynamic, single-amplitude response based on the first mode natural frequency and G, or G_{RMS}. G_{RMS} is the response of the PWB based on the transmissibility, Q, first mode natural frequency, and the power spectral density value where the first mode is. The transmissibility is then calculated through random analysis. These results are then compared for accuracy based on the FEM. Although the displacement calculated is accurate, it does not give you the worst case. Based on a Gaussian distribution shown by Figure 9.3, Steinberg (2000) uses a three-band method approach to determine maximum displacements of the PWB that is used for the high cycle fatigue life. This is the point where maximum damage will occur in the electrical components. Although based on this statistical approach, damage will only occur 4.33% of the time, it must be considered in the overall damage calculations.

The basic idea is that when the number of cycles reaches the number of allowed, or calculated cycles, the part will ideally fail. Traditionally, this calculation is used to ensure that this never happens for any electrical component; BGAs do not always follow the same "simple calculations" and have relatively short cycle lives. There is a lot that can go wrong with a single BGA—improper installation, solder joint failures due to thermal or structural inputs—and for each of these failures, every solder ball can be slightly different. Controlling this environment is crucial as well as the processes involved in assembling the BGAs to PWBs.

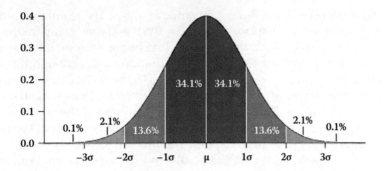

FIGURE 9.3
Gaussian distribution.

9.3 Finite Element Model: Random Vibration

BGA electronic packages as a function of components location, the number and size of the solder joints, mechanical clamping positions, and the vibration Power Spectral Density (PSD) during random vibration have been investigated. Assessing solder joint reliability under these conditions is especially important for harsh environments such as space, military, and automobile applications. A 3-D global/local FEM technique is used to simulate the random vibration responses of different-sized BGAs soldered onto a polyimide PCB and to determine the stresses/strains of BGA solder joints. A vibration fatigue life model based on Miner's cumulative damage index (CDI) and the derived solder effective strain is then used to predict BGA solder joint reliability. The FEM will be created in PATRAN with MD-enabled version 2010 and solved with MD NASTRAN 2008. The FEM is composed of a few different types of elements to try and decrease the model size. The PWB consists of 2-D Hex-8 shell elements, as well as the BGA body. The solder balls in the BGA will be modeled with 3D Hex-20 elements for detail and accuracy. All the elements are tied together using a Glue constraint for deformable bodies through PATRAN. The Glue constraint makes it much easier to connect multiple bodies and greatly reduces the number of user input REB2 or REB3 multi-point constraints.

The method that is used to solve the problem of BGAs and common aerospace random vibration environments consists of multiple independent steps. The first is to simply run a modal analysis of the PWB with no BGA attached. Having the BGA attached to the PWB will not affect the stiffness to any degree of concern and the model can now be much smaller. The modal analysis should line up with the calculated results; this is the success criterion for this step. The next step involves developing a model for random vibration cases. Again, the simple PWB model utilized for the modal analysis is used to keep the model size down during dynamic analysis. The result that is desired here is to find the max displacement of the PWB and then compare this to the manual calculations, which again should line up for each vibration case. The difference here is that for each random vibration case, although the natural frequencies are going to be the same as long as the boundary conditions are the same, the transmissibility will not be. This is the term that is calculated from the root-mean-square (RMS) of the vibration response of the board. From the modal analysis, it is easy to narrow down the area where Q will be the highest and thus the highest displacement. The manual calculations are modified and again compared with the FEM results to validate this step. With the idea of keeping the model as light as possible and the iterations as few as possible, it was not possible to run the full model with the BGA on the PWB. Instead, taking the results from the random acceleration analysis, and determining the max displacement, it is just as easy to run a static inertial analysis on the full FEM. This dramatically reduces the computation time for the analysis here. Although this deviates slightly from the original plan of running a dynamic simulation for each case, the results are very close to one another. This is a standard industry practice that is creative and very time forgiving.

Now that the thought process has been given for how the FEM will be solved, the boundary conditions must be analyzed. The most important boundary condition or load case is, of course, the random vibration environments that are analyzed. An example random vibration example is shown in Figure 9.4.

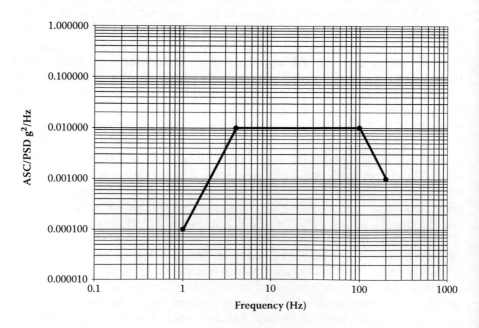

FIGURE 9.4
A random vibration example.

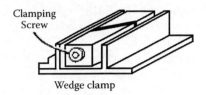

Clamping
Screw

Wedge clamp

FIGURE 9.5
Wedge lock card guides.

The PWB boundary conditions are simple as described in Steinberg (2000) for the 3 fixed edge case. This approach is chosen after a popular robust design in aerospace rack card assemblies. This design includes the use of wedge lock card guides. The wedge lock card guide essentially keeps the board edge fixed (all six degrees of freedom). In turn, this gives the designer much more stiffness that will have less displacement at the board center due to higher first modes. The last edge that is fixed comes from connectors on the daughterboard connecting to the motherboard as seen in Figure 9.5. Although this connection may not always be completely fixed due to tolerances in the rack assembly and improper mating with an interconnect, it is assumed that the mating conditions are fixed.

The last boundary condition for this model is the mating condition of the solder balls to the PWB and the solder balls to the BGA package. There are several ways to do this, including user-defined multi-point constraints (MPCs) (either REB2 or REB3), mesh matching, and the glue constraint. The REB2 MPC is used as a rigid or bolted connection that would be far too stiff for this application. The REB3 element would be ideal but very cumbersome, about 968 connections to match the solder balls to the PWB and the BGA body. Mesh matching is typically a very good option, although for this case, the mesh is very fine (0.008 inches), which results in a model too large to solve for either dynamic or static modeling with the computer hardware available. Finally, the glue boundary can be applied to the solid geometry to which the mesh is associated. This boundary sets up the model with REB3 connections that are ideal for this analysis case. These REB3 connections are created when the solution is solved in NASTRAN rather than PATRAN. The tolerances of these connections are set within PATRAN and are usually a percentage of the smallest shell or solid element.

Finally, the elements have all been chosen for a reason. The hex-8 shell elements are used for all shell elements. The hex-8 elements allow the mesh on the PWB and the BGA body to be much coarser due to the larger number of nodes. This is necessary to keep the model size down. For the 3-D solid solder balls, the mesh was done by taking a section view of the solder ball and performing a surface mesh seed. This 2-D mesh seed was then rotated to get the spherical shape of the solder ball. Hex-8 elements were used for the 2-D mesh seed so a more coarse mesh could be used and the element edges that are along the spherical edge actually resemble a sphere and map to the surface much better. A very fine hex-4 can be used to obtain the same result, but is much heavier in the model.

A noncontact laser holography exciting approach was adopted to conduct experimental modal analysis of a PCB assembly and find its dynamic characteristics. Then, its first-order natural frequency was used as the central frequency, and three different acceleration PSD amplitudes of narrow-band random vibration fatigue tests were respectively carried out. Subsequently, failed BGA solder joints were cross-sectioned and metallurgical analysis was done to investigate failure mechanisms of BGA lead-free solder joints under random vibration loading. The results showed that failure mechanisms of BGA lead-free solder joint vary as the PSD amplitude increases; solder joint failure locations change from the solder bump body of the PCB side to the solder ball neck, and finally to the Ni/intermetallic compound (IMC) interface of the package side; the corresponding failure modes change from fatigue fracture to brittle fracture.

9.4 FEM Model Optimization under High-Vibration Environment

FEA model optimization is crucial for this problem as the model size is a few hundred thousand elements and many more nodes. Because there are only three different bodies in this analysis—the PWB, BGA solder balls, and the BGA body—there are a few different mesh optimization ideas. First, the solder balls must be detailed and accurate, but cannot slow down the model and have such a fine mesh that the model will not even converge in a reasonable time. This will be done by looking at static cases of single solder balls under static loading conditions. Displacement and stress gradients are observed to get the most coarse mesh yet highly accurate results. The second option is to refine the mesh size of the PWB and the BGA body that best connect to the BGA balls with the very small tolerances applied to the glue boundary condition.

9.5 FEM Model Validation under High-Temperature Environment

Because there are no good experimental results available for this system, and the geometry and multiple materials present in the model make this fairly difficult to solve analytically, it is difficult to fully validate the results.

However, one can make a few observations. First of all, the curvature of the system agrees with intuition. The solder starts in an essentially stress-free molten state. Above the solder bumps is the Si semiconductor, which is fairly stiff and has a relatively small coefficient of thermal expansion; and below the solder bumps is the circuit board, which is fairly compliant and has a relatively large coefficient of thermal expansion (Figure 9.6). It is assumed that the solder bumps solidify just below their melting point. At this point, the solder bumps become rigidly connected to the Si semiconductor and the circuit board. To get an idea of curvature, imagine that the Si is much stiffer than the circuit board so that it has no curvature upon cooling to room temperature (obviously this is not entirely correct but Si is much stiffer than the circuit board; see Figure 9.6). Because the circuit board has a fairly large coefficient of thermal expansion, it will try to shrink significantly upon cooling. However, the solder bumps will resist this shrinking because they are now rigidly connected to the circuit board. Still, the bottom surface of the circuit board will be essentially free to shrink and hence will contract more than the top surface. This situation will create a concave downward curvature (see Figure 9.6).

Although it is difficult to develop an analytical model for the system model in Figure 9.6, one can easily develop an analytical solution for a thin layer of a material sandwiched between layers of another material that is cooled from an elevated temperature. A free-body diagram of this scenario is depicted in Figure 9.7. Free-body diagrams are diagrams used to show the relative magnitude and direction of all forces acting upon an object in a given situation. A free-body diagram is a sketch of an object of interest with all the surrounding objects stripped away and all of the forces acting on the body shown. The drawing of a free-body diagram is an important step in solving mechanics problems because it helps to visualize all the forces acting on a single object.

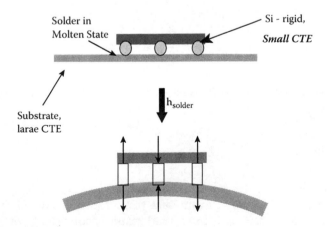

FIGURE 9.6
Predicted curvatures.

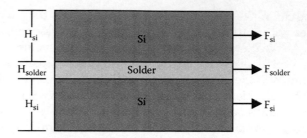

FIGURE 9.7
Free-body diagram of one material sandwiched between two others.

The net external force acting on the object must be obtained in order to apply Newton's Second Law to the motion of the object.

Deformation geometry (as solder and silicon are bonded):

$$\varepsilon_{Si} = \varepsilon_{Solder} \tag{9.5}$$

Force balance:

$$2F_{Si} + F_{Solder} = 0$$
$$\Rightarrow 2\sigma_{Si} h_{Si} + \sigma_{Solder} h_{Solder} = 0 \tag{9.6}$$

Material model:

$$\varepsilon_{Si} = \frac{\sigma_{Si}}{E_{Si}} - \nu_{Si}\frac{\sigma_{Si}}{E_{Si}} + \alpha_{Si}\left(T_{low} - T_{high}\right) \tag{9.7}$$

$$\varepsilon_{Solder} = \frac{\sigma_{Solder}}{E_{Solder}} - \nu_{Solder}\frac{\sigma_{Solder}}{E_{Solder}} + \alpha_{Solder}\left(T_{low} - T_{high}\right) \tag{9.8}$$

This is a system of four unknowns and four equations. The solution is

$$\sigma_{Solder} = \frac{(\alpha_{Solder} - \alpha_{Si})(T_{high} - T_{low})}{\dfrac{1 - \nu_{Solder}}{E_{Solder}} + \dfrac{1 - \nu_{Si}}{E_{Si}}\left(\dfrac{h_{Solder}}{2h_{Si}}\right)} \tag{9.9}$$

Substituting in the relevant parameters from Table 9.1 and Figure 9.7, one obtains $\sigma_{Solder} \approx 266.6 MPa$. This analytical result is consistent with the result from FEM. Hence, FEM is highly capable of handling this class of thermal problems with various materials.

TABLE 9.1

Materials Properties of Components

Material	Young's Modulus, E (GPa)	Poisson's Ratio (n)	Thermal Exp. Coefficient (10-6/K)
Sn-3.5Ag Solder	50	0.3	23
Underfill	6	0.35	30
Silicon chip	131	0.3	2.8
Bismaleimide triazene (BT) substrate	26	0.39	15

For example, although there are major differences with respect to scale and dimensions, the stress analysis of a chip component mounted at opposite ends has parallels with the stresses imposed by the environment on a bridge, and there are benefits from thinking in terms of the ability of both structures to withstand those stresses. In most mechanical structures, such thermal mismatch would be accommodated by elastic deformation, resulting sometimes in a high stress in the structure. With soldered assemblies, however, the situation is different, as the strength of the solder is low compared with that of the usual engineering materials. With leadless components, the materials of component and substrate are comparatively so rigid that a large part of the mismatch must be accommodated by plastic deformation in the solder joints. In this case, repeated movement due to temperature changes produces a cyclic stress, and fatigue failure may eventually follow. The shear strain experienced depends on the coefficient of thermal expansion (CTE) mismatch between the materials and the length:height ratio of the joint. As the CTE mismatch increases, so does the strain, and thus the thermal cycling life decreases. If rigid solder joints are to survive cycling during the specified life, the component size may have to be limited or the stand-off height increased to withstand large temperature fluctuations and CTE mismatch. The column grid array (CGA) is an example of a package where the stand-off height is deliberately made higher than a normal BGA (by using columns of high-melting solder) in order to accommodate CTE differences between its ceramic body and a PCB substrate.

Thermal mismatch, as a cause of plastic deformation in the solder, leading to fatigue fracture, finds its origin not only in differences in CTE, but also in differing rates of temperature change. During both soldering and operational life, the rates of heating and cooling of components and substrate are generally not the same, so that temperature differences are created even if the CTEs are matched, and these temperature differences generate stresses. In practice, the stresses fortunately remain fairly small, provided that no incorrect constructions have been used. However, if the rate of temperature change is very fast, as is the case in thermal shock testing, these stresses may become high.

Bibliography

Akay, H.U., Paydar, N.H., and Bilgic, A., 1997. Fatigue life predictions for thermally loaded solder joints using a volume-weighted averaging technique, *ASME Trans., J. Electronic Packaging*, 119 (December), 228–234.

Amagai, M., Watanabe, M., Omiya, M., Kishimoto, K., and Shibuya, T. (2002). Mechanical characterization of Sn–Ag based lead-free solders, *Micoelectronics Reliability*, 42, 951–966.

Antolovich, S.D., and Antolovich, B.F. (1996). *An Introduction to Fracture Mechanics in ASM Handbook 19. Fatigue and Fracture*, ASM International®.

Arora, N.D., Raol, K.V., Schumann, R., and Richardson, L.M. (1996). Modeling and extraction of interconnect capacitances for multilayer VLSI circuits, *IEEE Trans. on Computer Aided Design of Integrated Circuits and Systems*, 15(1), 58–66.

Bilotti, A.A. (1974). Static temperature distribution in IC chips with isothermal heat sources, *IEEE Trans. Electron Devices*, ED-21(March), 217–226.

Black, J.R. (1969). Electromigration failure models in aluminium metallization for semiconductor devices, *Proc. IEEE*, 57(9), 1587–1594.

Blech, I.A., and Herring, C. (1976). Stress generation by electromigration, *Appl. Phys. Lett.*, 29, 131–133.

Box, G.E.P., Hunter, W.G., and Hunter, J.S. (1978). *Statistics for Experimenters—An Introduction to Design, Data Analysis, and Model Building*. New York: John Wiley & Sons, Inc.

Brakke, K.A. (1994). *Surface Evolver Manual*. University of Minnesota, Geometry Center.

Chen, C., and Liang, S.W. (2007). Electromigration issues in lead-free solder joints, *J. Mater. Sci.*, 18, 259–268.

Coffin, L.F., Jr. (1954). A study of the effects of cyclic thermal stresses on a ductile metal, *ASME Trans.*, 76, 931–950.

Darveaux, R. (1996). How to use finite element analysis to predict solder joint fatigue life, *Proc. VIII International Congress on Experimental Mechanics*, Nashville, TN, June 10–13, 1996, pp. 41–42.

Darveaux, R. (2000). Effect of simulation methodology on solder joint crack growth correlation, *Proc. of 50th ECTC*, Las Vegas, NV, May 2000, pp. 1048–1058.

Dreezen, G., Deckx, E., and Luyckx, G. (2003). Solder alternative: Electrically conductive adhesives with stable contact resistance in combination with non-noble metallization, *CARTS Europe 2003*, pp. 223–227.

Gale, W.F., and Totemeier, T.C. (2004). *Smithells Metals Reference Book, (8th edition)*. Maryland Heights, MO: Elsevier.

Galyon, G.T. (2003). *Annotated Tin Whisker Bibliography*, Hearndon, VA: a NEMI Publication, July.

Guo, Q., Cuttiongco, E.C., Keer, L.M., and Fine, M.E. (1992). Thermomechanical fatigue life prediction of 63Sn/37Pb solder, *ASME Trans., J. Electronic Packaging*, 114, 145–150.

Hong, B.Z. (1997). Finite element modeling of thermal fatigue and damage of solder joints in a ceramic ball grid array package, *J. Electronic Materials*, 26(7), 814–820.

Hunter, W.R. (1997). Self-consistent solutions for allowed interconnect current density. I. Implication for technology evolution, *IEEE Trans. Electron Devices*, 44(2), 304–309.

Hunter, W.R. (1997). Self-consistent solutions for allowed interconnect current density. II. Application to design guidelines, *IEEE Trans. Electron Devices*, 44(2), 310–316.

Ju, T.H., Chan, Y.W., Hareb, S.A., and Lee, Y.C. (1995). An integrated model for ball grid array solder joint reliability, *Structural Analysis in Microelectronic and Fiber Optic Systems*, ASME, EEP-12, 83–89.

Jung, W., Lau, J.H., and Pao, Y.-H. (1997). Nonlinear analysis of full-matrix and perimeter plastic ball grid array solder joints, *ASME Trans., J. Electronic Packaging*, Vol. 119 (September) 163–170.

Lall, P., Islam, N., Suhling, J., and Darveaux, R. (2003). Model for BGA and CSP reliability in automotive underhood applications, *Proc. 53rd Electronic Components and Technology Conference*, May 27–30, 2003, pp. 189–196.

Lall, P., Pecht, M., and Hakim, E. (1997). *Influence of Temperature on Microelectronic and System Reliability*. Boca Raton, FL: CRC Press.

Lau, J.H., and Pao, Y.H. (1997). *Solder Joint Reliability of BGA, CSP, Flip Chip and Fine Pitch SMT Assemblies*. New York: McGraw-Hill.

Lee, S.-W.R. and Lau, J.H. (1998). Solder joint reliability of cavity-down plastic ball grid array assemblies, *Soldering & Surface Mount Technology*, 10(1), 26–31.

Lucic, M.J. (2010). "A Finite Element Study of Ball Grid Array Components in Common Aerospace Random Vibration Environments," Rensselaer Polytechnic Institute Thesis (Master of Science), Hartford, CT, December.

Manson, S.S. (1965). Fatigue: A complex subject—Some simple approximations, *Experimental Mechanics*, 5(7), 193–226.

Nagaraj, B., and Mahalingam, M. (1993). Package-to-board attachment reliability methodology and case study on OMPAC package, *ASME Advances in Electronic Packaging*, EEP-4-1, 537–543.

Pan, T.-Y. (1994). Critical accumulated strain energy (CASE) failure criterion for thermal cycling fatigue of solder joints, *ASME Trans., J. Electronic Packaging*, 116 (September), 163–170.

Pang, J.H.L., and Chong, D.Y.R. (2001). Flip Chip on Board Solder Joint Reliability Analysis Using 2-D and 3-D FEA Models, *IEEE Trans. Advanced Packaging*, 24(4), 499–506.

Pang, J.H.L., Xiong, B.S., and Low, H. (2004). Creep and fatigue characterization of lead free 95.5Sn-3.8Ag-0.7Cu solder, *Proc. 54th ECTC*, June 2004, pp. 1333–1337.

Pao, Y.-H., Jih, E., Adams, R., and Song, X. (1998). Bags in Automotive Applications, *SMT*, January, pp. 50–54.

Paydar, N., Tong, Y., and Akay, H. U. (1994), A finite element study of factors affecting fatigue life of solder joints, *ASME Trans., J. Electronic Packaging*, 116 (December), 265–273.

Razz, L. M., and Sleekly, J. (1993). An analysis of the applicability of wetting balance measurements of components with dissimilar surfaces, *Advances in Electronic Packaging*, ASME, EEP-4-2, 1103–1111.

Shi, E.X., Pang, O.K., Shout, W., and Wang, Z.A.P. (1999). A modified energy-based low cycle fatigue model for eutectic solder alloy, *Scripts Material*, 41(3), 289–296.

Steinberg, D.S. (2000). *Vibration Analysis for Electronic Equipment, (3rd edition)*. New York: John Wiley & Sons.

Strauss, R. (1998). *SMT Soldering Handbook, (Second edition)*. Maryland Height, MO: Elsevier/Newnes.

Suo, Z. (2004). A continuum theory that couples creep and self-diffusion, *J. Appl. Mechanics*, 71, 646–651.

Syed, A.R. (2004). Accumulated creep strain and energy density based thermal fatigue life prediction models for SnAgCu solder joints, *Proc. 54th ECTC,* June 2004, pp. 737–746.

Syed, A.R. (1995). Creep crack growth prediction of solder joints during temperature cycling: An engineering approach. *Trans. of the ASME,* 117, 116–122, June.

Teng, C.C., Cheng, Y.K., Rosenbaum, E., and Kang, S.M., iTEM: A temperature-dependent electromigration reliability diagnosis tool, *IEEE Trans. Computer-Aided Design of Integrated Circuits and Systems,* 16(8), 882–893.

Tu, K.N. (2003). *Solder Joint Technology: Materials, Properties, and Reliability.* Berlin, Germany: Springer.

Tu, K.N. (2003). Recent advances on electromigration in very-large-scale integration of interconnects, *J. Appl. Phys.,* 94, 5451–5473.

Tu, K.N. (1994). Irreversible processes of spontaneous whisker growth in bimetallic Cu–Sn thin-film reactions, *Phys. Rev. B,* 49(3), 2030–2034.

Tunga, K., Pyland, J., Pucha, R.V., and Sitaraman, S.K. (2003). Field-use conditions vs. thermal cycles: A physics-based mapping study, *Proc. 53rd Electronic Components and Technology Conference,* May 27–30, 2003, pp. 182–188.

Wiese, S., Schubert, A., Walter, H., Dudek, R., Feustel, F., Meusel E., and Michel, B. (2001). Constitutive behaviour of lead-free solders vs. lead-containing solders: Experiments on bulk specimens and flip-chip joints, *Proc. 51st Electronic Components and Technology Conference,* pp. 890–902.

Winter, P. R., and Wallach E. R. (1997). Microstructural modeling and electronic interconnect reliability, *Int. J. Microcircuits and Electronic Packaging,* 20(2), 124–129.

Yeh, C.-P., Zhou, W.X., and Wyatt, K. (1996). Parametric finite element analysis of flip chip reliability, *Int. J. Microcircuits and Electronic Packaging,* 19(2), 120–127.

Zahn, B.A. (2002). Finite element based solder joint fatigue life predictions for a same die size-stacked-chip scale-ball grid array package, *SEMICON West, International Electronics Manufacturing Technology (IEMT) Symposium,* pp. 274–284.

Zahn, B.A. (2003). Solder joint fatigue life model methodology for 63Sn37Pb and 95.5Sn4Ag0.5Cu materials, *Proc. 53rd Electronic Components and Technology Conference,* May 27–30, 2003, pp. 83–94.

10

Probabilistic Modeling of the Elastic-Plastic Behavior of 63Sn-37Pb Solder Alloys

The mechanism of solder joint failure under various mechanical loading conditions has emerged as one of the critical research subjects in the field of microelectronic packaging as its control has been proven to be key to the success of lead-free electronic packaging technology and also successful migration to lead-free technology. While recent studies address a few key failure mechanisms and their relation to loading conditions, it is not yet clear how fatigue failure interacts with the dynamic change in solder microstructure.

This chapter presents a comprehensive mechanics approach for solder reliability and integrity. This approach is capable of predicting the integrity and reliability of solder joint material under fatigue loading without viscoplastic damage considerations. It also presents the comprehensive damage model describing life prediction of the solder material under thermomechanical fatigue (TMF) loading. The method is based on the theory of damage mechanics, which makes possible a macroscopic description of the successive material deterioration caused by the presence of micro-cracks and -voids in engineering materials. A damage mechanics model based on the thermodynamic theory of irreversible processes with internal state variables is proposed and used to provide a unified approach in characterizing the cyclic behavior of a typical solder material. With the introduction of a damage effect tensor, the constitutive equations are derived to enable the formulation of a fatigue damage dissipative potential function and a fatigue damage criterion. The fatigue evolution is subsequently developed on the basis of the hypothesis that the overall damage is induced by the accumulation of fatigue and plastic damage. This damage mechanics approach offers a systematic and versatile means that is effective in modeling the entire process of material failure, ranging from damage initiation and propagation leading eventually to macro-crack initiation and growth. As the model takes into account the load history effect and the interaction between plasticity damage and fatigue damage, with the aid of a modified general-purpose finite element program, the method can readily be applied to estimate the fatigue life of solder joints under different loading conditions.

On a global scale, there is a continual movement to reduce the use of "toxic" materials in industrial processes and consumer products. This movement is present in electronics, an industry that has traditionally relied heavily on the use of eutectic tin–lead (63Sn-37Pb, wt.%) solder for the attachment of surface-mount electronics components to printed wiring boards. It is traditionally thought that leaded electronics devices are environmentally unsound, both from the perspective of industrial processes, as well as solid waste disposal (and subsequent leaching of lead waste into water resources). It appears, however, that the best approach to reduce the environmental impact of electronics products is to rework the entire product life cycle, particularly the recycling and disposal components, rather than create a blanket ban on leaded solders. Nonetheless, the pressure to eliminate lead in electronic interconnections will continue in the future from both the legislative and competitive sides.

The reliability concerns for solder joints are increasing exponentially with the increasing use of surface-mount technology (SMT) in the microelectronics industry. Solder alloys are most commonly used bonding materials in electronics packaging, which provide electrical and thermal interconnection, as well as mechanical support. The temperature fluctuations due to device internal heat dissipation and ambient temperature changes, along with the Coefficient of Thermal Expansion (CTE) mismatch between the soldered layers, result in thermomechanical fatigue of the solder joints. Progressive damage in solder balls eventually leads to device failure. Fatigue life prediction of solder joints is critical to the reliability assessment of electronic packaging.

The reliability of solder joints under the influence of thermomechanical loads in electronics packaging is influenced directly by the thermomechanical properties of the solder alloys. Most of the solder alloy creep constitutive relationships reported in the literature are based on a deterministic approach, which is known to be inadequate in representing the complex phenomena associated with the hot deformation of solder joints. This inadequacy, in turn, results in an additional uncertainty, especially in the mechanical behavior of near-eutectic (63Sn-37Pb) solders. It is well documented in the literature that measured thermomechanical properties of the near-eutectic solders exhibit large scatter even under relatively controlled laboratory measurement conditions. In complex loading conditions, creep and/or fatigue behavior usually occurs at high temperatures under cyclic loading, which results in material damage and hence shortening of the component's life. When the creep and fatigue operate simultaneously, it is necessary to consider not only the individual effects, but also the effect of their interaction, to obtain a more accurate prediction of component life.

Lead-free solders are soon to become the standard for use in electronic devices due largely to legislation prohibiting the use of lead and other toxic materials in electronics. Fortunately, there are inherent reliability advantages with certain lead-free solders. In the electronics industry, progress is marked by miniaturization, and increased integration and power density. In many applications, the Sn–Pb solder alloy (eutectic or otherwise) has reached its reliability performance limit. In terms of developing a new solder alloy that will allow for increased reliability in electronic devices, the alloy may as well be lead-free. Experts in the field of electronics generally agree that various versions of Sn–Ag–Cu solder are the best alternatives to the formerly ubiquitous eutectic Sn–Pb solder alloy. As such, two main motivations exist for a global shift to lead-free solder technologies for applications in electronics manufacture:

1. Environmental concerns
2. Reliability concerns

To minimize the impact of these uncertainties and concerns, Continuum Damage Mechanics (CDM) and Neural Networks Principles (NNP) have been applied to measured Sn–Pb solder property data to predict optimum elastic-plastic material model parameters. These parameters have been applied to conduct detailed nonlinear finite element analyses (FEA) of solder joints subjected to coupled thermomechanical loads.

10.1 Continuum Damage Mechanics

The state-of-the-art method for thermal fatigue life prediction is based on using empirical relations, such as the Coffin–Manson approach. Usually, the plastic strain range of a solder joint under thermal cycling is determined by finite element methods. Then Coffin–Manson curves, which are obtained from isothermal mechanical testing, are used to predict the fatigue life of solder joints. Usually, this approach yields very conservative results for thermal fatigue life prediction of BGA (ball grid array) packaging. Solder alloys as cast are thermodynamically unstable materials and microstructurally evolve into a stable equiaxed configuration over time under strain and heat. As a result, plastic strain accumulation in solder joints under thermal cycling is a nonlinear process, and the plastic strain range of just one or several cycles cannot appropriately reflect the physical mechanism of fatigue damage evolution. Continuum Damage Mechanics (CDM) has proved that this state-of-the-art practice underestimates fatigue life significantly. Moreover, the damage mechanism under mechanical loading is quite different from that under thermal loading. CDM quantitatively has shown that signal-to-noise (S-N) curves from isothermal loading cannot be used to predict thermal fatigue life.

It is well known that reliability and workability are the more important issues in the field of chip size package (CSP). Creep and fatigue behaviors are the main loads of the solder joints, the reliability of which should take account of those two main loads. The CDM focuses on damage evolution of interaction between the fatigue and creep. A new damage model of fatigue–creep interaction has been developed.

For CDM, the theoretical framework is based on the thermodynamic theory of energy and material dissipation and is described by a set of fundamental formulations of constitutive equations of damaged materials, the development equations of the damaged state, and evolution equations of microstructures. According to concepts of damage-dissipation of the material state and effective evolution of material properties, all these advanced equations, which take non-symmetrized effects of damage aspects into account, are developed and modified from the traditional general failure models so they are more easily applied and verified in a wide range of engineering practices by experimental testing.

- Use a state variable D to represent the effects of damage on stiffness, etc.
- Use the Second Law of Thermodynamics to develop the theory.
- Insights from experiments help us simplify the model; for example, we see only traverse and parallel cracks, even under shear, so we make D a diagonal second-order tensor.
- Derive explicit equations relating the thermodynamic forces to stress, thus greatly simplifying the theory and computations.
- Material behaviors are predicted up to initiation of a micro-crack.
- The model is incorporated in the finite element method (FEM).

CDM are widely used for the determination of material failure under creep, fatigue, and particularly creep–fatigue interaction processes. It is known that creep failure, as a function of time, usually occurs at high temperature by either ductile transgranular or brittle intergranular fractures. However, fatigue failure, as a function of the number of cycles during cyclic loading, takes place through the initiation and growth of micro-cracks. In addition, as the temperature increases, the interaction between the processes of creep and fatigue can lead to significant reductions in product life. Researches have shown that it may not be possible to get a reliable prediction of the processes by means of individual deformation behavior analysis. Unified constitutive models in terms of internal variables representing the damage caused by creep and fatigue, and the damage caused by their interaction, have thus been proposed. The main idea for the development of unified constitutive equations is based on the analysis of dislocation motion, and this concept has been considered by the inclusion of an inelastic strain term.

A thermodynamics-based damage mechanics rate-dependent constitutive model is used to simulate experiments conducted on thin-layer eutectic Sn–Ag–Cu (SAC) solder joints. The nondamage constitutive is measured by the bulk tensile test. The relationship between true stress and strain is $\sigma = 85.26\varepsilon^{0.3536}$. A damage evolution equation is proposed based on Lemaitre ductile damage theory and the constant in the equation is measured by the unloading elastic modulus method. The damage evolution equation is $D = 1.0689\varepsilon P - 0.0008$. FEM simulation of the shear test of the solder joint between Cu sticks employing damage mechanics rate-independent constitutives is uniform to practicable testing.

Much research has been done on SMT using the FEM. Little of this, however, has employed fracture mechanics and/or CDM. Here we present two finite element approaches incorporating fracture mechanics and continuum damage mechanics to predict the time-dependent and temperature-dependent fatigue life of solder joints. For fracture mechanics, the J-integral fatigue formula, $da/dN = C(J)m$, is used to quantify fatigue crack growth and the fatigue life of J-leaded solder joints. For CDM, the anisotropic creep–fatigue damage formula with partially reversible damage effects is used to find the initial crack, crack growth path, and fatigue life of solder joints. The concept of partially reversible damage is especially novel and, based on laboratory tests we have conducted, appears to be necessary for solder joints undergoing cyclic loading. Both of these methods are adequate to predict the fatigue life of solder joints. The advantage of the fracture mechanics approach is that little computer time is required. The disadvantage is that assumptions must be made on the initial crack position and the crack growth path. The advantage of CDM is that the initial crack and its growth path are automatically evaluated, with the temporary disadvantage of requiring a lot of computer time.

Researchers have investigated the crack growth path and creep rupture life of 63Sn-37Pb bulk solder experimentally by Moiré analysis and theoretically using CDM. In conventional fracture mechanics analysis, a crack growth path is assumed a priori. Fracture parameters such as the J-integral and stress intensity factor are subsequently calculated to predict the crack growth rate and structural life. However, it is often difficult to postulate the correct crack growth path for a complex structure that is subject to complex loading. CDM is an alternative method that can be used to compute structural life with the important feature that the crack growth path is computed automatically. A theory has been developed for partially reversible creep–fatigue damage. A finite element procedure incorporating this damage theory is developed and used to analyze the response of bulk solder plates with holes. The displacement fields obtained by finite element analysis are compared to the fringe patterns obtained from conventional Moiré experiments. Furthermore, the accuracy of the predicted crack growth paths was investigated by comparison between the maximum damage contours obtained by finite element simulation and the actual crack paths occurring in laboratory specimens. These comparisons indicate the new continuum damage theory developed herein can adequately predict creep displacements, crack growth paths, and structural life.

Fatigue damage is a progressive process of material degradation. The objective of this CDM is to experimentally qualify the damage mechanism in solder joints in electronic packaging under thermal fatigue loading.

Another objective of CDM is to show that the damage mechanisms under thermal cycling and mechanical cycling are very different. Elastic modulus degradation under thermal cycling, which is considered a physically detectable quantity of material degradation, was measured with a nanoindenter. It was compared with the tendency of inelastic strain accumulation of solder joints in a ball grid array (BGA) package under thermal cycling, which was measured by Moiré interferometry. Fatigue damage evolution in solder joints with a traditional load-drop criterion was also investigated by shear-strain hysteresis loops from strain-controlled cyclic shear testing of thin-layer solder joints. Load-drop behavior was compared with elastic modulus degradation of solder joints under thermal cycling. Following the conventional Coffin–Manson approach, the S-N curve was obtained from isothermal fatigue testing with load-drop criterion. Coffin–Manson curves obtained from strain-controlled mechanical tests have been used to predict the fatigue life of solder joints. CDM has shown that this approach underestimates the fatigue life such as the number of thermal fatigue cycles, by 10 times. Results obtained from CDM indicate that thermal fatigue and isothermal mechanical fatigue are completely different damage mechanisms for microstructurally evolving materials.

Recently, numerous damage mechanics-based models have been developed for the evaluation of thermomechanical fatigue reliability, which considers damage as an intrinsic material state. Most models use a set of internal state variables, known as damage variables, to describe the state of damage. With the introduction of damage variables at Gauss integration points in an FEM, damage distribution all over the structure can be characterized adequately as a function of time. On the other hand, traditional approaches just give a number-of-cycles-to-failure prediction, which cannot reflect the progressive process of fatigue damage evolution due to the growth and coalescence of micro-cracks. Furthermore, the traditional fatigue theory of Coffin–Manson, which is not an inherent intrinsic approach, cannot give the damage distribution of structure under fatigue loading.

Herein, damage is defined as "the gradual degradation of material strength due to growth and coalescence of smeared micro-voids or micro-cracks to initiate a single crack in the representative volume element (RVE) under continuous load application." As an intrinsic material property, the damage variable can be readily determined experimentally at the microscale, such as dislocation density or micro-crack density. Nevertheless, presently it is not feasible to directly quantify dislocation density or micro-crack density for use in a boundary value continuum mechanics problem. Elasticity is directly influenced by damage because the number of atomic bonds responsible for elasticity decreases with damage. For an actual engineering system, it is extremely difficult for the current state of material science to provide the level of theoretical guidance that is needed to develop a predictive model based solely on dislocation or crack density considerations. Instead, measurement of the degradation of global mechanical properties, such as elastic modulus, can be used to represent the evolution of dislocation density or micro-crack density.

10.2 Probabilistic Continuum Damage Mechanics Model

Solder joint structural damage accumulation is an intrinsically random phenomenon due to variations in the solder material microstructure, the joint loading history, and the operating environment of the electronic package. This is particularly true of 67Sn-37Pb solders used in electronic packaging, as the ratio of their operating temperature to the melting temperature is high and hence promotes hot deformation damage. When a material is deformed above 0.5Tm (where Tm is the material melting temperature), the resulting deformation is referred to as a hot deformation, where Tm is a function of the material density. Failed solder joints in electronic packaging data published in various journals revealed significant degradation in the microstructure resulting from hot deformation phenomena. Direct measurement of anisotropic material properties is a difficult task to accomplish, and therefore damage-related parameters have been used to quantify the cumulative damage in the laboratory setting. Extensive metallurgical interaction between the solder and the substrate metal induced by thermomechanical loads can also cause changes in the solder microstructure resulting in the onset of microcracking at the mesoscale. These microstructural defects may include voids, discontinuities, and the material inhomogenetics. Different damage methods have been proposed in the literature to quantify damage occurring in a material. This includes the plastic strain or maximum displacement method. Research has been ongoing to develop a Probabilistic Continuum Damage Mechanics (PCDM) model to predict the elastic-plastic constitutive parameters based on thermodynamic principles to argument finite element analysis.

10.2.1 Creep–Time-Dependent Deformation and Uncertainty

The near-eutectic solder alloy 63Sn-37Pb is the solder material used mainly for interconnection of components in electronic packaging. The eutectic composition 61.9Sn-38.1Pb exhibits the lowest temperature for total solidification of the alloy. The eutectic composition melting temperature is 183°C and, for this analysis, was assumed to be the melting temperature for the near-eutectic composition. The characterization of solders presents a challenge because the material behavior depends on temperature and strain rate. For thin solder layers commonly found in electronic packages, the problem becomes very significant because of the complication of simulating the thermomechanical loading history that may be considerably different for different layers even under controlled laboratory conditions. In light of this fact, CDM principles are applied to experimental data to develop probabilistic elastic-plastic constitutive parameters. From fundamental creep theory, the uniaxial creep equation is shown in Equation (10.1):

$$f(\varepsilon, \sigma, t, T) = 0 \tag{10.1}$$

where
  $\varepsilon(t)$ is strain history
  T is temperature
  $\sigma(t)$ is stress response in time
  t denotes time

Equation (10.1) can also be expressed in terms of creep strain as Equation (10.2):

$$\varepsilon^c = f(\sigma, t, T) \tag{10.2}$$

where
  ε^c is the creep strain.

Experimental studies of creep mechanisms found in the literature suggest that creep phenomena of Sn–Pb solders are mainly influenced by temperature in the following two ways:

1. The dependence of solder material constitutive constants on temperature
2. The structural changes of the solder alloy producing strain hardening or softening conditions

At an intermediate temperature range ($\sim 0.4 < T < 0.5 Tm$), the strain hardening effect in the material is diminished by the dislocation creep mechanism caused by thermal activity of the crystal lattice structure of the material. Also, at a higher temperature range ($\sim 0.5 Tm < T < 0.6 Tm$), the dislocation creep mechanism overcomes the strain hardening effect due to an increase in thermal activity, resulting in a secondary phase. This is mainly true for the steady-state creep regime. Under these conditions, the solder strength may be greatly reduced to a level such that it may not be adequate to resist the hot deformation caused by the creep mechanisms acting on the joint.

Several creep constitutive relationships have been proposed, including their respective inherent degrees of uncertainty. Mathematical models, in particular Sn–Pb eutectic creep models, are usually imperfect descriptors of complex physical phenomena; hence, predictions made from these models are plagued with some levels of uncertainties. To quantify these levels of uncertainties, a probabilistic approach to material damage behavior called *continuum damage mechanics* is presented, where the damage in the solder alloy is considered a random variable that must be predicted.

10.2.2 Thermodynamic Representation of Damage

The relationship between material microstructural disorder (damage) and entropy may be made by considering a deformable body R, receiving dilathermal Q from a heat reservoir. The rate of change in the body's entropy Ś, due to heat transfer Q under constant temperature θ, based on the Second Law of Thermodynamics, is given by

$$\dot{S} - \frac{\dot{Q}}{\theta} \geq 0 \tag{10.3}$$

Furthermore, Boltzmann, using statistical mechanics, proposed the connection between disorder and entropy as

$$\bar{s} = k \ln W \tag{10.4}$$

where
$\bar{s}$ *is* the entropy of the system
k is the Boltzmann constant
W is the disordered parameter

The disordered parameter W is considered stochastic in nature because it accounts for all the disorder or randomness in R, including void interactions, thermal fluctuations, and environmental effects. A function like the Helmholtz free energy, when defined with respect to entropy, is given as

$$\Phi = U - \theta S \tag{10.5}$$

where
U is the internal energy of R
θ is the absolute temperature
S is the entropy

If we consider the case of a unit mass in R, its entropy may be given by

$$S = \frac{N_o k}{m_S} \ln W \tag{10.6}$$

where
N_o is Avogadro's number
m_s is the specific mass

Substituting Equation (10.6) into Equation (10.4), we arrive at the following disorder expression:

$$W = e^{\left(\dfrac{(U - \Phi)}{\dfrac{N_o k\theta}{m_S}} \right)} \tag{10.7}$$

If we denote the initial damage in R by W_o, then the ratio of the change in disorder to initial disorder is given by

$$D = \frac{\Delta W}{W_o} \tag{10.8}$$

where D is referred to as the dynamic internal state variable of R. It measures the level of disorderness in R and can assume any value between 0 (undamaged) and 1.0 (fully damaged), depending on the degree of randomness or disorder in R and hence becomes a random variable. For a small temperature change θ, Equation (9.4) can be simplified to yield the damage parameter D for this analysis as

$$D = 1 - e^{\left(\dfrac{\dfrac{\Delta U - \Delta \Phi}{N_o k\theta}}{m_s} \right)} \tag{10.9}$$

In CDM, the damage is represented as a field variable. Inserting the damage models of discrete systems into continuum mechanics, one can get local state models without a length scale in the damage fields. However, a thermodynamic approach offers the possibility to extend the constitutive space beyond the local state assumption, including space derivatives of the damage variable. In this case, one can derive partial differential equations instead of ordinary ones for the evolution of damage. The weakly nonlocal thermodynamic theory gives a definite structure and a systematic method to get damage evolution equations. Evolution equations of CDM are usually expressed in terms of stresses; some researchers introduce other quantities, like strains or elastic energy.

10.2.3 Mechanical Representation of Damage

Damage is a progressive physical mechanism that leads to the initiation and growth of micro-voids or micro-cracks, leading to failure of a system. A proper understanding of damage growth is therefore essential in understanding the effect of the presence of voids and internal defects on the global response of mechanical and structural systems and also on the process that leads these internal defects to final fracture. CDM seeks to express the aggregate effect of microscopic defects present within the solder joint materials in terms of macroscopically defined quantities. These include quantities that can be measured at the macroscopic level such as tensile strength, shear strength, hardness, and Poisson's ratio. CDM defines damage as the density of defects/discontinuities on a cross-section in a given orientation, amplified by their stress-raising effect. For isotropic damage conditions, the concept of effective stress and strain equivalence may be used to describe relative to constitutive relationship under uniaxially loading as follows:

$$\bar{\sigma} = \frac{\sigma}{1-D} \tag{10.10}$$

By application of the principle of strain equivalence, the damage may alternatively be related to the fractional loss in material stiffness, given as

$$D = 1 - \frac{\bar{E}}{E} \tag{10.11}$$

where
σ is the normal stress
$\bar{\sigma}$ is the effective stress
D is the scalar damage variable
E is the elastic modulus of the undamaged and damaged materials
$\bar{E}$ is the elastic modulus of the undamaged and damaged materials

Equations (10.10) and (10.11) provide a means of experimentally measuring the extent of damage in a structural component using one of the many conventional and nondestructive methods; direct tension tests, ultrasonic pulse velocity, and electrical resistivity measurements have all been used to generate constitutive data. Here, failure in the solder material is assumed to occur when the damage variable equals a critical damage value D_c. It does not necessarily correspond to rupture but rather to the formation of macroscopic defects. The parameter D_c varies from 0.15 to 0.85 for engineering alloys, and hence is treated as a random variable and is used in this analysis for the purpose of providing the basis for stochastic CDM equation formulation.

CDM deals with elastic or inelastic materials that undergo structural weakening as a result of micro-crack formation or micro-void growth. In damage mechanics, scales are considered on three different levels: micro, meso, and macro. Micro-level damage is the accumulation of dislocations and micro-stresses near defects and interfaces, as well as bond tear-off. On the meso level, cracks begin to form due to unified motion and growth of micro-cracks or micro-voids inside a representative volume element. At the macro level, the cracks and voids grow. The first two scales can be examined using CDM variables while the macro scale is subject to crack mechanics. Considering micro scales, when the distance between atoms is at a critical value, the energy between these atoms reaches its maximum level. As a result of external loads and consequent changes in this distance between the atoms, a decrease in the interaction energy between the atoms occurs. This leads to a weakening in atomic bonds and a decrease in cohesion power. This situation creates micro-voids and discontinuity surfaces in a material. Thus, it can be stated that micro-level damage starts in the material. Continuum mechanics deals with quantities defined at a mathematical point. From a physical point of view, these quantities represent averages on a certain volume.

10.2.4 Stochastic Creep Damage Growth Prediction

With respect to the experimental observations, creep processes can be divided into three stages:

1. Primary creep, creep proceeds at a diminishing rate due to work hardening of the metal
2. Secondary or stationary creep, which reflects the equilibrium between softening and hardening
3. Tertiary creep, with a dominant increasing in material deterioration

These three stages can be obtained for all materials at elevated temperatures (in comparison with the melting temperature), but are independent of the temperature level, the loading rate, etc. These stages may be more or less significant.

Various constitutive relations for Sn–Pb eutectic solder joint alloys that account for both time-independent plasticity and steady-state creep have been developed. If we assume that the total strain rate consists of elastic strain and plastic strain rates, then in terms of pure shear, the total strain can be expressed as follows:

$$\varepsilon_{total} = \varepsilon_e + \varepsilon_p = \frac{\overline{\tau}}{G} + A\Phi\,\overline{\tau}t^{\Phi-1} \tag{10.12}$$

where
  G is the elastic shear modulus
  A is the material constant
  Φ is assumed to be 1.0 for steady-state conditions
  $\overline{\tau}$ is the effective shear stress
  t is time

The second part of Equation (10.12) is the Norton–Bailey Creep strain rate relationship under uniaxial loading conditions. The Norton–Bailey law is a suitable description for the stationary creep. This relation can be modified for the primary creep, introducing an explicit dependence on the time t, or for the tertiary creep, introducing a damage variable ω and defining a damage evolution law. An implicit, iterative semi-analytical integration scheme is used to integrate the Leckie–Hayhurst isotropic creep damage evolution equation as well as the Norton–Bailey creep constitutive equation. This scheme is incorporated into a finite element program dealing with thermal elastic-plastic creep problems. Creep damage evolution and rupture time in a high-temperature structure can be predicted. In recent years, several important advances in understanding the mechanics and mechanisms of creep deformation, damage, and fracture in polycrystalline alloys have been achieved through the synergistic efforts of materials scientists and engineers. Current understanding, while far from complete, nonetheless provides a physically based framework for developing computationally tractable continuum constitutive relationships that capture major features of the intrinsic mechanisms of creep damage and deformation. Such models, when applied to the analysis of scientifically and technologically relevant boundary value problems, provide a basis for a local approach to high-temperature fracture. It is concluded that an important internal damage parameter influencing macroscopic tertiary creep in conventional polycrystalline materials is the density of grain boundary facet cracks.

The development of a "damage" constitutive model can conceptually be broken into two parts. One aspect is to quantify the effect of an (instantaneous) state of damage on the mechanical behavior. The remaining aspect of the model is to provide evolution equations for the damage variable(s), and then develop a simplified model for the evolution of facet crack density under conditions of creep-constrained cavitation. A central feature of the model is the variability in cavity nucleation potency over the grain boundary population.

Under constant stress or steady-state creep, the increment in total strain is the same as that in creep strain and, therefore, when data noise is neglected, the following damage expression can be used to predict the critical damage parameter:

$$D_c(\tau_f) = 1 - \left[(1 - D_o)^{m+1} - \frac{4}{3} \left(\frac{A\tau_o^{m+1}}{\tau_f} \right) \right]^{\frac{1}{m+1}} \tag{10.13}$$

where

τ_o is the far-field stress acting on the surface approximately equal to residual stress

$D_c(\tau_f)$ is the critical value of D

A is the material constant determined from test data

τ_f is the true failure stress

m is the exponential material constant assumed to be unaffected by damage and determined by test data; assumed to be 1.2 for this work

Φ is taken to be 1 for steady-state condition

D_o is the initial damage (disorder in the material) assumed to equal to initial strain

Equations (10.9) and (10.11) both seek to define the disorder or damage parameter D, although from different perspectives. Also, Equation (10.13) assumes that, given an experimental dataset, the critical damage parameter $D_c(\tau_f)$, may be estimated indirectly by stochastic methods. For this work, the probabilistic neural networks (PNN) method was used and therefore a holistic aspect of neural networks is presented in the next section. The analysis of creep-damage processes is becoming more and more important in engineering practice due to the fact that exploitation conditions such as temperature and vibration are increasing while the weight of the structure should decrease. At the same time, safety standards are increasing too. The accuracy of mechanical state estimation (stresses, strains, and displacements) depends primarily on the introduced constitutive equations and on the chosen structural analysis model.

10.2.5 Neural Networks

Industry is highly interested in finding—quickly and accurately—the impact of various design or process parameters on the reliability of electronic packages under thermal cycling conditions. Based on the fact that performance of Pb-free solders in thermal cycling depends on and is sensitive to a large number of factors, a hybrid technique has been presented in research. The outcome of this research results in a (software-based) tool that can be used in design, evaluation, optimization, and trade-offs of electronic packages. The technique relies on combinations of mechanistic insights, trends identification, and neural network analysis to interpolate between accelerated test results for Pb-free soldered packages in a large dynamic database. Results show the effectiveness of combining those methods in achieving more accurate results. Such a tool serves as an aid for predicting and understanding the sensitivity of component reliability to working environments, material, package architectures, and board attributes. The advantage of this software-based tool is that it gives a designer the ability to make quick decisions about different packages or design scenarios in an offline, computer-based environment with minimal need for costly, time-consuming experimentation. The functionality and capabilities of this tool have been verified and validated against experimental test data.

Mechanical behavior is being determined by thermomechanical fatigue testing. A neural network model is also being developed concurrently with mechanical tests for predicting solder joint fatigue life. A neural network is a parallel distribution processor composed of simple processing units or nodes. It has a natural propensity for storing experimental knowledge and making it available for use. The network has the ability to learn a test dataset of the material by adaptation of the weights it assigns to each dataset or data point to achieve the desired objective. Figure 10.1 illustrates a typical neural network structure showing the input layer, the hidden layer, and an output layer.

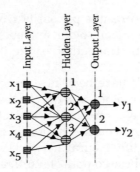

FIGURE 10.1
Partially connected feed-forward back-propagation neural network.

Similar connections exist for the other hidden neurons in the network. To satisfy the weight-sharing constraints for each of the neurons in the hidden layer of the network, the same set of synaptic weights is used. This weight-sharing constraint concept may be applied to Figure 10.1, where three local connections are connected to one neuron to make a total of three neurons. We can express the induced local field of hidden neuron j as given by Equation (10.14):

$$u_j = \sum_{j=1}^{3} w_i X_k \qquad j = 1, 2, 3 \qquad (10.14)$$

where $\{w_i\}^6_{i=1}$ constitutes the same set of weights shared by all four hidden neurons, and X_k is the signal picked up from the source node $k = i + j - 1$. $6 = 2 \times 3 =$ number of output layer nodes × number of hidden layer nodes. The mechanics of the solder joints are determined by thermomechanical fatigue testing. Once again, the potential for forming brittle intermetallics and degrading subsequent joint strength is a major concern. The data from these tests provides input to a neural network model for predicting solder joint lifetime. Test variables include strain rate, hold time, and test temperature, with two thermal cycling ranges of −55°C/125°C and 0°C/100°C as the test target. The test sample is a standard PWB thermomechanical fatigue tab with Au–Pt–Pd deposited on copper lands. Testing actual alumina substrates was not possible due to the inherent brittleness of the ceramic. Solders that demonstrated potential during the wetting and aging phase of the investigation were chosen for mechanical evaluation.

10.2.6 Neuron Models

A reliability model, based on neural networks, is developed for use on surface-mount devices. While a number of failure mechanisms are possible, the dominant failure mode for surface-mount reliability is fatigue resulting from cyclic differential thermal expansion.

This damage mechanism results primarily from thermal expansion differences between the component, solder, and substrate materials. These differences induce cyclic shear strains that lead to the accumulation of fatigue damage.

Several empirical models have been developed to describe the expected fatigue life of typical solder joints. Most of the resulting functions relate the different parameters to only a median life expectancy. This is of particular concern, as two solder alloys could exhibit different life characteristics, but the same median fatigue life. Neural networks, however, should increase the accuracy in modeling these complex failure relationships. The technique depends on "training" the network by processing experimental errors back through the network and adjusting the model until a minimum criterion is achieved. Experimental data consequently drives the life prediction, rather than arbitrary parametric functions.

The basic element (neuron) of neural networks was reported in the early 1940s. Consider the basic neuron model shown in Figure 10.2, consisting of a processing element with synaptic input connections and a single output. In this simple model, inputs are multiplied by their respective connection weights w_i, which will yield the effective input to the processing element given in Equation (10.2).

$$Z = \sum_{i=1}^{n} w_i X_i \tag{10.15}$$

This weight input is then passed to an activation function called the sigmoid function, which is expressed as

$$f(Z) = \frac{1}{1 + e^{-(Z+T)}} \tag{10.16}$$

where Z is the weighted input processing element and T is the bias parameter used to modulate the output element. The proper weight coefficients w_i and the bias parameter T are determined in the network training process.

A combination of experimental design, neural network modeling, and probabilistic design theory should therefore offer an improved approach

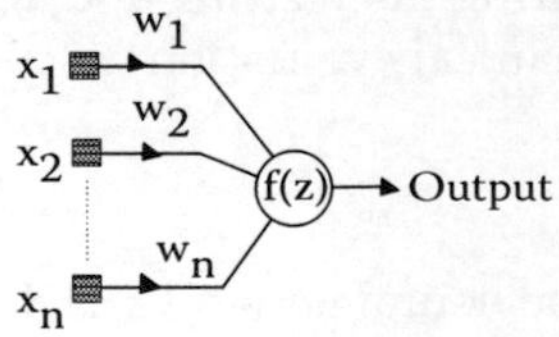

FIGURE 10.2
Simple neuron model.

for predicting solder joint life and providing a more direct link between the experimental results and solder joint reliability. Initial modeling has involved working with three input parameters: coefficients of thermal expansion for the solder and component/substrate materials, and cyclic temperature extremes. The output variable is fatigue life or cycles to failure. Thermomechanical fatigue test data is being collected for input to the model and for subsequent "tuning" of the neural network.

10.2.7 Network Training

Area-of-spread wetting experiments have been conducted with several fluxes and Pb-free solders on an Au–Pt–Pd thick film. Wetting was generally good with rosin and citric-acid-based fluxes. The low-solids flux yielded the poorest wetting results. Preliminary thermal aging data revealed substantial intermetallic growth in the solder joint with increasing aging temperature and time. Darkening of the low-solids and citric-acid flux residues occurred during environmental stressing (temperature-humidity-time), although no apparent corrosive reactions were observed between the residues and the solders or thick film. A neural network model is being developed with experimental mechanical inputs for predicting solder joint life.

A neural network requires training before it can be used to optimize an objective function. Weights are systematically modified so that the response of the network progressively approximates the desired response. Such a process may be interpreted as optimization of the objective function. The design variables of the optimization are the synaptic strengths (weights) and biases within each node, as well as the objective function. The output error function is defined by

$$F = \sum_n \left[\sum_k \|d_{kn} - Output_{kn}\|^2 \right] \qquad n = 1, N; \qquad k = 1, K \qquad (10.17)$$

where N is the number of training trials and K is the number of output nodes. The learning algorithm modifies the weights so that the error decreases during the learning process. Once the network is trained so that $error_{trained} \leq error_{max}$, it can be used as a function approximator for both continuous and discrete optimization problems.

10.2.8 Application to 63Sn-37Pb Solder Data

Probabilistic neural networks are feed-forward networks built with three layers. They are derived from Bayes decision networks. They train quickly because the training is done in one pass of each training vector, rather than several. Probabilistic neural networks estimate the probability density function for each class based on the training samples. The probabilistic neural network uses Parzen or a similar probability density function. This is calculated for each test vector. This is what is used in the dot product against the input vector as described below. Usually, a spherical Gaussian basis function is used, although many other functions work equally well.

Vectors must be normalized prior to input into the network. There is an input unit for each dimension in the vector. The input layer is fully connected to the hidden layer. The hidden layer has a node for each classification. Each hidden node calculates the dot product of the input vector with a test vector, subtracts 1 from it, and divides the result by the standard deviation squared (the standard deviation is the variance). The output layer has a node for each pattern classification. The sum for each hidden node is sent to the output layer and the highest values win. The probabilistic neural network trains immediately but execution time is slow and requires a large amount of space in memory. It only works for classifying data. The training set must be a thorough representation of the data. Probabilistic neural networks handle data that has spikes and points outside the norm better than other neural nets.

The prediction of the properties of the Pb-free solders has been focused on using the BP (back-propagation) neural network. Then different algorithms and parameters of the BP neural network were used in the training process, and the results have been analyzed for obtaining the optimum algorithm and parameter. To find the optimum solder alloy elastic-plastic parameters, a neural networks method was used to learn the mapping of the input variables (stresses) to the continuous output variable (strain). The dataset is divided into three data subsets: training, verification, and testing sets, respectively. For performance comparison with the standard regression analysis methods, a neural network computes the standard least squares fitting parameters, that is, the Pearson-R correlation coefficient between the actual and predicted outputs. A perfect prediction will have a correlation coefficient of unity.

Bibliography

Akay, H.U., Paydar, N.H., and Bilgic, A. (1997). Fatigue life predictions for thermally loaded solder joints using a volume-weighted averaging technique, *ASME Trans., J. Electronic Packaging*, 119 (December) 228–234.

Amagai, M., Watanabe, M., Omiya, M., Kishimoto, K., and Shibuya, T. (2002). Mechanical characterization of Sn–Ag based lead-free solders, *Micoelectronics Reliability*, 42, 951–966.

Antolovich, S.D., and Antolovich, B.F. (1996). *An Introduction to Fracture Mechanics in ASM Handbook 19 Fatigue and Fracture*, ASM International®, 1996.

Arora, N.D., Raol, K.V., Schumann, R., and Richardson, L.M. (1996). Modeling and extraction of interconnect capacitances for multilayer VLSI circuits, *IEEE Trans. Computer Aided Design of Integrated Circuits and Systems*, 15(1), 58–66.

Basaran, C., and Yan, C.Y. (1995). A thermodynamic framework of damage mechanics of solder joints, *J. Electronic Packaging*, 10(1), 365–376.

Bhattachaiya, B., and Ellingwood, B. (1998). Continuum damage mechanics-based model of stochastic damage growth, *J. Engineering Mechanics*, September, pp. 1000–1009.

Bilotti, A.A. (1974). Static temperature distribution in IC chips with isothermal heat sources, *IEEE Trans. Electron Devices*, ED-21(March), 217–226.

Black, J.R. (1969). Electromigration failure models in aluminium metallization for semiconductor devices, *Proc. IEEE*, 57(9), 1587–1594.

Blech, I.A., and Herring, C. (1976). Stress generation by electromigration, *Appl. Phys. Lett.*, 29, 131–133.

Boresi, A.P., et al. (1993). Advanced Mechanics of Materials, *(5th edition).* Canada: John Wiley & Sons.

Box, G.E.P., Hunter, W.G., and Hunter, J.S. (1978). *Statistics for Experimenters—An Introduction to Design, Data Analysis, and Model Building.* New York: John Wiley & Sons, Inc.

Brakke, K.A. (1994). *Surface Evolver Manual.* University of Minnesota, Geometry Center.

Chen, C., and Liang, S.W. (2007). Electromigration issues in lead-free solder joints, *J. Mater. Sci.*, 18, 259–268.

Coffin, L.F., Jr. (1954). A study of the effects of cyclic thermal stresses on a ductile metal, ASME Trans., 76, 931–950.

Darveaux, R. (1996). How to use finite element analysis to predict solder joint fatigue life, *Proc. VIII Int. Congress on Experimental Mechanics*, Nashville, TN, June 10–13, 1996, pp. 41–42.

Darveaux, R. (2000). Effect of simulation methodology on solder joint crack growth correlation, *Proc. 50th ECTC*, May 2000, pp. 1048–1058.

Dayhoff, J. (1990). *Neural Network Architectures – An Introduction.* New York: Van Nostrand Reinhold, pp. 217–243.

Dreezen G., Deckx, E., and Luyckx, G. (2003). Solder alternative: Electrically conductive adhesives with stable contact resistance in combination with non-noble metallization, *CARTS Europe 2003*, pp. 223–227.

Frear, D.R., and Kinsman, K.R. (1991). *Solder Mechanics—A State of the Art Assessment*, Minerals, Metals, and Material Society, PA: Warrendale.

Gale, W.F., and Totemeier, T.C. (2004). *Smithells Metals Reference Book, (8th edition).* Maryland Heights, MO: Elsevier.

Galyon, G.T. (2003). *Annotated Tin Whisker Bibliography*, a NEMI Publication, July.

Gilat, A., and Krisha, K. (1997). The effects of strain rate and thickness on the response of thin layers of solder loaded in pure shear, *J. Electronic Packaging*, 119(2), 81–84.

Grunwald, J., and Schnack, E. (1995). Models for shape optimization of dynamically loaded machine parts, *Proc. WCSM01*, Oxford: Pergamon, pp. 307–310.

Guo, Q., Cuttiongco, E.C., Keer, L.M., and Fine, M.E. (1992). Thermomechanical fatigue life prediction of 63Sn/37Pb Solder, *ASME Trans., J. Electronic Packaging*, 114, 145–150.

Haykin, S. (1997). *Neural Networks — A Comprehensive Foundation, (2nd edition)*, Upper Saddle River, NJ: Prentice-Hall, pp. 2–10.

Hong, B.Z. (1997). Finite element modeling of thermal fatigue and damage of solder joints in a ceramic ball grid array package, *J. Electronic Materials*, 26(7), 814–820.

Hunter, W.R. (1997). Self-consistent solutions for allowed interconnect current density. I. Implication for Technology Evolution, *IEEE Trans. Electron Devices*, 44(2), 304–309.

Hunter, W.R. (1997). Self-consistent solutions for allowed interconnect current density. II. Application to design guidelines, *IEEE Trans. Electron Devices*, 44(2), 310–316.

Ju, T.H., Chan, Y.W., Hareb, S.A., and Lee, Y.C. (1995). An integrated model for ball grid array solder joint reliability, Structural analysis in microelectronic and fiber optic systems, *ASME*, EEP-12, 83–89.

Jung, W., Lau, J.H., and Pao, Y.-H. (1997). Nonlinear analysis of full-matrix and perimeter plastic ball grid array solder joints, *ASME Trans., J. Electronic Packaging*, 119 (September), 163–170.

Lall, P., Islam, N., Suhling, J., and Darveaux, R. (2003). Model for BGA and CSP reliability in automotive underhood applications, *Proc. 53rd Electronic Components and Technology Conference*, May 27–30, 2003, pp. 189–196.

Lall, P., Pecht, M., and Hakim, E. (1997). *Influence of Temperature on Microelectronic and System Reliability*. Boca Raton, FL: CRC Press.

Lau, J.H., (Editor) (1991). *Solder Joint Reliability—Theory and Application*. New York: Van Nostrand Reinhold, p. 279.

Lau, J.H., and Pao, Y.H. (1997). *Solder Joint Reliability of BGA, CSP, Flip Chip and Fine Pitch SMT Assemblies*. New York: McGraw-Hill.

Lee, S.-W.R. and Lau, J.H. (1998). Solder joint reliability of cavity-down plastic ball grid array assemblies, *Soldering & Surface Mount Technology*, 10(1), 26–31.

Lemaitre, J. (1996). *A Course on Damage Mechanics*. Berlin: Springer-Verlag, pp. 11–36.

Manson, S.S. (1965). Fatigue: A complex subject—Some simple approximations, *Experimental Mechanics*, 5(7), 193–226.

Meekisho, L., and Nelson-Owusu, K. (1999). Stress analysis of solder joint with torsional eccentricity subjected to based excitation, Conference paper presented at the *12th Int. Conf. on Mathematical and Computer Modeling and Scientific Computing*, Chicago, IL, August 1999.

Muju, S., et al. (1999). Predicting durability, *Mechanical Engineering Magazine of ASME*, March, pp. 64–67.

Nagaraj, B., and Mahalingam, M. (1993). Package-to-board attachment reliability methodology and case study on OMPAC package, *ASME Advances in Electronic Packaging*, EEP-4-1, 537–543.

Ohring, M. (1998). *Reliability and Failure of Electronic Materials and Devices*. San Diego: Academic Press.

Pan, T.-Y. (1994). Critical accumulated strain energy (CASE) failure criterion for thermal cycling fatigue of solder joints, *ASME Trans., J. of Electronic Packaging*, 11 (September), 163–170.

Pang, J.H.L., and Chong, D.Y.R. (2001). Flip chip on board solder joint reliability analysis using 2-D and 3-D FEA models, *IEEE Trans. Advanced Packaging*, 24(4), 499–506.

Pang, J.H.L., Xiong, B.S., and Low, H. (2004). Creep and fatigue characterization of lead free 95.5Sn-3.8Ag-0.7Cu solder, *Proc. 54th ECTC*, June 2004, pp. 1333–1337.

Pao, Y.-H., Jih, E., Adams, R., and Song, X. (1998). BGAs in automotive applications, *SMT*, January, pp. 50–54.

Paydar, N., Tong, Y., and Akay, H.U. (1994). A finite element study of factors affecting fatigue life of solder joints, *ASME Trans., J. Electronic Packaging*, 116 (December), 265–273.

Racz, L.M., and Szekely, J. (1993). An analysis of the applicability of wetting balance measurements of components with dissimilar surfaces, *Advances in Electronic Packaging, ASME*, EEP-4-2, 1103–1111.

Shi, X.Q., Pang, H.L.J., Zhou, W., and Wang, Z.P. (1999). A modified energy-based low cycle fatigue model for eutectic solder alloy, *Scripts Material*, 41(3), 289–296.

Skrzypek, J.J., and Hetnarski, R.B. (1993). *Plasticity and Creep-Theory, Examples, and Problems*, Boca Raton, FL: CRC Press.

Steinberg, D.S. (2000). *Vibration Analysis for Electronic Equipment*. New York: John Wiley & Sons.

Strauss, R. (1998). *SMT Soldering Handbook, (Second edition)*. Maryland Heights, MO: Elsevier/Newnes.

Suo, Z. (2004). A continuum theory that couples creep and self-diffusion, *J. Appl. Mechanics*, 71, 646–651.

Syed, A.R. (2004). Accumulated creep strain and energy density based thermal fatigue life prediction models for SnAgCu solder joints, *Proc. 54th ECTC*, June 2004, pp. 737–746.

Syed, A.R. (1995). Creep crack growth prediction of solder joints during temperature cycling: An engineering approach, *Trans. ASME*, 117 (June), 116–122.

Teng, C.C., Cheng, Y.K., Rosenbaum, E., Kang, S.M. (1997). iTEM: A temperature-dependent electromigration reliability diagnosis tool, *IEEE Trans. Computer-Aided Design of Integrated Circuits and Systems*, 16(8), 882–893.

Tu, K.N. (2003). *Solder Joint Technology: Materials, Properties, and Reliability*. Berlin, Germany: Springer.

Tu, K.N. (2003). Recent advances on electromigration in very-large-scale integration of interconnects, *J. Appl. Phys.*, 94, 5451–5473.

Tu, K.N. (1994). Irreversible processes of spontaneous whisker growth in bimetallic Cu–Sn thin-film reactions, *Phys. Rev. B*, 49(3), 2030–2034.

Tunga, K., Pyland, J., Pucha, R.V., and Sitaraman, S.K. (2003). Field-use conditions vs. thermal cycles: A physics-based mapping study, *Proc. 53rd Electronic Components and Technology Conference*, May 27–30, 2003, pp. 182–188.

Wiese, S., Schubert, A., Walter, H., Dudek, R., Feustel, F., Meusel E., and Michel, B. (2001). Constitutive behaviour of lead-free solders vs. lead-containing solders experiments on bulk specimens and flip-chip joints, *Proc. 51st Electronic Components and Technology Conference*, pp. 890–902.

Winter, P.R., and Wallach, E.R. (1997). Microstructural modeling and electronic interconnect reliability, *Int. J. Microcircuits and Electronic Packaging*, 20(2), 124–129.

Yeh, C.-P., Zhou, W.X., and Wyatt, K. (1996). Parametric finite element analysis of flip chip reliability, *Int. J. Microcircuits and Electronic Packaging*, 19(2), 120–127.

Zahn, B.A. (2002). Finite element based solder joint fatigue life predictions for a same die size-stacked-chip scale-ball grid array package, *SEMICON West, International Electronics Manufacturing Technology (IEMT) Symposium*, pp. 274–284.

Zahn, B.A. (2003). Solder joint fatigue life model methodology for 63Sn37Pb and 95.5Sn4Ag0.5Cu materials, *Proc. 53rd Electronic Components and Technology Conference*, May 27–30, 2003, pp. 83–94.

11

Flip-Chip Assembly for Lead-Free Electronics

The fatigue crack growth in a solder joint has been monitored during dynamic cyclic loading induced by mechanical shock. Crack growth has been tracked via measurement of tiny resistance changes in the solder joint coupled with computational simulations of cracked joints. The unique fatigue characteristics observed are an insignificant crack nucleation period and a distinctive growth pattern for bulk solder cracks. Crack growth accelerates upon movement of the crack into intermetallic regions, a common occurrence for lead-free solders under high material strain rates.

Lead-free solder alloys are being investigated from both assembly and reliability perspectives. The lead-free solder bumps require more flux when using a dip flux application process and a nitrogen reflow atmosphere. The liquid-to-liquid thermal cycle reliability of the tin–silver–copper (Sn–Ag–Cu, SAC) eutectic bumps is slightly less than that of Sn–Pb bumps. Rapid innovation in packaging is evident from the introduction of new package formats, including area array packages (flip-chip ball grid array (BGA) and flip-chip chip-scale packages (CSP)); leadless packages, direct chip attach, wafer-level packaging (WLP), wirebond die stacking, flip chip–wirebond hybrid, PoP, PiP and other forms of 3-D integration and others. In addition, there are new packaging requirements emerging such as Cu/low-k materials, and interconnects to address the need for flexibility and expanding heat and speed requirements.

The introduction of low k and E-low k materials makes the low-k layer in the chip susceptible to mechanical stresses in the combined chip package structure. New environmental constraints such as Pb-free and halogen-free requirements enforced by law, and the use of electronics in extreme environments also force rapid changes. The introduction of these new materials and package architectures are posing new reliability challenges. For example, in the flip-chip package, the interaction of the stiffer Pb-free solder bump with the mechanically weaker low-k dielectric requires chip and design and materials selection to address reliability risks in chip-to-packaging interaction (CPI). This comes at a time when there must be substantially higher reliability on a per-transistor basis to meet market requirements.

The technology requirement for the Cost Performance Market has been the leader for package technology innovations in the past decade with the drive for performance in notebooks, game consoles, routers, and servers as the technology advances while keeping cost at bay. The leading package technologies are flip-chip BGA organic packages with large die and high density. The issues have been speed, heat dissipation, reliability, and cost. The rise of the mobile market with cell phones, smart phones, portable personal devices, and portable entertainment systems has brought up a different set of technology challenges in form factor and weight, functional diversification such as RF and video, system integration, reliability, time to market, and cost. The packaging community has responded with wafer-level packaging, new generations of flip-chip CSPs, various forms of 3-D stacked die and stacked packages, as well as fine pitch surface-mount and 3-D integrated circuits (ICs). They illustrate the dynamic nature of the packaging and assembly world in "More Moore" and "More than Moore." (More Moore means scaling as per Moore's Law. The "More than Moore's Law" movement focuses on system integration rather than transistor density.) As the decade closes, the transition from Pb-based solder, and the implementation of low=k and E low-k dielectric and finer bond pad pitch adds a new set of challenges for packaging technologists. Finally, there is the continuing rise in the price of gold works against consumer markets' expectation for cost reduction.

More and more electronics manufacturers are adopting the latest flip-chip packaging technology in their designs. To incorporate this technology successfully, manufacturers need to make some modifications to their SMT (surface-mount technology) assembly equipment, materials, and processes. A number of manufacturing-related issues must be addressed, including fluxing and underfill, and additional attention must be paid to yield- and quality-related issues. Since its development, and with the appropriate tweaks to assembly equipment and processes, flip-chip technology has been successfully integrated into many electronics products.

By implementing flip-chip technology, it is possible to reduce cost, increase yields, and reduce overall process steps. All these benefits can be achieved if a company converts to flip-chip design and selects appropriate equipment and materials for the assembly process. The use of low-k ILD (Inter-level dielectric) to reduce on-chip interconnect parasitic capacitance has exacerbated the difficultly of maintaining high thermomechanical reliability of die assembled on organic substrates in flip-chip packages. Due to the fragile nature of low-k ILDs in the die and their relatively poor adhesion to the surrounding materials, it is becoming progressively critical to minimize stresses imparted to the chip during thermal cycling and wafer-level probing. The large coefficient of thermal expansion (CTE) mismatch between the silicon die (3 ppm/°C) and the organic substrate (17 ppm/°C) has been shown to be destructive for ILD materials and their interfaces. This issue has motivated the investigation of new I/O interconnect technologies that minimize mechanical stresses on the chip. The pending replacement of lower-modulus Pb solder bump material by Pb-free solder bump material or copper pillar makes the problem more difficult. To this end, the device and package communities must collaborate together to address the chip package interaction issue in the design of under-bump metallurgy (UBM) structure, solder bump or Cu pillar, underfill materials, and surface finishes.

In addition, the use of solder bumps augmented with mechanically flexible electrical leads to replace underfill is a potential solution.

11.1 Flip-Chip Assembly Process

Introduced in 1964 by IBM as the C4 process (Controlled Collapse Chip Connection), flip chip now offers a viable and proven alternative to standard assembly technologies for products requiring enhanced performance. "Flip chip" is a term used to describe the multiplicity of mounting technologies that orient the face of the die toward the interconnecting substrate. Although flip-chip technology was inaugurated by IBM and Delco in the 1960s, it is now poised to become the interconnection method of choice for many die devices.

The term "flip chip" refers to an electronic component or semiconductor device that can be mounted directly onto a substrate, board, or carrier in a "face-down" manner. Electrical connection is achieved through conductive bumps built on the surface of the chips, which is why the mounting process is "face-down" in nature. During mounting, the chip is flipped on the substrate, board, or carrier (hence the term "flip chip"), with the bumps being precisely positioned on their target locations. Because flip chips do not require wire-bonds, their size is much smaller than that of their conventional counterparts.

The low parasitic electrical elements introduced by the bumps present the best interface to the board, the ability to place power and ground connections throughout the face of the die, and the fact that this is, in general, a "gang" bonding technique, in which are all playing a role in the conversion of wirebond devices to flip chip. To incorporate flip-chip technology into an existing product design, a number of changes to standard SMT processes are required. Typically, placement of the flip-chip device can be achieved using existing placement machines; in contrast, the underfill process required for flip-chip assembly does require special equipment. A typical flip-chip production line may be made up of a fluxing, placement and reflow segment, and a separate section with underfill and curing. The flip chip is structurally different from traditional semiconductor packages and therefore requires an assembly process that also differs from conventional semiconductor assembly. Flip-chip assembly consists of three major steps:

1. Bumping the chips
2. "Face-down" attachment of the bumped chips to the substrate or board
3. Underfilling, which is the process of filling the open spaces between the chip and the substrate or board with a nonconductive but mechanically protective material

Given the many different materials and technologies used in the bumping, attachment, and underfilling steps, the flip chip now comes in a vast array of variants. For flip-chip assembly, one of the most important advantages is the fact that the bonding pads are not required to be placed at the periphery of the die—and in fact, the preferred arrangement is an array configuration over the face of the die. Array I/O also improves power delivery and high-speed performance, and provides relaxed pitch for ease of die product assembly. Some of the technologies now finding favor are solder ball flip chip and adhesive flip chip. Assembly operations include handling, placing, fluxing, and solder joining. The assembly process is influenced by the bumped die packaging, the solder bump, the substrate or board material and size, the assembly equipment, the end product, and the cost.

Bumped die may be transported in waffle packs or tape and reel. Presentation may be either bump-up or bump-down, depending on the bumping and the equipment. Tape and reel requires special tapes designed for carrying flip chips. Placing the bumped die may be by fine-pitch surface-mount equipment, or by high-accuracy flip-chip placement equipment. In either case, the die must be aligned with the bond pads on the board before placement. Fluxes may be conventional or no-clean fluxes, with differing application and cleaning requirements. Soldering may be in a belt furnace or by hot gas or other local means.

In general, flip-chip assembly tends to be rather sensitive to variations in the small dimensions of the solder bumps and contact pads, as well as to substrate warpage in reflow. Because of the large numbers involved, particularly in high-density applications, manufacturing defect levels of concern are often related to the "tails" of statistical distributions of these parameters. To make matters worse, the importance of even a relatively low assembly defect level is enhanced by the difficulty (at best) of a subsequent repair.

- One type of assembly defect is proving sensitive, not only to the substrate tolerances but also to the accuracy of the placement machine. However, the risk of placing one of the solder bumps on top of the solder mask, or otherwise not in contact with the corresponding pad, can often be substantially reduced by the proper substrate technology selection and design.

- Another potential source of defects is bridging or openings due to the combined effects of solder bump height variations, substrate pad size variations, and substrate warpage in reflow. These effects depend, to some extent in contrasting fashions, on substrate technology and design, that is, whether contact pads are mask- or pad-defined, as well as on pad size, thickness, and shape.

Programs have been developed to predict the expected yield of flip-chip assemblies, based on substrate design and the statistics of actual manufactured boards (e.g., in pad sizes and locations, mask registration, etc.), as well as placement machine accuracy, variations in bump sizes, and possible substrate warpage. These predictions and the trends they reveal can be used to direct changes in design so that defect levels will fall below the acceptable limits. Shapes of joints are calculated analytically or, when this is not possible, numerically by means of a public domain program called Surface Evolver. The method is illustrated with an example involving the substrate for a flip-chip BGA. It was found that the original design would lead to unacceptably high defect levels, but alternative designs significantly improved the yield without creating other significant problems such as bridging.

11.2 Placement Stage

For placement, vision systems typically recognize SMD (surface-mounted device) components by their shape, but these systems do not work well with flip chips because size varies with the accuracy of the dicing saw. Flip chips are usually recognized by the bump pattern, requiring a technique that uses a high-resolution camera. Because the bumps are on the face of the die, the camera can mistake some non-bump features, such as large metal structures, for bumps. To compensate, an engineer may choose to program the vision system to exclude specific bumps from the recognition pattern.

Equipment for the placement of flip chips is in its third generation. Equipment is available to place devices with bump pitches down to 150 μm on 20-mm die. The ability to accomplish such high-accuracy placement in a high-volume manufacturing environment depends more on the vision system and the dimensional tolerance of the package than on the mechanical accuracy of the placement tool. The accuracy of placement is only as good as what the vision system "sees."

Vision and lighting systems must be adjusted to account for changes in substrate coloring, surface textures, reflectivity, and substrate transparency. These variables all affect what the vision system sees and therefore the calculated position of the die and package in space. The correct use of fiducials can simplify the vision process. The assembler should be included early on in the substrate design process to identify vision requirements and include them in the design or the assembly yield may suffer. Also, the importance of the dimensional tolerances of pad positions and size and their effects on bump yields cannot be overemphasized.

Increasing throughput for flip-chip placement is a practical challenge for future generations of equipment. Practical throughputs are currently approximately 1,000 CPH (chips per hour) when working with large complex die, and 25% to 50% slower if die are fluxed on the system. Technology is available for fluxing upstream of the flip chip attachment equipment on less expensive equipment. Fluxing options include dispensing, brushing, pad stamping, and non-contact jet spray fluxing.

The flip-chip components themselves may be handled out of trays. There are several configuration choices that must be made when considering flip-chip assembly on a pick-and-place machine. They include optical resolution, lighting geometry, processing capability, pick-and-place tooling, and substrate lifter tooling. Because the interconnect medium of a flip chip is a solder bump, simple die edge techniques are inadequate for locating and placing flip chips.

To avoid the possibility of placing flip chips in the wrong orientation, the programming bump pattern used for recognition should be asymmetric. The pick-and-place machine must have the optical resolution, lighting geometry, and processing capability to locate a pattern of bumps on the bottom side of the flip chip. The pick-and-place machine must then possess the intrinsic accuracy to place the die precisely on a substrate. In addition, pickup tooling must be properly sized and of the correct material such that when a die is dipped in flux (thin-film flux application), it is not dropped in the flux applicator and there is a clean release when it is placed.

Finally, the under-board support must rigidly capture the substrate so that there is no movement in the substrates due to die placement contact. A well-chosen support mechanism will simplify the handling while supporting the substrate during printing, and will protect it from warpage and deformation due to sagging in the reflow and curing ovens. As always, there will be other issues that are specific to a given process that must be considered when choosing a pick-and-place configuration.

The yield of the placement process depends on a number of parameters, including board tolerance (solder mask, copper layer), pad design, and placement accuracy. Solder mask tolerance is one of the major factors affecting flip-chip assembly yields; because of this, it may be necessary to experiment with different parameters to minimize the solder mask effect. For example, smaller pitches may introduce the effect of solder mask registration due to different required pad designs and therefore require drastically better placement accuracy and solder mask tolerances.

11.3 Underfill Stage

One function of the bump is to provide a space between the chip and the board. In the final stage of assembly, this under-chip space is usually filled with a nonconductive "underfill" adhesive joining the entire surface of the chip to the substrate. The underfill protects the bumps from moisture or other environmental hazards, and provides additional mechanical strength to the assembly. However, its most important purpose is to compensate for any thermal expansion difference between the chip and the substrate. Underfill mechanically "locks together" chip and substrate so that differences in thermal expansion do not break or damage the electrical connection of the bumps.

Underfill may be needle-dispensed along the edges of each chip. It is drawn into the under-chip space by capillary action, and heat-cured to form a permanent bond. Various parameters have an impact on the underfill process. The multiple interactions between the various different materials involved clearly show that the materials used must be treated as a system; this should be reevaluated as soon as any one of the materials or the application changes.

Specifically, the process windows for flow temperatures, dispense pattern, and time must be established for the individual application and materials. Even more importantly, these parameters, as well as the choice of materials, have a tremendous impact on reliability performance.

Contamination of board and die during handling can have a major impact on the underfill process and overall reliability. Because of the robustness of the soldering process, a clean room is not required, although handling boards with gloves and protection of the boards from contamination, dust particles, and chemicals are mandatory before underfill curing.

Board cleaning after misprints or contamination of the board during repair must be avoided. Exposure of assemblies to moisture before underfill will impact the reliability performance. Unfortunately, the tolerable exposure to moisture depends on substrate design and materials, as well as the reliability requirements.

Inspection of materials and testing in the production are necessary to ensure high yield in production and acceptable reliability. The assembly yields and necessary process parameters depend, as discussed, on the substrate and die quality. Consequently, the quality of these materials must be monitored.

11.3.1 Capillary Flow Underfills

Underfills are used in flip-chip packaging to help mitigate the effects of the large coefficient of thermal expansion (CTE) mismatch between the silicon chip and the laminate circuit board. Underfills reduce the strain on solder joints to improve interconnect fatigue life. The conventional capillary flow underfill process involves fluxing, placing, and reflowing the flip chip, and dispensing the underfill along the sides of the chip. The underfill flows by capillary action to fill the area underneath the chip. Finally, a cure must be completed in an oven. Capillary flow underfill is needed today due to the large CTE mismatch between the silicon chip, laminate board, and solder interconnects. Underfills are used to reduce the strain on solder joints, resulting in significant improvements in interconnect fatigue life. Past studies, using cost analysis software, have indicated that the industry may lose at least $1 to 2 billion in lost production with the current SMT process. It has been estimated that the typical flip-chip assembly for cellular phones slows down the cycle time by more than 30%. However, no flow processing has shown that it can be done quicker and easier, while still meeting today's electronics on-board requirements.

Capillary materials are currently used in the manufacture of flip-chip assemblies. Researchers have examined the manufacturing steps required for flip-chip-on-laminate and flip-chip-in-package assembly and the impact of these additional steps on the standard SMT production cycle time. The three key process steps of concern are substrate dehydration, underfill dispense and flow, and underfill cure. Heating during the reflow cycle just prior to underfill dispense is shown to significantly decrease the moisture content of the substrate, potentially eliminating the need for a dehydration bake prior to assembly. Substrate temperature during underfill dispense is shown to be a major factor in underflow speed. Fast flow materials are typically self-filleting, thus eliminating the need for an additional fillet dispense. Snap cure underfill materials can be cured in-line using a standard reflow oven with a modified temperature profile.

Numerical methods for simulating the underfill flow of conventional capillary flow and no-flow types in flip-chip packaging have been developed. Analytical models for the two types of underfill encapsulation processes have been proposed. In the capillary flow type, the underfill material is driven into the cavity with solder bump by the surface tension with an effect of contact angle as the capillary action. In the no-flow type, the movement of the IC chip during the reflow attachment is controlled by an appropriate loading to get the proper interconnect between the IC chip and substrate. In both types, the flow behavior and filling time of underfill material in the underfilling encapsulation process have been investigated, taking the fluid dynamic force acting on the solder bump into account. It was found that the proposed analytical models have a considerable potential for predicting the underfill flow.

11.3.2 Fluxing Underfills

Flip-chip-on-laminate (FCOL) provides advantages in size, weight, and performance for portable products. However, the industry has been slow in adopting FCOL technology. One issue often cited is the equipment and time associated with capillary underfill dispense, flow, and cure. Fast flow, snap cure underfills are one solution to decreasing dispense, flow, and cure times in a high-volume production environment.

Fluxing underfills eliminate the post-reflow underfill dispense and cure steps. The fluxing underfill is applied to the board prior to die placement, typically by dispensing. The fluxing underfill serves as the soldering flux, and then cures to function as the underfill. The material may cure during the reflow process or require post-reflow curing. Issues with these materials have included shelf life, placement voids, repeatability of fluxing activity, profile sensitivity, post-reflow curing, and thermal cycle performance. Researchers are evaluating a new generation of fluxing underfill that demonstrates good fluxing activity and cures during the reflow cycle. Assembly process optimization and reliability studies are under investigation.

Fluxing underfill eliminates process steps in the assembly of FCOL when compared to conventional capillary flow underfill processing. In the fluxing underfill process, the underfill is dispensed onto the board prior to die placement. During placement, the underfill flows in a "squeeze flow" process until the solder balls contact the pads on the board. The material properties, dispense pattern, and resulting shape, solder mask design pattern, placement force, placement speed, and hold time all impact the placement process and the potential for void formation. A design of experiments (DOE) was used to optimize the placement process to minimize placement-induced voids. The major factor identified was board design, followed by placement acceleration.

Fluxing underfills are polymer systems that incorporate fluxing activity into an underfill. Fluxing underfills contain three basic components: epoxy resins to provide the cured structure, fluxing materials such as organic acids in sufficient quantity to remove the oxides, and a catalyst to cure the material. The catalyst is chosen to provide latency until the solder starts to melt. The material and method of flux application have a significant impact on reliability and yield. Dip fluxing and spray fluxing are the most commonly used methods. Most of the fluxes used for spray fluxing are so-called "no residue" fluxes with an alcohol content of 98% to 99%. The fluxes are sprayed on the substrate before placement, and the alcohol evaporates rapidly at room temperature. The remaining flux does not provide sufficient tackiness to hold the flip chip in place during board handling and reflow, and frequently results in movement of the die. As a result, a dip-flux process may be adopted to apply a tacky flux to the flip-chip components. Typically, no-clean flux may be applied in a thick film on a rotating plate using a doctor blade. The die bumps may then be dipped in the flux and placed onto the substrate.

Prior to assembly, the boards must be dehydrated to remove absorbed moisture and to ensure that the solder mask has been fully cured. Moisture and volatiles from under-cured solder mask can cause bubbles (voids) in the fluxing underfill during the reflow temperature cycle.

The fluxing underfill is dispensed onto the dehydrated printed wire board (PWB) at the flip-chip site, and the die is placed through the fluxing underfill. During placement, the underfill flows in a "squeeze flow" process until the solder balls contact the pads on the board. Ideally, the fluxing underfill will make initial contact with the center of the die and flow radially outward as the die is placed. This would happen if the bottom of the flip chip were a flat plate, as is usually assumed in squeeze flow models.

However, flip-chip die have solder balls that interfere with the ideal flow pattern. Voids (trapped air) can be created in the underfill near the solder balls during placement. The viscosity, surface tension, and wetting characteristics of the underfill along with the placement parameters of force, velocity, or acceleration and hold time affect the formation of these placement voids. The assembly is then sent through a reflow oven. The fluxing underfill provides the necessary fluxing activity for good solder wetting. Depending on the cure kinetics, the underfill may cure during the reflow cycle or a post-reflow cure may be required.

During the reflow cycle, the fluxing underfill provides the fluxing action required for good wetting and then cures by the end of the reflow cycle. With small, homogeneous circuit boards, it is relatively easy to develop a reflow profile to achieve good solder wetting. However, with complex SMT assemblies involving components with significant thermal mass, this is more challenging. To get the large thermal mass components to temperature, the small flip-chip die will be at higher temperatures for longer periods of time. Use of predictive software tools to optimize the reflow profile and minimize temperature differences across the board is required.

A series of experiments was performed using these tools to optimize the reflow profile of a complex FCOL/SMT assembly. The profile obtained was used to successfully assemble flip-chip die with fluxing underfill. The reflow profile and cure kinetics are important for high-yield assembly. As the underfill temperature increases during the reflow cycle, the viscosity of the underfill decreases. However, as cure is initiated, the viscosity of the under-fill will increase. When the solder melts and begins to wet to the substrate metallization, the viscosity of the underfill must be low enough to allow collapse of the chip. Failure of the chip to collapse will result in poor or open solder joints.

Compared to liquid fluxes, which provide no tackiness for the die, tacky paste fluxes will hold the die in place during handling. As a result, problems due to handling are unlikely. The choice of flux and encapsulant is influenced by the die passivation, bump metallurgy, substrate, solder mask, and pad metallurgy. A good knowledge of bump height distributions and possible bump defects is essential in determining the process window for the dip fluxing process.

The minimum flux film thickness depends on bump height variations within a die. To ensure good soldering of the eutectic bumps on the die, all the bumps must be dipped in the flux. To establish a suitable process window for the fluxing, experiments may be performed using bare laminate.

11.3.3 Reworkable Underfills

Filling the interspace between the package and the PWB, namely under-filling, was demonstrated to yield dramatic reliability improvement in the mechanical shock and bending (flexing) of most CSP assemblies in mobile phone applications. Underfills protect the active surface of the die of flip chips, BGA, and CSP package types while improving their reliability by distributing stress away from the solder interconnects. This increases the performance of products in meeting drop, shock, and bend criteria.

Newer underfills are specifically designed to minimize the need to scrap entire boards with high-cost devices bonded on them because testing has determined that a device is defective. However, the ability of these devices to be reworked once they have been underfilled is challenging and time consuming. Studies have shown that underfill encapsulation dramatically improves the solder joint fatigue reliability of flip-chip-on-board (FCOB) assemblies. Traditional underfills do not allow easy removal and replacement of defective die after the underfill is dispensed and cured. A family of thermally reworkable materials is being studied. Key research activities center on developing a manufacturable rework process and assessing the reliability of flip-chip assemblies using reworkable underfills.

In response to the development of these reworkable underfills, some manufacturers of surface-mount rework stations are converting their machines for flip-chip placement capability in order to capitalize on the new market potential for flip-chip rework. These machines suffer from being designed to handle large circuit boards and large components, and thus lack the fine precision capability needed when working with flip chips. They do not provide the finely controlled spot heating, viewing magnifications, and precision bond load that are needed. The equipment also tends to have large footprints and is very expensive. In essence, these designs have been influenced by soldering rather than microelectronic considerations.

11.3.4 Wafer-Applied Underfills

Direct chip attach (DCA) on laminate is being used by only a few OEMs in high-volume manufacturing, mostly because it offers the advantages of reduced product size and weight, improved electrical performance, and higher input/output (I/O) counts. These are critical enabling characteristics for future high-performance, portable products and are equally important in other market segments. However, DCA use has been limited by the need to underfill after assembly, which adds process steps to the surface-mount assembly process. Wafer-applied underfill is applied and b-staged on the wafer, eliminating the underfill dispense and cure processes at the point of assembly. Key research areas include wafer coating, assembly process development, and reliability.

While the application of underfill materials directly to the wafer seems straightforward, many of the material requirements are incompatible with each other. For instance, the necessity of dicing the wafer using water is not compatible with the use of uncured epoxy materials. In addition, the incorporation of fluxing materials into the bulk underfill is known to degrade long-term stability at room temperature. This must be addressed, as the stated goal of the program is to provide at least 6 months of on-part life prior to use.

In the placement process, vision recognition of the coated die is critical. Contrast between the solder ball and the underfill as well as the partial coating over the ball can impact the vision system's ability to recognize the die. There are three distinct parts to the development of wafer-applied materials:

1. First, the coating process must be defined and then the chosen process must be further defined to provide a robust process. This process includes choosing the coating thickness, the uniformity on the wafer, the reproducibility of wafer formation, and dicing the wafer into individual die.

2. Second, the underfill material must be developed. Factors that must be considered include generating proper material rheology to provide uniform coating as noted above, as well as proper coloration to provide visual inspection capabilities of the solder balls.

3. Finally, the reliability of the wafer-applied underfill on the part must be assessed. This can be accomplished initially by evaluating various aspects of the material, such as adhesion as initially cured, after reflow cycles, after moisture exposure, after JEDEC (Joint Electron Devices Engineering Council) preconditioning, etc. Physical properties of the wafer-applied material, such as modulus, T_g (glass transition temperature), fracture toughness, and CTE, can all be measured without actually using flip-chip assemblies. The final conclusion is, of course, the assembly of the flip chips with the wafer-applied underfill into functional interconnect units and then evaluating their performance through thermal cycling, JEDEC preconditioning, etc.

For wafer-applied underfills, one of the key requirements is extended on part shelf life prior to use. Current technology used with liquid fluxing underfills falls far short in providing this stability as these systems struggle to have hours of stability at room temperature, let alone months. Fortunately, because the wafer-applied underfill is preapplied, a simple yet elegant solution is achieved. The materials of the fluxing layer are separated from the bulk underfill layer so that on-die life is not compromised, thus providing an avenue to provide the desired stability.

Underfills are used to improve the reliability of flip-chip assemblies. Reliability studies are an underlying element of flip-chip assessments. Liquid-to-liquid thermal shock is used as a rapid screening method to evaluate the relative reliability performance of different material and assembly process combinations. Thermal cycling is used as a second method of determining reliability, but requires longer test times. In both cases, the test vehicles are continuously scanning for continuity during the test. It has been observed that intermittent electrical failures occur at the high temperature extreme 100 to 750 cycles before the failure becomes permanent at room temperature. Thus, thermal cycle or thermal shock data generated by periodic measurements at room temperature significantly overestimate the reliability.

Underfill cracking, solder extrusion, and electrical shorting between adjacent, fine-pitched bumps have been observed by x-ray and specially designed test vehicles. Additional tests are underway with modified underfills to further assess this failure mode.

11.4 Finite Element Modeling of Die Stress

Failure induced by die cracking is one of the concerns in flip-chip packaging design and reliability analysis. In this section, a thermal stress model called the bi-material plate (BMP) model for analyzing flip-chip packages is presented. This analytical model, which has a closed-form solution, is validated by finite element modeling (FEM) and extensive experimental measurements for applications in flip-chip packages. Using this model, die stress and curvature can be determined effectively. It offers a significant advantage in estimating die stress and package reliability in the process of selecting and evaluating the design and material parameters for flip-chip packages. From this model, it is evident that the curvature and the bending stress are independent of die size if the edge effect is neglected. Furthermore, the bending stress is independent of absolute die thickness if the substrate-to-die thickness ratio is kept the same. The die curvature and the bending stress have a simple correlation in a certain range of thickness ratio.

The device performance of microelectromechanical system (MEMS) inertial sensors such as accelerometers and gyroscopes is strongly influenced by the stress developed in the silicon die during packaging processes. This is due to the die warpage in the presence of stress. It has previously been shown that most of the stress is generated during a die-attach process. Researchers have employed both experimental and theoretical approaches to gain a better understanding of stress development induced during the packaging processes of small silicon die (3.5×3.5 mm^2). The former approach is accompanied by an optical profilometer, the latter by a finite element analysis and an analytical model. Specific emphasis is given to the effects of structural parameters such as the die-attach adhesive thickness and material properties on stress development. The results from all three approaches show good agreement, in that more compliant and thicker adhesives offer greater relief in stress development, as well as bending the die convex-downward from its central location. A stress model not only provides a diagnostic tool for very small stress-sensitive devices, but it also presents a design tool for low-stress MEMS packaging systems.

Research efforts have focused on the FEM prediction of vertical die crack stresses in a flip-chip configuration, induced in major package assembly processes and subsequent thermomechanical loading. An extended Maxwell model is used to describe the time-dependent inelastic behavior of the solder

bumps. Two types of viscoelastic models, describing the mechanical properties of underfill resin during and after the curing process, are used. The die
stresses caused by both the soldering and the underfill curing processes are
obtained. These stresses are used as the initial stress-state for the further
modeling of subsequent thermal cycling. Using this methodology, the complete die stress evolution in a selected flip chip can be obtained; the physics
of thermal stress-induced vertical die cracks can be better understood and
the possible die cracks can be reliably predicted.

Mechanical stress distributions in packaged silicon die that have
resulted during assembly or environmental testing can be accurately characterized using test chips incorporating integral piezoresistive sensors.
Measurements of thermally induced stresses in FCOL assemblies are presented. In addition, stress variations have been monitored in the assembled
flip-chip die as the test boards were subjected to slow temperature changes
from −40°C to +150°C. Using these measurements and ongoing numerical
simulations, valuable insight has been gained as to the effects of assembly
variables and underfill material properties on the reliability of flip-chip
packages.

A series of stress die has been developed based on piezoresistive sensors
fabricated in oriented silicon. These can be fabricated into test die using standard semiconductor processing. Test die have been fabricated to measure
the stress on the backside of a flip-chip die. Backside die stress has been
observed in the industry to lead to die fracture. A second test die has been
fabricated that can measure the stress on the silicon due to the deposition of
UBM, solder ball formation, and underfill application and cure. The stress
during thermal cycling can also be measured. A key feature of oriented silicon is the ability to measure out-of-plane shear stresses. These stresses are
responsible for delamination of the underfill at the die surface.

The flip-chip plastic ball grid array (FC-PBGA) package is widely used in
high-performance components. However, its die back is normally under tensile stress at low temperatures. A probabilistic mechanics approach has been
used to predict the die failure rate in the FC-PBGA qualification process. The
methodology consists of three parts:

1. *Die strength test using four-point bending (4PB) method.* A specially
   modified three-parameter Weibull function is used to fit the 4PB
   die strength data. The three parameters of the Weibull distribution
   are used as the sole description of the cracking characteristics for a
   specific die process.

2. *The radius of curvature (ROC) measurement of the assembled FC-PBGA
   at room temperature.* The measured ROC of FC-PBGA at room temperature is used as a calibration input to determine the effective
   stress-free temperature of the FC-PBGA. It is used to overcome the
   difficulty caused by process-induced residual stress and unknown

material properties, for example, the underfill's viscoelasticity, that are normally unavailable. This effective stress-free temperature can be used in the stress analysis in the third part of the mentioned methodology.

3. *Finite element method (FEM) stress analysis.* FEM is used to calculate the die stress distribution under the most critical stage of a certain qualification process. The calculated stress distribution is combined with the Weibull distribution parameters of the die strength test to predict the die failure percentage.

A probabilistic mechanics approach is a general description of a probabilistic formalism of mechanics, that is, an extension of the Newtonian mechanics principles to the systems undergoing random motion. From an analysis of the induction procedure from experimental data to the Newtonian laws, it has been shown that the experimental verification of Newton law in a random motion implies a stochastic extension of the virtual work principle and the least action principle, that is, <dW> = 0 and <dA> = 0 averaged over all the random paths instead of dW = 0 and dA = 0 for a single path in regular dynamics. A probabilistic mechanics was formulated and applied to thermodynamic systems. Several known results, rules, and principles can be reproduced and justified from this new point of view. To mention some, we have obtained the entropy variation of the free expansion of gas and heat conduction without considering local equilibrium, and a violation of the Liouville theorem and the Poincaré recurrence theorem, which allows us to relate the entropy production to the work performed by random forces in a nonequilibrium process.

11.5 Gold Stud Bump Bonding

Stud bump bonding is a modified wire bonding process. As in wire bonding, there is bonding of the ball to the die pad. Unlike wire bonding, there is no second wire bond to a lead. The wire is terminated after the first bond, so there is only a bump on the die pad. To complete the interconnect, the die is flip-chipped onto a substrate using a thermosonic or thermocompression process.

Gold stud bumps are placed on the die bond pads through a modification of the "ball bonding" process used in conventional wire bonding. In ball bonding, the tip of the gold bond wire is melted to form a sphere. The wire bonding tool presses this sphere against the aluminum bond pad, applying mechanical force, heat, and ultrasonic energy to create a metallic connection. The wire bonding tool next extends the gold wire to the connection pad on the board, substrate, or lead frame, and makes a "stitch" bond to that pad, finishing by breaking off the bond wire to begin another cycle. The gold stud bump is summarized as follows:

- For gold stud bumping, the first ball bond is made as described, but the wire is then broken close above the ball. The resulting gold ball, or "stud bump," remaining on the bond pad provides a permanent, reliable connection through the aluminum oxide to the underlying metal.

- After the stud bumps are placed on a chip, they may be flattened (or "coined") by mechanical pressure to provide a flatter top surface and more uniform bump heights, while pressing any remaining wire tail into the ball. Each bump may be coined by a tool immediately after forming, or all bumps on the die may be simultaneously coined by pressure against a flat surface in a separate operation following bumping.

Gold stud bumping of semiconductor die provides an alternative to solder bumping for low-volume/low-I/O density applications. In this process, a thermosonic gold wire bonder is used to place gold balls on the semiconductor I/O pads. For assembly, the die is inverted and dipped into a thin layer of silver-filled epoxy. The die is then aligned and placed on the circuit board. Underfill is dispensed and flows under the die by capillary action. Curing the underfill completes the assembly process.

The assembly process and reliability of gold stud bump bonding is being evaluated. An alternate assembly process being developed involves thermo-compression bonding the gold stud bumped die onto Sn-plated pads on the PWB. The need for underfill with this assembly process for small MEMS die is under evaluation. Gold stud bumped die may be attached by conductive or nonconductive adhesives, or by ultrasonic assembly without adhesive. Conductive adhesive may be isotropic, conducting in all directions, or aniso-tropic, conducting in a preferred direction only.

Gold stud bumping forms gold bumps using a process very similar to gold ball wire bonding. Like wire bonding, it forms a gold ball (stud). However, the wire is terminated after the first bond, so there is only a bump on the die. Gold stud bumping serves over 97% of the present stud bump market. Gold stud bumps require no special die surface treatment or under-bump metal (UBM). They are formed with a modified wire bonder on any surface that is wire-bondable. After placing the ball portion of a gold wire bond, the wire is severed, leaving behind a gold stud. Gold stud bumps can be planarized by coining to ensure uniform bump heights. Gold stud bump reliability is well documented with many years of data. Unfortunately, gold stud bumps have several limitations for single-chip bumping. Most flip chips are intended for solder assembly onto boards or in packages. The electrical, mechanical, and thermal characteristics of gold stud bumps differ from those of solder bumps, adding a risk factor to prototype device performance evaluation. Gold stud bump connections use specialized attachment methods, such as conductive adhesives, requiring different materials, equipment, and process flows from solder attachments in a standard board line. An alternative is to use a stud bump for attachment to untreated die pads, and cap it with solder for the board connection.

Gold stud bumping requires no UBM or special wafer preparation, unlike the requirements for solder bumping. It also offers finer bump spacing than most solder bump technology without the added expense of a solder redistribution layer. Gold stud bumps with a gold–gold (Au–Au) interconnect (GGI) have developed as a niche segment of the flip-chip market. Gold stud bumping uses a variation of traditional wire bond technology to generate gold bumps on a wafer. After bumping, a wafer is diced and flipped, then thermosonically welded to the gold-plated substrate. Metallurgically, a monometallic thermosonic weld has higher strength and reliability than a solder joint produced by conventional flip-chip methods.

Gold stud bumps can be used on individual die or wafers and typically have much lower set-up costs than a solder bump approach. The ability to bump individual die makes Au stud bumping an extremely valuable tool in the prototyping phase, as well as a viable option for volume manufacturing. Joint development of the stud bump and flip chip die attach process, with optimization of all processes, provides a faster development path than a single-party development. These partnerships advance the capabilities of the industry by providing a complete solution. For many sensitive devices such as lasers, MEMS, and sensors, the use of flux or adhesives is not allowed, and a thermosonic or thermocompression Au–Au attach process offers a reliable flux-free process to improve device reliability.

11.6 Impacts on the Process

Contamination of board and die during handling can have a major impact on the underfill process and overall reliability. Because of the robustness of the soldering process, a clean room is not required, although handling boards with gloves and protection of the boards from contamination, dust particles, and chemicals is mandatory before underfill curing.

Board cleaning after misprints or contamination of the board during repair must be avoided. Exposure of assemblies to moisture before underfill will impact the reliability performance. Unfortunately, the tolerable exposure to moisture depends on substrate design and materials, as well as reliability requirements.

Inspection of materials and testing in production is necessary to ensure high yield in production and acceptable reliability. The assembly yields and necessary process parameters depend, as discussed, on the substrate and die quality. Consequently, the quality of these materials must be monitored.

Specifically, the impact upon the process of any changes such as new cleaning procedures or process flows for the substrate, as well as any changes in the bumping technique and underfill material, must be considered in order to evaluate the impact on the process.

Additionally, substrate tolerance must be monitored closely to ensure high yields in assembly. To ensure quality, it is necessary to understand the bumping process and monitor the bump height distributions. In addition, shear tests of the bumps can disclose systematic interface weaknesses of the bump. Depending on bumping technology, a change in failure mode or shear force can indicate systematic problems with the wafer lot. A change could indicate a weakness and possible interface problem of UBM with the die or solder bump.

11.7 Materials and Process Variations

Lot-to-lot changes of the encapsulants' properties are often difficult to detect. Underfill needs storage temperatures below −40°C and rapidly ages at room temperature. Temperature variations that occur during shipping or storage can substantially alter the performance of underfill materials. Simple standardized tests can be performed to monitor changes in encapsulant batches. Possible tests include a flow test and a test measuring the wetting angle to the substrate.

Material selection becomes one of the most essential parameters for reliability as well as processability. For example, materials with a low thermal mass will make uniform heating easier in reflow, during underfill, and curing.

Off-line x-ray test equipment is being used to do regular testing of soldering quality. Because of the circular pad design with attached traces, a two-dimensional x-ray is, in this case, sufficient to determine solder joint quality. Non-wetted pads on the substrate are easy to detect because the bumps will not deform during reflow.

Other observable failures are solder voids and solder bridging. Solder bridging can be caused by solder extrusions filling the voids touching the soldered bumps. Depending on the bump pitch, these voids can link two solder joints and cause bridging during thermocycling or additional reflow.

The underfill quality of the process was monitored with acoustic microscopy. Because submersion of all assemblies in water for testing is not desirable, spot-checks were performed in order to detect voids and incomplete flow. Alternatively, a destructive test can be performed by grinding the die off the assembly in order to inspect the underfill layer.

11.8 Integrating Flip Chip into a Standard SMT Process Flow

Various issues must be addressed for the integration of flip chip into a standard SMT process flow. The material selection of flux and encapsulant is influenced by the die passivation, bump metallurgy, substrate, solder mask, and pad metallurgy. Process windows and robustness make a dip fluxing process preferable. Board design and component pitches will significantly impact the assembly yield. For the underfill process, the process windows for flow temperatures, dispense pattern, and time must be established for the individual application and materials. Monitoring of the incoming materials such as bump height distribution and underfill flow performance as well as board quality is important because of the number of interactions in the process, specifically the underfill process. Contamination of the board and die during handling can have a major impact on the underfill process and overall reliability. Additional board baking may be required to ensure sufficient reliability and good underfill performance.

New materials and processes make it very difficult to maintain high yields, acceptable quality, and a "competitive edge." Ongoing support from and partnership with a large, application-relevant R&D organization is required. Nevertheless, standard solutions for conservative applications in terms of pitch and reliability requirements are becoming possible.

Selection of the material system is the key to required reliability as well as assembly yield. Different from other technologies, it is possible to implement flip-chip assembly in a step-by-step fashion as far as investments in facilities, equipment, and training are concerned. The decision to go with flip chip in a specific application can then be made on a case-by-case basis.

Bibliography

Akay, H.U., Paydar, N.H., and Bilgic, A. (1997) Fatigue life predictions for thermally loaded solder joints using a volume-weighted averaging technique, *ASME Trans., J. Electronic Packaging,* 119 (December), 228–234.

Amagai, M., Watanabe, M., Omiya, M., Kishimoto, K., and Shibuya, T. (2002). Mechanical characterization of Sn–Ag based lead-free solders, *Micoelectronics Reliability,* 42, 951–966.

Antolovich, S. D., and Bruce F. Antolovich (1996), *An Introduction to Fracture Mechanics in ASM Handbook 19 Fatigue and Fracture,* ASM International®, 1996.

Arora, N. D., Raol, K.V., Schumann, R., and Richardson, L.M. (1996), Modeling and extraction of interconnect capacitances for multilayer VLSI circuits, *IEEE Trans. Computer Aided Design of Integrated Circuits and Systems,* 15(1), 58–66.

Bilotti, A.A. (1974). Static temperature distribution in IC chips with isothermal heat sources, *IEEE Trans. Electron Devices,* ED-21 (March), 217–226.

Black, J.R. (1969). Electromigration failure models in aluminium metallization for semiconductor devices, *Proc. IEEE*, 57(9), 1587–1594.

Blech, I.A., and Herring, C. (1976). Stress Generation by Electromigration, *Appl. Phys. Lett.*, 29, 131–133.

Brakke, K.A. (1994). *Surface Evolver Manual*. University of Minnesota, Geometry Center.

Box, G.E.P., Hunter, W.G., and Hunter, J.S. (1978). *Statistics for Experimenters—An Introduction to Design, Data Analysis, and Model Building*. New York: John Wiley & Sons, Inc.

Chen, C., and Liang, S.W. (2007). Electromigration issues in lead-free solder joints, *J. Mater. Sci.*, 18, 259–268.

Coffin, L.F., Jr. (1954). A study of the effects of cyclic thermal stresses on a ductile metal, *ASME Trans.*, 76, 931–950.

Darveaux, R. (1996). How to use finite element analysis to predict solder joint fatigue life, *Proc. VIII International Congress on Experimental Mechanics*, Nashville, TN, June 10–13, 1996, pp. 41–42.

Darveaux, R. (2000). Effect of simulation methodology on solder joint crack growth correlation, *Proc. 50th ECTC*, May 2000, pp. 1048–1058.

Dreezen, G., Deckx, E., and Luyckx, G. (2003). Solder alternative: Electrically conductive adhesives with stable contact resistance in combination with non-noble metallization, *CARTS Europe 2003*, pp. 223–227.

Gale, W.F., and Totemeier, T.C. (2004). *Smithells Metals Reference Book, (8th edition)*. Maryland Heights, MO: Elsevier.

Galyon, G.T. (2003). *Annotated Tin Whisker Bibliography*, Hearndon, VA: a NEMI Publication, July.

Guo, Q., Cuttiongco, E.C., Keer, L.M., and Fine, M.E. (1992). Thermomechanical fatigue life prediction of 63Sn/37Pb solder, *ASME Trans., J. Electronic Packaging*, 114, 145–150.

Hong, B.Z. (1997). Finite element modeling of thermal fatigue and damage of solder joints in a ceramic ball grid array package, *J. Electronic Materials*, 27(7), 814–820.

Hunter, W.R. (1997). Self-consistent solutions for allowed interconnect current density. I. Implication for technology evolution, *IEEE Trans. Electron Devices*, 44(2), 304–309.

Hunter, W.R. (1997). Self-consistent solutions for allowed interconnect current density. II. Application to design guidelines, *IEEE Trans. Electron Devices*, 44(2), 310–316.

Ju, T.H., Chan, Y.W., Hareb, S.A., and Lee, Y.C. (1995). An integrated model for ball grid array solder joint reliability, *Structural Analysis in Microelectronic and Fiber Optic Systems, ASME*, EEP-12, 83–89.

Jung, W., Lau, J.H., and Pao, Y.-H. (1997). Nonlinear analysis of full-matrix and perimeter plastic ball grid array solder joints, *ASME Trans. J. Electronic Packaging*, 119 (September), 163–170.

Lall, P., Islam, N., Suhling, J., and Darveaux, R. (2003). Model for BGA and CSP reliability in automotive underhood applications, *Proc. 53rd Electronic Components and Technology Conference*, May 27–30, 2003, pp. 189–196.

Lall, P., Pecht, M., and Hakim, E. (1997). *Influence of Temperature on Microelectronic and System Reliability*. Boca Raton, FL: CRC Press.

Lau, J.H., and Pao, Y.H. (1997). *Solder Joint Reliability of BGA, CSP, Flip Chip and Fine Pitch SMT Assemblies*. New York: McGraw-Hill.

Lee, S.-W.R. and Lau, J.H. (1998). Solder joint reliability of cavity-down plastic ball grid array assemblies, *Soldering & Surface Mount Technology*, 10(1), 26–31.

Manson, S.S. (1965). Fatigue: A complex subject—Some simple approximations, *Experimental Mechanics*, 5(7), 193–226.

Nagaraj, B., and Mahalingam, M. (1993). Package-to-board attachment reliability methodology and case study on OMPAC package, *ASME Advances in Electronic Packaging*, EEP-4-1, 537–543.

Pan, T.-Y. (1994). Critical accumulated strain energy (CASE) failure criterion for thermal cycling fatigue of solder joints, *ASME Trans., J. Electronic Packaging*, 116 (September), 163–170.

Pang, J.H.L., and Chong, D.Y.R. (2001). Flip chip on board solder joint reliability analysis using 2-D and 3-D FEA models, *IEEE Trans. Advanced Packaging*, 24(4), 499–506.

Pang, J.H.L., Xiong, B.S., and Low, H. (2004). Creep and fatigue characterization of lead free 95.5Sn-3.8Ag-0.7Cu solder, *Proc. 54th ECTC*, June 2004, pp. 1333–1337.

Pao, Y.-H., Jih, E., Adams, R., and Song, X. (1998). BGAs in Automotive Applications, *SMT*, January, pp. 50–54.

Paydar, N., Tong, Y., and Akay, H.U. (1994). A finite element study of factors affecting fatigue life of solder joints, *ASME Trans., J. Electronic Packaging*, 116(December), 265–273.

Racz, L.M., and Szekely, J. (1993). An analysis of the applicability of wetting balance measurements of components with dissimilar surfaces, *Advances in Electronic Packaging, ASME*, EEP-4-2, 1103–1111.

Shi, X.Q., Pang, H.L.J., Zhou, W., and Wang, Z.P. (1999). A modified energy-based low cycle fatigue model for eutectic solder alloy, *Scripts Material*, 41(3), 289–296.

Steinberg, D.S. (2000). *Vibration Analysis for Electronic Equipment*. New York: John Wiley & Sons.

Strauss, R. (1998). *SMT Soldering Handbook, (Second edition)*. Maryland Heights, MO: Elsevier/Newnes.

Suo, Z. (2004). A continuum theory that couples creep and self diffusion, *J. Appl. Mechanics*, 71, 646–651.

Syed, A.R. (2004). Accumulated creep strain and energy density based thermal fatigue life prediction models for SnAgCu solder joints, *Proc. 54th ECTC*, June 2004, pp. 737–746.

Syed, A.R. (1995). Creep crack growth prediction of solder joints during temperature cycling: An engineering approach, *Trans. ASME*, 117 (June), 116–122.

Teng, C.C., Cheng, Y.K., Rosenbaum, E., and Kang, S.M. (1997). iTEM: A temperature-dependent electromigration reliability diagnosis tool, *IEEE Trans. Computer-Aided Design of Integrated Circuits and Systems*, 16(8), 882–893.

Tu, K.N. (2003). *Solder Joint Technology: Materials, Properties, and Reliability*. Berlin, Germany: Springer.

Tu, K.N. (2003). Recent advances on electromigration in very-large-scale integration of interconnects, *J. Appl. Phys.*, 94, 5451–5473.

Tu, K.N. (1994). Irreversible processes of spontaneous whisker growth in bimetallic Cu–Sn thin-film reactions, *Phys. Rev. B*, 49(3), 2030–2034.

Tunga, K., Pyland, J., Pucha, R.V., and Sitaraman, S.K. (2003). Field-use conditions vs. thermal cycles: A physics-based mapping study, *Proc. 53rd Electronic Components and Technology Conference*, May 27–30, 2003, pp. 182–188.

Wiese, S., Schubert, A., Walter, H., Dudek, R., Feustel, F., Meusel E., and Michel, B. (2001). Constitutive behaviour of lead-free solders vs. lead-containing solders experiments on bulk specimens and flip-chip joints, *Proc. 51st Electronic Components and Technology Conference*, 2001, pp. 890–902.

Winter, P.R., and Wallach E.R. (1997). Microstructural modeling and electronic interconnect reliability, *Int. J. Microcircuits and Electronic Packaging*, 20(2), 124–129.

Yeh, C.-P., Zhou, W.X., and Wyatt, K. (1996). Parametric finite element analysis of flip chip reliability, *Int. J. Microcircuits and Electronic Packaging*, 19(2), 120–127.

Zahn, B.A. (2002). Finite element based solder joint fatigue life predictions for a same die size-stacked-chip scale-ball grid array package, *SEMICON West, International Electronics Manufacturing Technology (IEMT) Symposium*, pp. 274–284.

Zahn, B.A. (2003). Solder joint fatigue life model methodology for 63Sn37Pb and 95.5Sn4Ag0.5Cu materials, *Proc. 53rd Electronic Components and Technology Conference*, May 27–30, 2003, pp. 83–94.

12

Flip-Chip Bonding Technique for Lead-Free Electronics

Current printed circuit board assemblies are empirically tested for reliability certification. A more rigorous approach is to use fracture mechanics. To apply fracture mechanics and crack growth models to reliability testing of microelectronics, such as ball grid array solder joints on microelectronic circuit boards, fundamental properties must be known. Fracture toughness and the stress intensity factor for the specific metallurgical system and geometry must also be known.

Emerging environmental regulations worldwide, most commonly in Japan and Europe, have targeted the elimination of lead in electronic products. Pure tin (Sn) and eutectic tin–copper (Sn–Cu) alloy are two possible candidates to replace tin–lead (Sn–Pb) as the solder bumping materials for flip-chip bonding. Without Pb, the stability of the Sn–Cu interface during reflow suffers from the rapid dissolution of Cu into molten Sn. The requirements of multiple reflow during the flip-chip manufacturing steps further complicate the issue. In some severe cases, some intermetallic droplets can move to the top surface of the bumps, which can cause failures during the flip-chip bonding process.

With the increased awareness of the lead-free solder in the electronic packaging industry, the development of the Pb-free solder on the low-cost FR-4 substrate is a necessity for flip-chip applications. However, it is found that the high reflow temperature of the Pb-free solders (normally >250°C) caused substrate burnt-related defects in the FR-4 substrates (T_g < 200°C) during the bonding process in the traditional reflow oven. To solve this problem, bonding was performed on the flip-chip bonder. Using this approach, no substrate burned was observed. Using the flip-chip bonder instead of the reflow oven, it is possible to solve the low T_g of the FR-4 substrate for lead-free solders' flip-chip applications. However, the drawback is that solder reflow data provided by solder paste suppliers must be adjusted for the flip-chip bonder.

In this chapter, a survey of flip-chip techniques, giving their main technological features, and the area of their applications are presented. Particular emphasis is given to micro-jet techniques as the only method that enables, in one technological process, the production of multilayer passive circuits with resistive or dielectric and some opto-electronic components and solder bumps for flip-chip bonding. The advantages and drawbacks of micro-jet technology and the range of its applications are also emphasized. The main factors affecting the bonding process by the flip-chip bonder in the Sn0.7Cu lead-free solder have been evaluated. These main factors include:

- Stage temperature
- Bonding pressure
- Head temperature
- Bonding time
- Moving distance

Sn0.7Cu solder is chosen due to its high melting temperature of up to 227°C. If we can demonstrate the solution for this high reflow temperature solder, we can apply the same approach to the other lower melting temperature solders such as Sn–Ag–Cu, Sn–Ag, and Sn–Bi on FR-4 substrates. The experiments were designed based on the Taguchi method. A die shear test of the bonded samples is selected as the tool to verify the die shear strength of the samples under different experiments. The signal-to-noise ratio (S/N ratio) is used to determine the optimum conditions for bonding by the flip-chip bonder. Based on the Analysis of the Variance (ANOVA), the degree of distribution of the five bonding factors to the die shear strength is determined.

12.1 Lead-Free Reflow Soldering Techniques and Analytical Methods

The assessment of Pb-free solder has been become a hot subject over the past few years. Governments want to exercise stricter control over lead-containing waste disposal and recycling requirements because of lead's potential hazards. Recently, a great number of Pb-containing electronic products such as televisions, radios, and mobile phones are being disposed of in landfills. Pb can potentially leach from the solder into underground water or municipal water supplies and cause serious effects to the environment. In the United States, a legislative ban has been proposed since 1990. At the time, there was no Pb-containing alloy replacement being identified, and legislation was dropped under strong pressure from the electronics industry. However, the history of government regulation suggests that Pb is being eliminated in products such as paint, plumbing, and gasoline. Therefore, it is believed that lead will be prohibited in the electronics industry in the near future.

Pb-free reflow soldering techniques applying Au–Sn as well as Sn–Ag electroplated bumps were chosen for the evaluation of the flip-chip bonding process for an x-ray pixel detector. Both can be used in pick-and-place processes with a subsequent batch reflow suitable for high-volume production. Au–Sn solder was selected due to its fluxless bondability, good wettability, and self-alignment process capability; Sn–Ag solder was selected due to its more ductile behavior and lower yield stress compared to Au–Sn. Ga–As test chips with daisy chain and four-point Kelvin probe structures, together with appropriate Si test substrates, were designed, manufactured, and bumped. Test chips with 55- and 170-μm pitch and different chip sizes (maximum 16.3 mm down to 4 mm square) were used. Au–Sn bumps were deposited by electroplating Au first and Sn on top. Au bumps were also formed on the substrate side. Two under-bump metallizations (UBM) were used for the Sn–Ag samples: Cu and Ni.

12.1.1 Finite Element Modeling

The road to the Pb-free alloy faces many technical challenges in electronics. One of them is the bonding method. FR-4 substrate is generally used as chip carrier in electronic packaging, but it is not compatible with the Pb-free solder process due to its low glass transition temperature. The reflow temperature of the Pb-free solder is generally more than 250°C. FR-4 substrate would burn during solder bonding in the traditional reflow oven. The flip-chip bonder can eliminate this problem, as the bumped die and the FR-4 substrate are bonded by the localized heating source in the substrate pads rather than the whole samples; in addition, a much shorter cycle time is needed as compared to a 4- to 6-minute reflow time in the oven.

Finite element modeling (FEM) simulation was performed for Au–Sn and Sn–Ag interconnections and for different chip sizes. A local model was designed for the bump interconnection and a global octant model for the whole assembly. Very high values were calculated for the peel stress using Au–Sn bumps. Sn–Ag bumps, on the other hand, showed a 3 to 5 times reduced peel stress dependent on chip size.

12.1.2 Taguchi Methodology

Experiments are often conducted to determine if changing the values of certain variables leads to worthwhile improvements in the mean yield of a process or system. Another common goal is estimation of the mean yield at given experimental conditions. In practice, it is difficult to fit an accurate and interpretable model to the data that can achieve both goals. A certain manufacturer of assembled electronic circuit boards was suffering from severe quality problems in terms of high percentage of solder defects. A traditional method of improving the quality of a product is to adjust one factor at a time to observe the results. The major disadvantage of this method is that it is costly and unreliable due to the interaction of other factors.

Design of experiments (DoE) is widely used in research and development, where a large proportion of the resources go toward solving optimization problems. The key to minimizing optimization costs is to conduct as few experiments as possible. DoE requires only a small set of experiments and thus helps to reduce costs. It is now widely accepted that using designed experiments is the most effective way to optimize surface-mount technology (SMT) processes. This situation begs the question, "What is an effective strategy in implementing this powerful tool?"

The recognized analysis approach known as the Taguchi method integrates an innovative approach to quality with traditional methods for the design of experiments, in which a series of interrelated techniques are developed that in concert help minimize unwanted variability, reduce production waste, and provide greater customer satisfaction. The Taguchi approach for reducing variation in production is a two-step process:

1. Manufacture the product in the best manner most of the time (less deviation from the target).
2. Produce all products as identically as possible (less variation between the products).

The Taguchi method employs experiments using a specially constructed "orthogonal array" table to influence the design process, such that quality is built into a product during its design stage. An orthogonal array is an experimental design constructed to allow a mathematically independent assessment of the effect of the different factors affecting the experiment.

The process of Pb-free soldering has been closely monitored since July 2006, when the RoHS (Restriction of Hazardous Substances) directive was activated. The Taguchi method has been used to analyze the influence of four control parameters of a process of lead-free soldering on bridging and filling of through-holes on a test board. The following control parameters (factors) have been taken into account:

- Solder temperature
- Time of a contact between the soldered area and solder
- Preheating temperature and flux wettability

Every factor was used in three levels. A Taguchi orthogonal array of the type L9 was used for process analysis. The analysis showed that bridging is influenced predominantly by the preheating temperature and by the time of contact between solder and a pad, and that the influence of the solder temperature is minimal. The filling of the through-hole depends, above all, on the preheating temperature; the influence of other control parameters is low. This has also confirmed the effectiveness of the use of the Taguchi orthogonal array in comparison with full factorial experiments.

Solder paste suppliers only suggest a temperature profile for the conventional reflow oven. They do not provide any information about reflow temperature in the flip-chip bonder. The influence of other factors such as pressure is unknown in solder reflow. Research has investigated the effect of the five critical factors followed in the Taguchi methodology. In addition, the interaction of the bonding factors in the flip-chip bonding process is also investigated. These five critical factors are

- Stage temperature
- Bonding pressure
- Head temperature
- Bonding time and moving distance to the die shear strength

The Taguchi method refers to the techniques of quality engineering that embody both statistical process control (SPC) and new quality-related management techniques. The objectives are to improve the process and the product design through the identification of controllable factors and their settings. It emphasizes the attainment of the specified target value and the elimination of variation. In addition, it emphasizes that control factors must be optimized through the design of experiment rather than by trial and error. The Taguchi method can change many factors simultaneously in a systematic way (thus ensuring an independent assessment of the product factors). Once the factors have been adequately characterized, the quality of a product can be improved. The S/N ratio is used to identify the optimum condition for the bonding process by the flip-chip bonder. ANOVA is used to determine the most influential factors in the flip-chip bonding process. In statistics, ANOVA is a collection of statistical models, and their associated procedures, in which the observed variance in a particular variable is partitioned into components attributable to different sources of variation. ANOVA is a general technique that can be used to test the hypothesis that the means among two or more groups are equal, under the assumption that the sampled populations are normally distributed.

12.2 Electromigration Analysis for Mean-Time-to-Failure Calculations

The trend in flip-chip and ball grid array (BGA) packaging is to increase the input–output (I/O) count. This trend drives the interconnecting solder joints to be smaller in size and, thus, they have a higher current density. The current densities further increase as chip voltage decreases and as the absolute current level increases.

There is also a similar drive in flip-chip power semiconductors and evolving system-on-package power processors to increase the current densities. A physical limit to increasing current density in both microelectronics and power electronics is electromigration. Electromigration of interconnecting metal lines is the major source of failure in integrated circuits, but it is seldom recognized as a reliability concern for solder joints. Most of the published literature on electromigration focuses on thin, pure metal lines, and there is little published on present-day solder interconnects.

Electromigration is caused by high current density stress in metallization patterns and is a major source of breakdown in electronic devices. It is therefore an important reliability issue to verify current densities within all stressed metallization patterns. Both electromigration and Joule heating are used in self-consistent solutions for maximum allowed interconnect peak current density. The maximum allowed temperature and current density solutions monotonically increase as the duty cycle r decreases (Amagai et al., 2002). With the help of layout parameters, the peak current density is calculated and analyzed with the estimated values obtained from various interconnect nodes of the circuit. Using these analyses, peak current density solutions can be used to generate adequately safe current density design guidelines.

12.2.1 Electromigration

Electromigration (EM) is the current-induced transport of the conducting material. In the presence of high current stresses, electron momentum is transferred to atoms in the conducting material, yielding a net atomic flux. This net flux causes the conducting material to be depleted "upwind" and accumulated "downwind." Regions where the interconnect material has been depleted will form a *void*, leading to interconnect open-circuit failure. Likewise, interconnect material can also accumulate and extrude to make electrical contact with neighboring interconnect segments, potentially leading to circuit failure due to the formation of a "short-circuit." Either outcome can contribute to the gradual "wearing out" of a current-stressed interconnect over time.

EM is atomic diffusion caused by a combination of high temperature and high current conditions where the momentum of the electrons is transferred to the atoms. This phenomenon, with respect to solder balls, correlates to several failure mechanisms, such as solder/silicon interface voiding as a result of current crowding, as well as the formation of intermetallic compounds and under-bump metallization consumption caused by migrated phases.

The formation of voids and accumulations depends on the underlying microstructure of the metal film from which the interconnect has been patterned, as discussed previously. Once deposited, the metal film has a distribution of grain sizes. This metal film is then etched to produce the desired interconnect layer.

EM is a diffusion-controlled process that results in the mass transport of metal atoms in the presence of a current. In the past four decades, extensive efforts have been made to understand EM in aluminum and copper lines in integrated circuits so as to better control it. However, the continuing drive to improve device performance has been accompanied by an increased I/O packing density, coupled with a reduction in the pitch and size of flip-chip solder interconnections. This phenomenon has emerged as a critical concern for the EM reliability of solder bumps carrying an increasing current density with shrinking dimensions.

12.2.2 Modern Approach to the Electromigration of Aluminum

Electromigration in aluminum interconnects has been a key and persistent reliability problem in microelectronics technology. With further dimensional scaling down of very-large-scale integration (VLSI) circuits and increasing current density, EM in copper interconnects continues to be a subject of concern, but a new reliability issue is EM in flip-chip solder joints. Owing to the need for a higher on-chip input/output (I/O) interconnect density and better performance, the trend of miniaturization of VLSI circuits demands flip-chip packaging technology. In flip-chip technology, a high density of area arrays of tiny solder bumps is used to join the circuits on a chip to its packaging module or board.

With the aid of micro-fabrication techniques, sub-micrometer relief of a large number of parallel grooves in silica is made. Aluminum (Al) is sputter-deposited onto the groove pattern. If molten, it flows into the grooves. If the correct amount of Al is deposited, all grooves are filled with Al, while the ridges are left uncovered. After solidification, the Al in the grooves is single-crystalline and in neighboring grooves it often has the same crystallographic orientation. These single-crystalline structures are used as a starting point for a number of EM tests.

The EM behavior of a single-crystalline Al line in a groove includes an exceptionally long lifetime for the single-crystalline Al lines. Moreover, the noise—related to movement of defects in the metal—in the resistance is very low. Until now, no one has unraveled the exact nature of this resistance noise. Test lines with more complicated grain structures are being developed by varying the geometry. Alternatively, continued growth (after melting and recrystallization) will be used to grow a thin film with very large grains in which eventually test lines will be defined. The Al in the groove pattern acts, therefore, as a seed layer.

One of the most significant differences between EM in aluminum lines and EM in solder balls is the nonuniform current density that occurs within the solder ball. The uneven distribution can be attributed to current crowding, in which a large portion of the current enters the solder bump at the nearest corner. The concentration of high current and high temperature at this corner causes the formation of voids. This continues across the entire solder/silicon interface and leads to failure. Electromigration estimation is separated into the following two steps:

1. Check for violations of the current density limits.
2. Assess the mean-time-to-failure (MTTF) for all wire segments.

While most interconnect segments exhibit AC current behavior, almost every signal interconnect line on a chip includes interconnect segments that exhibit DC current behavior. Therefore, signal wires must be checked for both peak and RMS current density violations. Hence, the cumulative probability of failure for a projected lifetime must be determined.

12.3 Electromigration Analysis

Advanced electro-thermo coupling models have been developed to investigate the EM and electrothermomechanical effects on electronic packaging, especially on package-on-package (POP). POP packaging involves an ultra-thin gold wire (phi = 1 mil) on wirebonding and Sn4.0Ag0.6Cu (SAC405) solder ball on the package. The current density arising in the aluminum pad (wirebonding) and in the Copper trace above the SAC405 solder ball imply the hot-spot that results in electromigration along the current direction. Finite element predictions reveal that the maximum electrothermomechanical effective stress is located at the regions where electromigration potentially occurred. Reliability of electrothermomechanical effects for wirebonding and SAC405 solder ball is evaluated. Current crowding, temperature distribution, and electrothermal-induced effective stress distribution are predicted. A series of comprehensive parametric studies were conducted.

Here we describe a system for reliability analysis of VLSI CMOS circuits with emphasis on electromigration analysis for MTTF calculations. This process does not restrict itself to the power and ground lines but includes all the metal lines in a circuit at the layout level. The procedure consists of three main steps:

1. Computation of peak current densities for power, ground, and other metal lines in a circuit
2. Extraction of RC parameters from layout designed, using circuit netlist
3. Verification of the estimated peak values with the values extracted from the layout parameters

The electromigration behavior of an 80-μm pitch C2 (Chip Connection) interconnection has been studied. C2 is a peripheral ultra-fine pitch flip-chip interconnection technique with Cu pillars and Sn/Ag-capped solder bumps formed on Al pads for wirebonding. The technique was reported in Orii et al., 2009. It allows for easy control of the space between dies and substrates just by varying the Cu pillar height. The control of the collapse of solder bumps is not necessary, and hence the technology is called the "C2 (Chip Connection)." C2 bumps are connected to OSP (organic solderability preservative) surface-treated Cu substrate pads on an organic substrate by a reflow and no-clean process. C2 is a low-cost, ultra-fine pitch flip-chip interconnection. However, the EM behavior for such a small flip-chip interconnection is still an open issue. The electromigration tests were performed on an 80-μm pitch C2 flip-chip interconnection. Interconnections with two different solder materials were tested: Sn-2.5Ag and pure Sn. The effect of the Ni barrier layer on the test was also studied. The tests showed that the presence of IMC (intermetallic compound) layers reduce the atomic migration of Cu. The test also showed that the Ni barrier is also effective in reducing the migration of Cu atoms into Sn solder. The under-bump metals (UBMs) are formed by sputtered Ti–Cu layers. The electroplated Cu pillar height is 45 μm and the solder height is 25 μm for an 80-μm pitch. The die size is 7.3 mm square, and the organic substrate is 20 mm square with four layers laminated prepreg with 310-μm thickness. The electromigration test condition is 7 to 10 kA/cm² at 125 to 170°C. IMCs were formed prior to the test by an aging process, which is 2,000 hr at 150°C and then the electromigration tests were performed. We have studied the effect of IMC thickness on the EM-induced failure mechanism in C2 flip-chip interconnection on an organic substrate.

12.3.1 Computation of Peak Current Densities

Electromigration in solder joints under high direct current density is a known reliability concern for future high-density microelectronic packaging and power electronic packaging. The trend in flip chip to increase the I/O count drives the interconnecting solder joints to be smaller in size and thus carry a higher current density. The current density will increase further as chip voltages decrease and absolute current levels increase. The research on electromigration and thermomigration in solder joints is still in its early stages, and hence the literature is very poor in publicly available data. The failure modes of flip-chip solder joints under high electric current stress are not yet well understood.

For the maximum allowed current density, the J_{peak} value, self-consistent solutions are obtained for the maximum allowed peak current density J_{peak} as a function of waveform (Amagai et al., 2002), which simultaneously comprehends both relevant temperature-dependent mechanisms: electromigration (EM) and Joule heating (JH). It was shown (Akay et al., 1997) that solutions for maximum allowed temperature and peak current density J_{peak} depend on the duty cycle r of the waveforms. One of the unique behaviors of these solutions is that J_{peak} has constant-temperature EM-like behavior near r = 1, but constant-temperature Joule heating-like behavior for smaller r. We consider only the case of unipolar (and rectangular) pulsed dc operation in an isolated single level of metal. Examples of the parametric dependence of J_{peak} versus r on lead width, underlying oxide thickness, and EM current density specification were given. Here, we focus on the application of these solutions to current density design guidelines that ensure that reliability requirements are met.

The failure modes of flip-chip solder joints under high electrical current density are studied experimentally. Three different failure modes are reported. Only one of the failure modes is caused by the combined effect of electromigration and thermomigration, where void nucleation and growth contribute to the ultimate failure of the module. The Ni under-bump metallization/solder joint interface is found to be the favorite site for void nucleation and growth. The effect of preexisting voids on the failure mechanism of a solder joint is also investigated. For a unipolar (and rectangular) pulsed dc waveform with duty cycle r and peak current density J_{peak}, the standard definition of J_{rms} results in Equation (12.1):

$$J_{rms} = r^{0.5} J_{peak} \tag{12.1}$$

Pb-free solders have replaced Pb-containing Sn–Pb solders in the electronic packaging industry due to environmental concerns. Both electromigration (EM) and thermomigration (TM) have serious reliability issues for fine-pitch Pb-free solder bumps in the flip-chip technology used in consumer electronic products. We review the unique features of EM and TM in flip-chip solder bumps, emphasizing the effects of current crowding and Joule heating. In addition, the challenges to a better understanding of EM and TM in Pb-free solders are discussed. For example, the anisotropic nature of the Sn microstructure in Pb-free solders can enhance the dissolution rates of Ni and Cu in solders driven by EM and TM.

The reason to consider the unipolar pulsed dc is that the maximum allowed J_{peak} for a symmetrical pure ac is greater than for the pulsed dc case, making the latter a worst-case scenario. We use the following relations:

$$J_{peak} = J_{rms}/r^{0.5} \tag{12.2}$$

and

$$J_{peak} = J_{avg}/r \tag{12.3}$$

The peak current density in the bump is found to have a significant effect on the EM lifetime of the tested structures and thus impacts the maximum allowable bump current. Finite element models (FEMs) have been developed that accurately predict the maximum allowable bump current as a function of the current crowding. The FEMs make it possible to predict the maximum bump current for a new current-distribution scheme based on results from EM tests on an existing current-distribution scheme. Here we eliminate J_{peak} to obtain the self-consistent equation between mean metal lead temperature T_m and r:

$$r = \frac{J_{EM}\left(T_m\right)^2}{J_{rms}\left(T_m\right)^2} \tag{12.4}$$

As flip-chip solder joints become smaller, the joints must endure an ever higher current density. Because of this increase, failures caused by electromigration become a real reliability concern (Lin, 2008).

Depending on the experimental conditions, such as the temperature, current density, UBM and surface finish design, and circuit design, several failure mechanisms have been identified, including void formation and propagation, the local melting mechanism, and the UBM dissolution mechanism. Recently, it was pointed out that the asymmetrical consumption of the Ni layer in UBM could also cause failures. To extend the lifetime of those solder joints that are prone to fail through the asymmetrical consumption of the Ni layer, the use of a thicker Ni UBM layer is a logical avenue.

Nevertheless, there is no published report to verify this strategy. Here, we focus on the effect of Ni UBM thickness on the lifetime of flip-chip solder joints. Specifically, three different Ni thicknesses—0.3, 0.5, and 0.8 lm—were used to assess the effect of Ni thickness on the mean-time-to-failure (MTTF) of flip-chip solder joints under current stressing. For Joule heating (JH), the steady-state equation for a quasi one-dimensional (1D) heat transport equation is given by Equation (12.5):

$$J_{rms^2} = \frac{\left(T_m - T_{ref}\right) K_{ox} \cdot w_{eff}}{t_{ox} \cdot t_m \cdot w_m \cdot \rho_m \left(T_m\right)} \tag{12.5}$$

where
J_{rms} is the root-mean-square (rms) current density
T_m is the mean metal lead temperature
T_{ref} is the maximum allowed junction reference temperature in the silicon (chosen to be 100°C)
K_{ox} is the underlying oxide thermal conductivity
t_{ox} is the underlying oxide thickness
t_m is the metal thickness
w_m is the metal width
ρ_m is the temperature-dependent metal resistivity

Researchers have employed three-dimensional (3-D) simulation to investigate the Joule heating effect under accelerated electromigration tests in flip-chip solder joints. It was found that the Joule heating effect was very serious during high current stressing, and a hot-spot exists in the solder bump.

The hot-spot may play an important role in void formation and thermomigration in solder bumps during electromigration.

When a voltage is applied to the boundary of a device and current flows through it, the current flow produces heating, which induces thermal stresses into the device. In many electromagnetic systems, electromagnetic energy is lost in the form of heat in dielectrics and resistive conductor materials. This is often referred to as Joule loss or Joule heating. One aspect of Joule heating is shown in Figure 12.1, where Sn-3.5Ag solder was

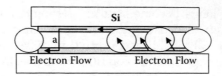

FIGURE 12.1
Schematic diagram of a sample electromigration test structure.

electroplated on the chip side using a conventional manufacturing process and then reflowed. The chip with the solder bump was then bonded onto the substrate with a Cu pad to form the flip-chip solder joints in the second reflow. The underfill material was filled into the space between the chip and the substrate. The sample electromigration test structure is characterized as follows:

- To induce failure in a solder bump at the anode side (bump (a) in Figure 12.1) intentionally, the electric current was dispersed by parallel connection of three solder bumps on the cathode side.
- Here, the three parallel bumps on the right-hand side of the specimen, as shown in Figure 12.1, lead to 3 times higher current density in one bump (a) on the anode side than in the three bumps on the cathode side.
- The top-to-bottom direction of electron flow in solder bump (a) as shown in Figure 12.1 is defined as the forward direction, and the reverse electron flow direction is defined as the reverse direction in this sample electromigration test structure.

To meet the miniaturization trend for portable devices, flip-chip technology has been adopted for high-density packaging due to its excellent electrical performance and better heat dissipation capability. As the required performance in microelectronics devices becomes higher, the current design rule requires that the current each bump needs to carry is 0.2 A, and it will increase to 0.4 A in the near future. Therefore, EM has become an important reliability issue. During the EM test, the applied current may be as high as 2.0 A (Antolovich and Antolovich, 1996; Arora et al., 1996) and thus the Joule heating effect in the solder bumps became a very serious issue. We assume that interconnect reliability is dominated by lead failure, rather than by contact or via failure. Black's equation for the dependence of EM lifetime on current density and temperature leads to the relation that must be satisfied by the current density to maintain equal EM reliability lifetimes:

$$\frac{\left(e^{Em/kB.T_m}\right)}{J_{EM^2}} = \frac{\left(e^{Em/kB.T_{ref}}\right)}{J_{EM^2,dc,ref}} \tag{12.6}$$

where

J_{EM} is the dc current density at temperature T_m

E_M is the activation energy for the EM mechanism

$J_{EM,dc,ref}$ is the dc EM current density specification at temperature T_{ref}

Solder bumps serve as electrical paths as well as structural support in a flip-chip package assembly. Owing to the differences in feature sizes and electric resistivities between a solder bump and its adjacent traces, current densities around the regions where traces connect the solder bump increase by a significant amount. This *current crowding* effect, along with the induced Joule heating, would accelerate fatigue failure due to electromigration. The 3-D electrothermal coupling analysis has been applied to investigate current crowding and Joule heating in a flip-chip package assembly carrying different constant electric currents under different ambient temperatures. Experiments were conducted to calibrate temperature-dependent electric resistivities of solder alloy, Al trace, and Cu trace, and to verify the numerical model by comparing calculated and measured maximum temperatures on the die surface. Through electrothermal coupling analysis, the effects of current crowding and Joule heating induced by different solder bump structures were examined and compared. One very important consequence of Equation (12.2) is that if the metal temperature increases, the activation energy for EM requires that J_{EM} decrease. From Equation (12.4) we define the function $J_{EM}(T_m)$:

$$J_{EM}(T_m) = J_{EM,dc,ref} \cdot \left(\frac{e^{EM/2.kB.T_m}}{e^{EM/2.kB.Tref}} \right) \tag{12.7}$$

Throughout this chapter, the material values in Table 12.1 were considered for all calculations. While these values are reasonable for illustrative purposes, they may not be the best values available in the literature. Thermomigration due to the thermal gradient (which is caused by Joule heating) within the flip-chip solder joint is significant and may be a leading cause of diffusion. The Joule heating effect in solder joints was investigated using thermal infrared microscopy and modeling in this analysis. With the increase in applied current, the temperature increased rapidly due to Joule heating. Furthermore, modeling results indicated that a hot-spot existed in the solder near the entrance point of the Al trace, and it became more pronounced as the applied current increased. The temperature difference between the hot-spot and the solder was as large as 9.4°C when the solder joint was powered by 0.8A. This hot-spot may play an important role in the initial void formation during electromigration.

12.3.2 Extraction of Interconnect Area from Layout

Flip-chip assembly of die onto a substrate has been in existence since the 1960s. Today there is a great deal of interest in flip-chip technology, especially

TABLE 12.1

Values of Material Parameters Used

Parameter	Value	Units
Kox	1.52	$W.m^{-1}.K^{-1}$
Km	243	$W.m^{-1}.K^{-1}$
$\rho_m\,(T_m)$	4.2918 E–8	$\Omega.m$
E_m	7 E–1	e.V
$J_{em,\,dc,\,ref}$	6 E–9	$A.m^{-2}$
Tox	3 E–6	.m
T_m	0.5 E–6	.m
W_m	3 E–6	.m
k_B	1.38 E–23	J/k

Source: From Hunter, W.R. (1997). *IEEE Transactions on Electron Devices*, 44(2), 304–309; Hunter, W.R. (1997). *IEEE Transactions on Electron Devices*, 44(2), 310–316.

its use in chip-scale packaging (CSP), where it has seen dramatic take-up in the mobile phone and display markets. Due to the continued drive to add further functionality to these products, the trend in flip-chip interconnects is toward an ever-finer pitch providing more I/O per square area of die. This trend is posing a number of challenges for package designers and board assemblers in terms of reliability. Researchers have investigated the manufacture and reliability of flip-chip interconnects at sub-100μm pitch. Interconnects with an insufficient width may be subject to electromigration and eventually cause the failure of the circuit at any time during its lifetime. This problem has becme worse over the past couple of years due to the ongoing reduction in circuit feature sizes. For this reason, it is becoming crucial to address the problems of current densities and electromigration during layout generation.

With advancements in integrated circuit process technology, feature dimensions below 0.35 μm are currently used by the semiconductor industry. As the physical size decreases, the delay of electrical signals traveling in the interconnections is equivalent to or greater than the gate delay. For example, the interconnect capacitances between aluminum and silicon dioxide dielectric represent 50% of the total delay in a 0.25-μm technology. In a 0.18-μm technology, this capacitance can represent up to 70%, and it is expected to contribute up to 80% in a 0.15-μm technology.

The parasitic capacitance of each net has two components: area and perimeter. The relationship between wire height and width increases in deep-submicron technology (1.8 for 0.25-μm technology and will reach 2.7 for 0.07-μm technology), resulting in a major contribution for perimeter capacitances. Also, the number of interconnect layers is increasing to six or seven layers. These two facts make the coupling capacitances as important as ground capacitances.

12.3.3 Capacitance Extraction for Area Estimation

An accurate computation of the parasitic capacitance requires the surfacic charge obtained by the normal derivative of the potential on the surface of conductors. The potential in the dielectric media is the solution of the Laplace equation with boundary conditions on the surfaces of conductors. For these simulations on a complex 3-D domain, large computing resources, both in CPU time and memory, are needed. The finite elements (as in, for example, Clever from Silvaco) can take into account complicated geometries and inhomogeneous dielectrics, but with large CPU times and memory needs. The other methods are more efficient for some problems, but less robust when the complexity increases—for instance, in the case of multiple dielectric media.

Here, the interconnect capacitance in each node in a circuit is calculated using the model shown in Figure 12.2. It consists of two conduction layers over the substrate, considered as the reference plane (ground plane). There are three capacitance components at any node:

1. Overlap capacitance (C_{over}): due to the overlap between two conductors in different planes. They are C21a and C23a in Figure 12.2.
2. Lateral capacitance (C_{lat}): the capacitance between two conductors in the same plane. This is C22lat in Figure 12.2.
3. Fringing capacitance (Cfr): due to the coupling between two conductors of different planes. They are C21fr and C23fr in Figure 12.2.

The intrinsic capacitance is the capacitance when no outside forces perturb the charge distribution. We will always assume that we are dealing with the intrinsic capacitance, which is usually a good approximation for real capacitors. The intrinsic capacitance is the capacitance between one conductor layer and the ground plane. It has two components: overlap capacitance and fringing

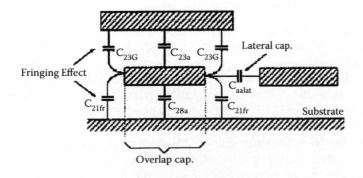

FIGURE 12.2
Capacitance model showing different capacitance components.

capacitance. Two parallel plates model the overlap capacitance. It is calculated using the traditional formulation based on the overlap area (Arora, 1996)

$$C_{over} = C_{area}\ W.L \qquad (12.8)$$

where C_{area} is the capacitance per unit area (fF/μm^2) and $W.L$ is the overlap area (.m^2).

The fringing capacitance is due to the edge of one conductor and the surface of the other one (in this case, the ground plane). It is calculated from Equation (12.9):

$$C_{fr} = 2.C_{length}.L \qquad (12.9)$$

where C_{length} is the capacitance per unit length (fF/μm). Thus, the intrinsic capacitance is the sum of these two components:

$$C_{int} = (C_{area}\ W + 2.C_{length}).L \qquad (12.10)$$

This modeling approach is not restricted to the structure shown here but is applicable to any arbitrary geometry. However, structures such as vias are not modeled using this approach. In order to extract the interconnect area, the overlap capacitance is observed using the Spectre simulator from Cadence tools, and the capacitance per unit area, C_{area}, is estimated using Advanced Design Systems (ADS).

The interconnect width extraction for the electromigration analysis was performed in two different ways. In the first approach, the parasitic capacitances were used to arrive at the metal interconnect area from which the width was calculated with some assumptions for constant length. So, for the parasitic capacitance extraction, some of the Cadence RCX tools such as Diva and the layout tool Virtuoso were used.

12.3.4 Mean-Time-to-Failure Calculations

Electromigration failure of contacts and vias in deep sub-micron IC technologies is the key concern for interconnect reliability. Electromigration failure of contacts and vias in advanced interconnect systems, where the low-resistivity conductors such as Al are clad by refractory layers of Ti or TiN, occurs by the drift of the conductor away from the contact, leaving the refractory materials intact. The reliability of a VLSI chip is ultimately limited by the failure characteristics of its basic building materials, under the stress imposed by the operating environment. Among all the IC technological trends, *scaling* is an important method for reducing die size and thus increasing circuit performance and complexity. Common wear-out processes in ICs are highly influenced by scaling the device dimensions, as this usually leads to increased electrical stress.

If the reliability of a system can be expressed in terms of a failure parameter, then it should be possible to express it as a numerical index so that it could be seen as a fitness of the design created. The mean-time-to-failure (MTTF) of a system is the expected time a system will operate before the first failure occurs. It turns out that the presence of dormant faults can drastically reduce the MTTF of a system. The effect of electromigration on the time to failure was investigated. The MTTF of a conductor under a constant current stress is expressed by the following equation (Goel and Au-Yeung, 1990):

$$\text{MTTF} = AJ^{-n} \exp \{Ea/kT\} \tag{12.11}$$

where
  Ea is the activation energy
  J is the current density
  T is the temperature, in degrees Kelvin
  A is a constant depending on geometry and material parameters—scaling
      factor
  K is Boltzmann's constant,
  n is a constant ranging from 1 to 7

Activation energy, usually denoted by Ea, is defined as the minimum amount of energy required to initiate a particular process. It is usually used in the context of chemical reactions, that is, as the minimum amount of energy that chemical reactants must possess before they can undergo a chemical reaction. In the context of semiconductor device reliability, however, activation energy refers to the minimum amount of energy required to trigger a temperature-accelerated failure mechanism.

Electromigration refers to the gradual displacement of the metal atoms of a conductor as a result of the current flowing through that conductor. The process of electromigration is analogous to the movement of small pebbles in a stream from one point to another as a result of the water gushing through the pebbles. The electromigration, Vm, can be expressed as

$$Vm = G J \exp \{v(Ea/kT)\} \tag{12.12}$$

where G is a proportionality constant. Combining these two equations yields

$$\text{MTTF}_{dc} = G A/(Vm)^n \exp \{-(n-1) Ea/kT\} \tag{12.13}$$

So, the various parameters involved in the calculation of the MTTF are activation energy, temperature, and the electromigration effects. One problem caused by electromigration is a reduction in the effective operating dimensions of interconnects to a micron or sub-micron. The other kind is material related, which is basically caused by high current densities. Because of the mass transport of metal atoms from one point to another during electromigration, this mechanism leads to the formation of voids at some points in the metal line, and hillocks or extrusions at other points. It can therefore result in either (1) an open circuit if the void(s) formed in the metal line become big enough to sever it, or (2) a short-circuit if the extrusions become long enough to serve as a bridge between the affected metal and another one adjacent to it.

Electromigration is actually not a function of current, but rather a function of current density. It is also accelerated by elevated temperature. Thus, electromigration is easily observed in Al metal lines that are subjected to high current densities at high temperature over time. Three associated problems in electromigration include

1. Joule heating
2. Current crowding
3. Material reactions

The effect of each of these parameters is required to arrive at a reliability parameter in terms of the MTTF, with which the fitness of the design can be evaluated. Electromigration is widely believed to be the effect of momentum transfer from the electrons of the metal, which move according to the applied electric field, to the ions that constitute the lattice of the metal.

12.3.5 Current Density, Current Crowding

There are two major driving factors that make electromigration happen:

1. Direct action of the electric field on the charged atoms or ions of the metal
2. Frictional force or momentum exchange between the flowing electrons and these ions

The total driving force is the sum of the effects of these two factors. Scaling down the geometry of ICs increases the current density and associated Joule heating in interconnect tracks and vias, leading to a greater incidence of thermal stress and electromigration failure mechanisms. The current density peaks for geometries containing sharp corners, such as those found at a track–via junction. The peak current densities will therefore be scaled according to finite-element discretization. Maximum current density was simulated by the Finite Element Method (FEM), which provides a better understanding of local heat as well as current crowding. Figure 12.3 shows the simulation results, indicating the serious current crowding distribution in solder bumps. This study found that the current crowding phenomenon was the main reason to hasten the solder bump electromigration failure. There is a very large current density change at the contact between the bump

e⁻ cathode/Chip   Anode/Chip

Anode/Substrate   →   Cathode/Substrate

•   High current density distribution

FIGURE 12.3
Current crowding occurs in the vicinity of the junction of the bumps and the under-bump metallization (UBM).

and the UBM. This change leads to current crowding, and also higher current density at the entry points and exit points of the solder bump. The current crowding phenomenon enhances void formation at the entry points of the cathode side of the solder bumps. This phenomenon also enhances the atomic tilted and clustered at the exit points of the anode side.

The increase in local current density is referred to as current crowding. Because Joule heating is proportional to the square of the current density, the current crowding effect leads to a local temperature rise around the void, which in turn further accelerates void growth. All metal films have imperfections or microstructural variations that cause the atomic flow rates through them to be nonuniformly distributed. These nonuniform atomic flow rates (or flux divergence) through different sections of the conductor result in mass depletion (which causes voids) and mass accumulation (which causes hillocks) as the mass transport mechanism occurs during electromigration.

12.3.6 Temperature's Impact

Current crowding generates electromigration damage through the effects of Joule heating. This accelerating effect of raised temperature is observed once a void has started to form in the metallization line. As ICs become progressively more complex, the individual components must become increasingly more reliable if the reliability of the whole is to be acceptable. However, due to continuing miniaturization of very-large-scale integrated (VLSI) circuits, thin-film metallic conductors or interconnects are subject to increasingly high current densities. Under these conditions, electromigration can lead to the electrical failure of interconnects in a relatively short time frame, thus reducing the circuit lifetime to an unacceptable level. It is therefore of great technological importance to understand and control electromigration failure in thin-film interconnects.

Electromigration is generally considered to be the result of momentum transfer from the electrons, which move in the applied electric field, to the ions that make up the lattice of the interconnect material. When electrons are conducted through a metal, they interact with imperfections in the lattice and scatter. Scattering occurs whenever an atom is out of place for any reason. Thermal energy produces scattering by causing atoms to vibrate. This is the source of the resistance of metals. The higher the temperature, the more out of place the atom is, the greater the scattering, and the greater the resistivity.

Electromigration needs a lot of electrons, and also needs electron scattering. Electromigration does not occur in semiconductors, but may in some semiconductor materials if they are so heavily doped that they exhibit metallic conduction. The identification of peak temperatures in multilevel interconnects is important because electromigration failure is temperature dependent and because it can lead to stress gradients being set up inside the structure or even melting. In general, the peak temperature does not coincide with the peak current density but is found to be geometry dependent. As the track width increases, the current density in the track decreases so that Joule heating is greater in a via, where the current density is largest.

As shown in Figure 12.4, the growth of a void causes a reduction in cross-sectional area, giving current crowding (i.e., an increase in current density) and therefore an increase in local temperature. The temperature increase causes the local rate of atomic migration to increase, thus increasing flux divergences and accelerating the rate of electromigration damage. Accelerating void growth continues until the void is large enough to cause a significant problem within the IC, through the increase of resistance, or ultimately, melting and the formation of an open circuit. Given the track and via dimensions, an approximate value for the current density can be determined by supposing that it is constant across the cross-sectional area of the track. At bends in the track and at contact vias, the current density often exceeds these estimates by significant amounts and can lead to preferential failure at these points. The FEM is used here for one-dimensional approximation. The temperature profile and peak temperature depends on the aspect ratio of the via and track widths. For the case where the track and via widths are equal,

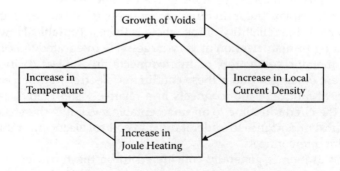

FIGURE 12.4
Growth of a void causes a reduction in cross-sectional area, giving current crowding and, therefore, an increase in local temperature.

the peak temperature coincides with the volume of current crowding. This suggests that such structures are most susceptible to failure.

Methods of failure in narrow interconnects include

- Void failures along the length of the line—internal failures
- Diffusive displacements at the terminals of the line that destroy the contact
- Joule heating and alloy composition

Electromigration causes several different kinds of failure in narrow interconnects. The most familiar are void failures along the length of the line (called internal failures) and diffusive displacements at the terminals of the line that destroy electrical contact. Recent research has shown that both of these failure modes are strongly affected by the microstructure of the line and can, therefore, be delayed or overcome by metallurgical changes that alter the microstructure.

Analysis must account for electromigration caused due to momentum transfer and scattering due to imperfections in the lattice; errors due to lattice defects are not noticeable at the design level. These are post-process defects that must be eliminated during fabrication. The three predominant mechanisms in electromigration failure processes discussed here include those associated with

1. Metallurgical-statistical properties of the interconnect
2. Thermal accelerating process
3. Healing effects

The *metallurgical-statistical properties* of a conductor film refer to the microstructure parameters of the conductor material, such as the grain size distribution, the distribution of grain boundary misorientation angles, and the inclinations of grain boundaries with respect to electron flow. These metallurgical parameters can only be dealt with statistically because they usually appear to be random.

The variation in all these microstructural parameters over a film causes a nonuniform distribution of atomic flow rate. Therefore, non-zero atomic flux divergence exists at places where the number of atoms flowing into the area is not equal to the number of atoms flowing out of that area per unit time. With the non-zero atomic flux divergence, there will be either a mass depletion (divergence > 0) or accumulation (divergence < 0), leading to the formation of voids and hillocks.

The *thermal accelerating process* refers to the acceleration process of electromigration damage due to the local increase in temperature. A uniform temperature distribution along an interconnect is possible only before any electromigration damage occurs. Once a void is initiated, it causes the current density to increase in the vicinity around itself because it reduces the cross-sectional area of the conductor.

The *healing effects* refer to those caused by the atomic flow in the direction opposite to the electron wind force, the back-flow, during or after electromigration. This back-flow of mass begins to take place once a redistribution of mass has begun to form. It tends to reduce the failure rate during EM and partially heals the damage after current is removed. The cause of this back-flow of mass is the inhomogenities, such as temperature and/or concentration gradients, resulting from electromigration damage.

Bibliography

Akay, H.U., Paydar, N.H., and Bilgic, A., 1997. Fatigue life predictions for thermally loaded solder joints using a volume-weighted averaging technique, *ASME Trans., J. Electronic Packaging*, 119 (December), 228–234.

Amagai, M., Watanabe, M., Omiya, M., Kishimoto, K., and Shibuya, T. (2002). Mechanical characterization of Sn–Ag based lead-free solders, *Micoelectronics Reliability*, 42, 951–966.

Antolovich, S.D., and Antolovich, B.F. (1996). *An Introduction to Fracture Mechanics in ASM Handbook. 19. Fatigue and Fracture*, Materials Park, OH: ASM International®, 1996.

Arora, N.D., Raol, K.V., Schumann, R., and Richardson, L.M. (1996). "Modeling and extraction of interconnect capacitances for multilayer VLSI circuits," *IEEE Trans. Computer Aided Design of Integrated Circuits and Systems*, January, 15(1), 58–67.

Basaran, C., and Yan, C.Y. (1995). A thermodynamic framework of damage mechanics of solder joints. *J. Electronic Packaging*, (10-1), 365–376.

Bhattachaiya, B., and Ellingwood, B. (1998). Continuum damage mechanics-based model of stochastic damage growth, *J. Engineering Mechanics*, September, 1000–1009.

Bilotti, A.A. (1974). Static temperature distribution in IC chips with isothermal heat sources, *IEEE Trans. Electron Devices*, ED-21 (March), 217–226.

Black, J.R. (1969). Electromigration failure models in aluminium metallization for semiconductor devices, *Proc. IEEE*, 57(9), 1587–1594.

Blech, I.A., and Herring, C. (1976). Stress generation by electromigration, *Appl. Phys. Lett.*, 29, 131–133.

Boresi, A.P., et al. (1993). *Advanced Mechanics of Materials, (5th edition)*. Canada: John Wiley & Sons.

Box, G.E.P., Hunter, W.G., and Hunter, J.S. (1978). *Statistics for Experimenters: An Introduction to Design, Data Analysis, and Model Building*. New York: John Wiley & Sons, Inc.

Brakke, K.A. (1994). *Surface Evolver Manual*. University of Minnesota, Geometry Center.

Chen, C., and Liang, S.W. (2007). Electromigration issues in lead-free solder joints, *J. Mater. Sci.*, 18, pp. 259–268.

Coffin, L.F., Jr. (1954). A study of the effects of cyclic thermal stresses on a ductile metal, *ASME Transactions,* 76, 931–950.

Darveaux, R. (2000). Effect of simulation methodology on solder joint crack growth correlation, *Proc. 50th ECTC,* May 2000, pp. 1048–1058.

Darveaux, R. (1996). How to use finite element analysis to predict solder joint fatigue life, *Proc. VIII International Congress on Experimental Mechanics,* Nashville, TN, June 10–13, 1996, pp. 41–42.

Dayhoff, J. (1990). *Neural Network Architectures—An Introduction.* New York: Van Nostrand Reinhold, pp. 217–243.

Dreezen, G., Deckx, E., and Luyckx, G. (2003). Solder alternative: Electrically conductive adhesives with stable contact resistance in combination with non-noble metallization, *CARTS Europe 2003,* pp. 223–227.

Ferreira, F.K., Moraes, F., and Reis, R. (2000). LASCA-interconnect parasitic extraction tool for deep-submicron IC design, *Proc. 13th Symp. Integrated Circuits and Systems Design,* 2000, pp. 327–332.

Frear, D.R., and Kinsman, K.R. (1991). *Solder Mechanics—A State of the Art Assessment,* Warrendale, PA: Minerals, Metals, and Material Society.

Gale, W.F., and Totemeier, T.C. (2004). *Smithells Metals Reference Book, (8th edition).* Maryland Heights, MO: Elsevier.

Galyon, G.T. (2003). *Annotated Tin Whisker Bibliography,* Hearndon, VA: a NEMI Publication, July.

Gilat, A., and Krisha, K. (1997). The effects of strain rate and thickness on the response of thin layers of solder loaded in pure shear, *J. Electronic Packaging,* 119, 81.

Goel, A.K., and Au-Yeung, Y.T. (1990). Electro migration in the VLSI interconnect metallizations, 1989, *Proc. 32nd Midwest Symp. on Circuits and Systems,* 2, 821–824.

Grunwald, J., and Schnack, E. (1995). Models for shape optimization of dynamically loaded machine parts, *Proc. WCSM01,* Germany, Pergamon Press, Oxford, pp. 307–310.

Guo, Q., Cuttiongco, E.C., Keer, L.M., and Fine, M.E. (1992). Thermomechanical fatigue life prediction of 63Sn/37Pb solder, *ASME Trans., J. Electronic Packaging,* 114, 145–150.

Haykin, S. (1997). *Neural Networks – A Comprehensive Foundation, (2nd edition).* Upper Saddle River, NJ: Prentice-Hall, pp. 2–10.

Hong, B.Z. (1997). Finite element modeling of thermal fatigue and damage of solder joints in a ceramic ball grid array package, *J. Electronic Materials,* 26(7), 814–820.

Hunter, W.R. (1997). Self-consistent solutions for allowed interconnect current density. I. Implication for technology evolution, *IEEE Trans. Electron Devices,* 44(2), 304–309.

Hunter, W.R. (1997). Self-consistent solutions for allowed interconnect current density. II. Application to design guidelines, *IEEE Transactions on Electron Devices,* 44(2), 310–316.

Hunter, W.R. (1995). The implications of self-consistent current density design guidelines comprehending electromigration and Joule heating for interconnect technology evolution, *Int. Electron Devices Meeting,* 1995, pp. 483–486.

Jerke, G., and Lienig, J. (2002). Hierarchical current density verification for electromigration analysis in arbitrarily shaped metallization patterns of analog circuits design, *Proc. Automation and Test in Europe Conference and Exhibition,* 2002, pp. 464–469.

Ju, T.H., Chan, Y.W., Hareb, S.A., and Lee, Y.C. (1995). An integrated model for ball grid array solder joint reliability, *Structural Analysis in Microelectronic and Fiber Optic Systems, ASME*, EEP-12, 83–89.

Jung, W., Lau, J.H., and Pao, Y.-H. (1997). Nonlinear analysis of full-matrix and perimeter plastic ball grid array solder joints, *ASME Trans., J. Electronic Packaging*, 119 (September), 163–170.

Lall, P., Islam, N., Suhling, J., and Darveaux, R. (2003). Model for BGA and CSP reliability in automotive underhood applications, *Proc. 53rd Electronic Components and Technology Conference*, May 27–30, 2003, pp. 189–196.

Lall, P., Pecht, M., and Hakim, E. (1997). *Influence of Temperature on Microelectronic and System Reliability*. Boca Raton, FL: CRC Press.

Lau, J.H., (Editor) (1991). *Solder Joint Reliability—Theory and Application*. New York: Van Nostrand Reinhold, p. 279.

Lau, J.H., and Pao, Y.H. (1997). *Solder Joint Reliability of BGA, CSP, Flip Chip and Fine Pitch SMT Assemblies*. New York: McGraw-Hill.

Lee, S.-W.R. and Lau, J.H. (1998). Solder joint reliability of cavity-down plastic ball grid array assemblies, *Soldering & Surface Mount Technology*, 10(1), pp. 26–31.

Lemaitre, J. (1996). *A Course on Damage Mechanics*. Berlin: Springer-Verlag, pp. 11–36.

Lienig, J., Jerke, G., and Adler, T. (2002). Electromigration avoidance in analog circuits: two methodologies for current-driven routing. *Proc. ASP-DAC 2002, Design Automation Conference; 7th Asia and South Pacific and the 15th International Conference on VLSI Design. Proceedings*.

Lin Y. L., Lai Y. S., Lin Y. W., and Kao C. R. (2008). "Effect of UBM thickness on the mean time to failure of flip-chip solder joints under electromigration," *Journal of Electronic Materials*, 37, 1, pp. 96–101.

Manson, S.S. (1965). Fatigue: A complex subject—Some simple approximations, *Experimental Mechanics*, 5(7), 193–226.

Meekisho, L., and Nelson-Owusu, K. (1999). Stress analysis of solder joint with torsional eccentricity subjected to based excitation, Conference paper presented at *12th Int. Conf. on Mathematical and Computer Modeling and Scientific Computing*, Chicago, IL, August.

Muju, S., et al. (1999). Predicting durability, *Mechanical Eng. Mag. of ASME*, March, pp. 64–67.

Nagaraj, B., and Mahalingam, M. (1993). Package-to-board attachment reliability methodology and case study on OMPAC package, *ASME Advances in Electronic Packaging*, EEP-4-1, 537–543.

Ohring, M. (1998). *Reliability and Failure of Electronic Materials and Devices*. San Diego: Academic Press.

Orii, Y., Toriyama, K., Noma, H., Oyama, Y., Nishiwaki, H., Ishida, M., Nishio, T., LaBianca, N.C., and Feger, C. (2009). "Ultrafine-pitch C2 flip chip interconnections with solder-capped Cu pillar bumps," *Electronic Components and Technology Conference* (ECTC 2009), San Diego, CA, May 26–29, 948–953.

Pan, T.-Y. (1994). Critical accumulated strain energy (CASE) failure criterion for thermal cycling fatigue of solder joints, *ASME Trans., J. Electronic Packaging*, 116 (September), 163–170.

Pang, J.H.L., and Chong, D.Y.R. (2001). Flip chip on board solder joint reliability analysis using 2-D and 3-D FEA models, *IEEE Trans. Adv. Packaging*, 24(4), 499–506.

Pang, J.H.L., Xiong, B.S., and Low, H. (2004). Creep and fatigue characterization of lead free 95.5Sn-3.8Ag-0.7Cu solder, *Proc. 54th ECTC*, June 2004, pp. 1333–1337.

Pao, Y.-H., Jih, E., Adams, R., and Song, X. (1998). BGAs in automotive applications, *SMT,* January, pp. 50–54.

Paydar, N., Tong, Y., and Akay, H.U. (1994). A finite element study of factors affecting fatigue life of solder joints, *ASME Trans., J. Electronic Packaging,* 116 (December), 265–273.

Racz, L.M., and Szekely, J. (1993). An analysis of the applicability of wetting balance measurements of components with dissimilar surfaces, *Adv. Electronic Packaging, ASME,* EEP-4-2, 1103–1111.

Sakimoto, M., Itoo, T.,Fujii, T., Yamaguchi, H., and Eguchi, K. (1995). Temperature measurement of Al metallization and the study of Black's model in high current density, *IEEE Int. Reliability Physics Symp., 33rd Annual Proceedings,* pp. 333–341.

Sampath, Barath. K. (2001). *Electromigration dependent MTTF Calculations,* Analog IC Research Group, University of Texas, Arlington.

Setlik, B., Eskett, D., Aubin, K., and Briere, M. (1997). Electromigration investigations of aluminum alloy interconnects, *University/Government/Industry Microelectronics Symposium, 1997, Proc. Twelfth Biennial,* pp. 159–160.

Shi, X.Q., Pang, H.L.J., Zhou, W., and Wang, Z.P. (1999). A modified energy-based low cycle fatigue model for eutectic solder alloy, *Scripts Material,* 41(3), 289–296.

Skrzypek, J.J., and Hetnarski, R.B. (1993). *Plasticity and Creep-Theory, Examples, and Problems,* Boca Raton, FL: CRC Press.

Steinberg, D.S. (2000). *Vibration Analysis for Electronic Equipment.* New York: John Wiley & Sons.

Strauss, R. (1998). *SMT Soldering Handbook, (Second edition).* Maryland Heights, MO: Elsevier/Newnes.

Suo, Z. (2004). A continuum theory that couples creep and self-diffusion, *J. Appl. Mechanics,* 71, 646–651.

Syed, A.R. (2004). Accumulated creep strain and energy density based thermal fatigue life prediction models for SnAgCu solder joints, *Proc. 54th ECTC,* June 2004, pp. 737–746.

Syed, A.R. (1995). Creep crack growth prediction of solder joints during temperature cycling: An engineering approach, *Trans. ASME,* 117 (June), 116–122.

Teng, C.C., Cheng, Y.K., Rosenbaum, E., and Kang, S.M. (1997). iTEM: A temperature-dependent electromigration reliability diagnosis tool, *IEEE Trans. Computer-Aided Design of Integrated Circuits and Systems,* 16(8), 882–893.

Tu, K.N. (2003). *Solder Joint Technology: Materials, Properties, and Reliability.* Berlin, Germany: Springer.

Tu, K.N. (2003). Recent advances on electromigration in very-large-scale integration of interconnects, *J. Appl. Phys.,* 94, 5451–5473.

Tu, K.N. (1994). Irreversible processes of spontaneous whisker growth in bimetallic Cu–Sn thin-film reactions, *Phys. Rev. B,* 49(3).

Tunga, K., Pyland, J., Pucha, R.V., and Sitaraman, S.K. (2003). Field-use conditions vs. thermal cycles: A physics-based mapping study, *Proc. 53rd Electronic Components and Technology Conference,* May 27–30, 2003, pp. 182–188.

Wiese, S., Schubert, A., Walter, H., Dudek, R., Feustel, F., Meusel, E., and Michel, B. (2001). Constitutive behaviour of lead-free solders vs. lead-containing solders experiments on bulk specimens and flip-chip joints, *Proc. 51st Electronic Components and Technology Conf.,* 2001, pp. 890–902.

Winter, P.R., and Wallach E.R. (1997). Microstructural modeling and electronic interconnect reliability, *Int. J. Microcircuits and Electronic Packaging,* 20(2), 124–129.

Wu, W., Kang, S.H., Yuan, J.S., and Oates, A.S. (2000). Electromigration performance for Al/SiO2, Cu/SiO2 and Cu/low-k interconnect systems including Joule heating effect, *2000 IEEE Int. Integrated Reliability Workshop Final Report*, pp. 165–166.

Yeh, C.-P., Zhou, W.X., and Wyatt, K. (1996). Parametric finite element analysis of flip chip reliability, *Int. J. Microcircuits and Electronic Packaging*, 19(2), 120–127.

Zahn, B.A. (2002). Finite element based solder joint fatigue life predictions for a same die size stacked chip scale ball grid array package, *SEMICON West, International Electronics Manufacturing Technology (IEMT) Symposium*, pp. 274–284.

Zahn, B.A. (2003). Solder joint fatigue life model methodology for 63Sn37Pb and 95.5Sn4Ag0.5Cu materials, *Proc. 53rd Electronic Components and Technology Conference*, May 27–30, pp. 83–94.

13

Flip-Chip Bonding of Opto-Electronic Integrated Circuits

The mechanics of solder joint failure under various loading conditions have been investigated in numerous recent studies, yet these studies often show several inconsistencies, especially regarding the location of failure site. While some report the solder/IMC (intermetallic compound) to be the predominant fracture site, others identify the IMC or solder matrix to be the crack growth path. Structural similarities in the solder joint used in these studies yet varying locations of cracking site suggest that fracture in the solder joint is affected greatly by a subtle change in microstructure and geometry of the solder.

Solders such as the 48Sn-52In (wt%) have been studied as interconnect materials in a vertical-cavity surface-emitting laser (VCSEL). The 48Sn-52In alloy is a low-temperature (melting point = 118°C) lead-free (Pb-free) solder and a potential candidate material for this device. In recent years, the integration of VCSELs onto integrated circuits has become a topic of intense research. An important focus of this research has been to develop and adapt methodologies to electrically connect VCSELs to electronic chips, including wire bonding, bridge bonding, and flip-chip bonding. Of these, the latter-most (i.e., flip-chip bonding) has proven to be the most promising chip-level packaging for this application.

A growing requirement to include optical devices such as lasers and high-brightness LEDs (light-emitting diodes) in microelectronic assemblies presents new challenges to the manufacturer. Maintaining position and alignment of optical components to other circuit elements for the lifetime of the assembly may be crucial for maximum performance and high reliability. Accurate initial placement and bonding of sensitive circuit elements depends on equipment capability and control. Maintaining lifetime accuracy depends on assembly materials.

In this chapter, thermal-structural analysis evaluates the solder alloys in the context of in-service operating conditions. Specifically, thermal analysis determines the in-service temperature distributions, as influenced by each solder alloy, due to power generation within the VCSEL and due to convective boundary conditions. Subsequently, these thermal profiles are used as the thermal loads in an evaluation of the stress and creep response in the laser pads and solder joints. Emphasis is placed on the relaxation of stresses at the laser pads and on the creep strains developed at the solder joints during a 24-hour power-on condition.

13.1 Gold–Tin Solder

Electronics are ubiquitous in modern life, permeating virtually every aspect of the human experience, from medicine, to transportation, to entertainment, to communications. Many would argue that the lattermost experience—communication—which enables the interaction of humans through barriers (physical, linguistic, cultural, and otherwise) has been the primary motivation for the rapid development of the electronics industry. As issues of scaling and cost dominate, the integration of the world of conventional (analog and digital) electronics with the optical realm appears imminent. Many would consider the development of the so-called "smart pixel" (or opto-electronic integrated circuit), a term coined for the integration of conventional circuits and optical devices, to be the focus of electronics research. The accomplishment of such a task would enable several key possibilities, including lower-cost manufacturing and optical interconnects.

One approach to improving the performance of large processing or telecommunications switching systems is to interconnect integrated circuits using optics. Smart pixels, with integrated optical detectors, modulators, and electronic logic, could potentially be used in these systems. The FET-SEED (field effect transistor-self-electro-optic-effect device) consisting of the monolithic integration of multiple quantum well (MQW) optical modulators and detectors with GaAs field effect transistors, is one design platform for these smart pixels. Another potential design platform uses the hybrid integration of MQW modulators and detectors with commercial electronic circuits. This latter approach allows one to design circuits with greater complexity and circuit yield, because it uses available established VLSI (very large scale integration) processes.

Solder attachment of components remains the most-used assembly method with solder-bumped flip chip in high-density assemblies. Gold–tin (Au–Sn) solder preforms have long been used for fluxless hermetic lid sealing, die attach, and heat sink attach. Now Au–Sn solder has become the material of choice for flip-chip mounting of precision optical die. Gold–tin solder has the following advantages:

- *Fluxless.* Unlike most common solders, Au–Sn does not require a chemical flux to remove oxides and prepare the surface. Eliminating flux and flux cleanup shortens the assembly process, while avoiding contamination of the optical surfaces with flux residues.

- *Hardness.* Au–Sn forms a very hard solder joint, with no creep, relaxation, or deformation in the joint, so that alignment remains unchanged over time.

- *Good wetting.* Au–Sn solder readily wets bond pads for strong, uniform, void-free joints, a deficiency with some newer lead-free substitutes.

- *Corrosion resistant.* Au–Sn solder joints are highly resistant to corrosion, without the additional protection required by some substitutes.

- *Excellent thermal properties.* The high thermal conductivity of gold–tin rapidly carries heat away without creating excessive stresses, a major concern in high-density packaging.

- *High electrical conductivity.* The high electrical conductivity of gold provides low-resistance connections important for high-power devices.

- *Long-term stability.* Intermetallic growth rates are low when deposited over nickel, palladium, or platinum.

- *Lead-free proven.* Au–Sn device mounting was used for decades before the political lead-free solder mandate created a flood of lesser-known lead-free solders.

Integration of high-speed Internet access, telephone, and cable is a major driving force for fiber-optic network and related technologies. Cost reduction is critical to accelerate the market growth of opto-electronic modules and systems. It is estimated that packaging contributes 60% to 90% of the overall cost of an opto-electronic module, while alignment can contribute up to 90% of the packaging cost. As a result, it is necessary to connect and maintain hundreds of optical precision alignments through a batch assembly process that is compatible with the existing manufacturing infrastructure. Soldering is the technology of choice for such a cost-effective assembly process. In addition to providing electrical connections, solder is useful in the formation of passive precision alignments for opto-electronic packaging. It can be used to couple optical fibers or waveguides to devices such as lasers, LEDs, or photodetectors. The alignments can vary from sub-millimeter to millimeter levels for single- or multi-mode fiber applications. Different designs have demonstrated precision alignments, and the aligned structures are becoming more and more complex.

Gold–tin eutectic solders (30% Sn) are used for packaging microelectronic and opto-electronic devices because of their excellent thermal and mechanical properties and relatively low melting or reflow temperatures (280°C). Electroplating is a cost-effective alternative to current commercial solder deposition processes, such as solder preforms and evaporation. A co-electroplating process for depositing Au–Sn alloys, from a slightly acidic, chloride-based solution using pulsed currents, onto metallized ceramic and semiconductor substrates has been developed. Two separate Au–Sn compositions, 15% Sn and 50% Sn, can be deposited under appropriate plating conditions, that is, current density, and pulse on/off time and duration. These compositions, according to the Au–Sn phase diagram, correspond to Au–5Sn and Au–Sn, respectively. Using multiple current pulses and varying their duration, it is possible to deposit a composite solder structure with an overall composition ranging from 15% to 50% Sn, including the important eutectic composition. Gold–tin trade-offs include:

- *Active atmosphere.* Eliminating flux also eliminates flux removal of surface oxides. Instead, Au–Sn assembly avoids oxidation by proceeding in a controlled, active atmosphere.
- *Narrow process window.* Obtaining the desired characteristics of Au–Sn requires careful process control in bumping and assembly. For example, a 1% increase in gold in the composition increases the melting temperature by 30°C. Accurate composition control is the essence of Au80-Sn20 bumping.
- *Material cost.* The worldwide financial crisis has recently driven the price of gold to new highs.

Eutectic Au–Sn solder is increasingly used in high-reliability and high-temperature applications where conventional Sn–Pb and Pb-free solders exhibit insufficient strength, creep resistance, and other issues. These applications include hybrid microelectronics (particularly flip chips), MEMS (micro-electro-mechanical systems), optical switches, LEDs, laser diodes, RF devices, and hermetic packaging for commercial, industrial, military, and telecommunications applications. For most of these applications, Au–Sn provides the additional benefit of not requiring flux during reflow, thereby significantly reducing the potential for contamination and pad corrosion. However, the materials and processing considerations are substantially different from those for conventional solders. Many companies struggle with issues such as poor solder flow, excessive void formation, variable reflow temperature (arising from off-eutectic compositions), heterogeneous phase distribution, and others, all contributing to development delays, process yield loss, and field reliability issues. Au–Sn opto-electronic assembly requires highly accurate alignment and placing of the device, as well as close control of bonding temperature, pressure, and time. A program-controlled single-assembly machine is the best way to automate the process while maintaining control.

The assembly of opto-electronic components requires fluxless processes to avoid the contamination of optical interfaces. Au–Sn solder is suitable for fluxless processing. In addition, the solder is very hard, and therefore no plastic deformation, creep, or relaxation is expected. Flip chip is advantageous for several reasons:

- Very low electrical parasitics due to the short interconnect from chip to module
- The potential self-alignment of the component in the soldering step

Therefore, it seems very likely to use Au–Sn solder bumps for flip-chip interconnects. One proven bump-forming approach is to electroplate a thick gold layer capped with a thin tin layer in the correct proportion. The bump is reflowed to obtain the eutectic composition. Other bumping methods include sequential evaporation of alternating gold and tin layers in the correct ratio, placing and reflowing preformed solder spheres of the proper composition, stenciling or jetting Au–Sn solder paste, or alloy electroplating with a single composition-controlled Au–Sn alloy solution. Gold–tin eutectic solder has many established advantages over other solders for accurate, stable mounting of opto-electronic devices, but the process requires close control. An automated machine can place and solder these devices to better than 1-μm accuracy.

13.2 Integrating Vertical-Cavity Surface-Emitting Lasers onto Integrated Circuits

In recent years, the integration of vertical-cavity surface-emitting lasers (VCSELs) onto integrated circuits has become a topic of intense research. An important focus of this research has been to develop and adapt methodologies to electrically connect VCSELs to electronic chips, including:

- *Wire bonding.* Wire bonding has been used in integrated circuit (IC) packaging for many years. However, there are many challenges in wire bonding for opto-electronics packaging. These challenges include bonding on sensitive devices, bonding over cavity, bonding over cantilever leads, and bonding temperature limitations. The opto-electronics package design brings another challenge, which requires wire bonding to have deep access capability.

- *Bridge bonding.* Large broadband asynchronous transfer mode (ATM) switching nodes require novel hardware solutions that could benefit from the inclusion of optical interconnect technology, as electronic solutions are limited by pin-out and by the capacitance/inductance of the interconnections.

- *Flip-chip bonding.* The system requires two-dimensional arrays of surface emitters that emit through the semiconductor substrate, thus making devices suitable for flip-chip bonding. Both VCSELs and resonant cavity LEDs (RCLEDs) are suitable for this application.

Of these, the lattermost, flip-chip bonding, has proven to be the most promising chip-level packaging for this application. There are many different alternative processes used for flip-chip joining. A common feature of the joined structures is that the chip is lying face-down to the substrate, and the connections between the chip and the substrate are made using bumps of electrically conducting material. Figure 13.1 shows flip-chip joints without underfill material.

FIGURE 13.1
Flip-chip joints without underfill material.

Flip-chip bonding has certain inherent electrical properties that make its use advantageous over other bonding methods, ranging from electromagnetic interference (EMI) immunity to lower parasitic impedance. Furthermore, the electrical interconnects of flip-chip bonding simultaneously act as mechanical bonds, with high mechanical reliability and efficient heat conduction. Flip-chip bonding, in addition to having advantageous electrical and mechanical properties, also possesses the capability for passive self-alignment. Surface tension in the solder balls achieves self-alignment during solder reflow.

Opto-electronic bonding has been performed on sample multi-chip modules (MCMs), which are to be used for optical testing on the HOLMS demonstrator. The photodiodes and VCSELs were successfully bonded. Ultra-thin silicon-on-sapphire complementary metal oxide semiconductor (UTSi CMOS) technology is a commercial, high-yield silicon-on-sapphire technology that yields circuitry well suited for optical communication functions on a transparent substrate. This characteristic, unique to the silicon-on-sapphire configuration, allows flip-chip bonding of opto-electronic (OE) devices onto CMOS circuitry to build flipped opto-electronic chips on UTSi (FOCUTS) optical transmit and receive modules. Flip-chip bonding eliminates the wire-bond inductance between driving/receiving circuits and the OE devices, which becomes problematic at data rates greater than about 2.5 Gbps. Such flip-chip integration also reduces the number of discrete components that must be handled, packaged, and aligned in a module, thereby improving reliability and reducing costs. Additional functions, such as electrically erasable programmable read-only memory (EEPROM) and self-aligned automatic power control (APC) photodetectors and control circuits are discussed. We describe measured results of flip-chip bonding of arrayed OE devices (VCSELs and photodetectors) and test results at 3 Gbps, as well as recent integrating and testing of phototransistors in UTSi circuits. We also describe the radiation sensitivity of all components and the applicability of this technique to remote sensing applications. These devices, operating at 850 nm, are aimed at multimode, short-reach optical fiber networks.

The integration and packaging of OE devices with electronic circuits and systems has increasing application in many fields, ranging from long- to micro-haul links. The opportunities, integration technologies, and some recent results using thin-film device heterogeneous integration with Si CMOS VLSI and GaAs MESFET (metal semiconductor field-effect transistors) circuit technologies have been explored. Applications explored herein include alignment tolerant OE links for network interconnections, smart pixel focal plane array processing through the integration of imaging arrays with sigma-delta analog-to-digital converters underneath each pixel, and three-dimensional (3-D) computational systems using vertical through-Si optical interconnections. A great deal of focused and proactive research has been dedicated to flip-chip bonding in the greater context of researching OE interconnects and OEICs. While some problems exist with the techniques studied thus far, most notably the incompatibility of materials and devices used, the potential benefits of effective flip-chip bonding has spurred a great deal of enthusiasm for this technique.

The VCSEL has many potential advantages over edge-emitting lasers. Its design allows the chips to be manufactured and tested on a single wafer. Large arrays of devices can be created exploiting methods such as flip-chip optical interconnects and optical neural network applications. In the telecommunications industry, the VCSEL's uniform, single-mode beam profile is desirable for coupling into optical fibers. However, concomitant with these advantages come a number of problems, particularly with regard to fabrication and operation at high powers. In this section, we look at the structure and operation of these devices and discuss the problems facing the designer of such devices.

VCSELs are semiconductor lasers, more specifically laser diodes with a monolithic laser resonator, where the emitted light leaves the device in a direction perpendicular to the chip surface. The resonator (cavity) is realized with two semiconductor Bragg mirrors ($\rightarrow$ distributed Bragg reflector lasers). Between those, there is an active region (gain structure) with (typically) several quantum wells and a total thickness of only a few micrometers. In most cases, the active region is electrically pumped with a few tens of milliwatts and generates an output power in the range from 0.5 to 5 mW, or higher powers for multimode devices (see below). The current is often applied through a ring electrode, through which the output beam can be extracted, and the current is confined to the region of the resonator mode using electrically conductive (doped) mirror layers with isolating material around them. Due to the short resonator round-trip time, VCSELs can be modulated with frequencies well into the gigahertz range. This makes them useful as transmitters for optical fiber communications. For short-range communications, 850-nm VCSELs are used in combination with multimode fibers. A data rate of, for example, 10 Gbps can be reached over a distance of a few hundred meters.

13.3 Flip-Chip Bonding and Opto-Electronic Integration

As the size of semiconductor devices rapidly scales down with each successive generation, the speed with which a signal travels through metallic interconnects becomes an important bottleneck limiting circuit performance. Similarly, the push toward higher throughput and lower cost in communications systems has called for higher levels of component integration. At the heart of both these issues is the full integration of optical and electrical devices, allowing for the creation of optical interconnects and OEICs. Intense research has focused on OE integration for well over a decade. Numerous attempts have been made, with significant progress made in all areas of research. Emerging heterogeneous integration techniques for integrating OE devices, analog interface circuitry, RF circuitry, and digital logic into mixed multisignal systems holds great promise for new packaged (SOP) and chip-based (SOC) microsystems. System design for high-yield, low-cost, alignment-tolerant mixed multisignal microsystems is paramount for opto-electronics to achieve pervasive implementation in lower cost, shorter-haul electronic systems. High-volume product opportunities are emerging for new optical interconnection lengths ranging from micro-haul SOC to milli-haul SOP to short-haul due to the societal thrust toward ubiquitous high-bandwidth data access, but cost is a critical factor in the development of these products. There will be products that address these needs; will they contain OE components? The current packaging solution for OE interfaces in electronic systems is through packaged discretes. The high-volume product solutions will contain integrated communication links, whether they be optical or electrical. Thus, integration of OE components with signal processing technology (current Si CMOS VLSI) is one significant step toward OE integration into electronic systems.

In the early 1990s, researchers at the Lockheed Palo Alto Research Laboratory succeeded in the monolithic integration of MESFETs and VCSELs. In their design, the epitaxial structures of both the VCSELs and MESFETs were simultaneously grown by molecular beam epitaxy on an n-type gallium-arsenide (GaAs) substrate. Processing was done in parallel, with an undoped epitaxial layer serving both as a mirror for the VCSELs and as a highly resistive material to prevent leakage currents in the MESFETs and isolate the two devices. Similarly, the Honeywell Technology Center succeeded in the monolithic integration of VCSELs and metal-semiconductor-metal (MSM) photodetectors on a GaAs substrate. The processing was done sequentially, with VCSEL layers grown first, followed by the photodetector layers.

The monolithic integration of micromechanical devices with their controlling electronics offers potential increases in performance as well as decreased cost for these devices. Analog devices have demonstrated the commercial viability of this integration by interleaving the micromechanical fabrication steps of an accelerometer with the microelectronic fabrication steps of its controlling electronics. Sandia's Microelectronics Development Laboratory has integrated the micromechanical and microelectronic processing sequences in a segregated fashion. In this CMOS-first, micromechanics-last approach, conventional aluminum metallization is replaced by tungsten metallization to allow the CMOS to withstand subsequent high-temperature processing

FIGURE 13.2
Flip-chip joints with underfill material.

during the micromechanical fabrication. Researchers at the University of California–Berkeley have developed a modular integrated approach in which the aluminum metallization of CMOS is replaced with tungsten to enable the CMOS to withstand subsequent micromechanical processing. Figure 13.2 shows flip-chip joints with underfill material.

While monolithic integration shows promise, the high cost of III–V semiconductors such as GaAs, as well as the lack of design flexibility with monolithic techniques, restricts its widespread application. In lieu of this, hybrid integration of OE devices and circuitry has grown in popularity. Advanced OE interconnects are rapidly being developed for high-bandwidth, high-density optical data communication, network switching, and signal processing. In free-space configured board-to-board, chip-to-chip, and even multi-stacked optical interconnects, two-dimensional (2-D) VCSEL arrays are especially desirable for their capabilities of providing parallel data links with aggregated ultra-high bandwidth. To be used as OE transmitters, VCSEL arrays must be hybridized on hosting driver substrates that contain either the CMOS circuits or another fan-out driving scheme. During recent years, high-speed, low-power consumption CMOS driver circuits were fabricated on sapphire substrates using ultra-thin silicon-on-sapphire (UTSi®) technology. The sapphire substrate provides superior electrical isolation between electrical signals and good thermal conductance for heat dissipation. The low-parasitic nature of UTSi circuitries allows their bandwidths to reach 40 GHz under a 0.13-μm processing feature.

13.4 Case Study: A VCSEL Bonded to a Driver Chip

13.4.1 Phase 1: A VCSEL Bonded to a Driver Chip

Phase 1 of the case study will select metals used for solder in the flip-chip bonding of VCSELs and other OE devices onto a foreign substrate. An example of a VCSEL bonded to a driver chip is illustrated in Figure 13.3, showing a specific application of flip-chip bonding. We present here the design, fabrication, and high-speed performance of a parallel optical transceiver based on a single CMOS amplifier chip incorporating sixteen transmitter and sixteen receiver channels. The optical interfaces to the chip

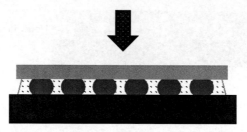

Thermo-compression Bonding

FIGURE 13.3
Principles of flip-chip joining by a thermo-compression process.

are provided by sixteen-channel photodiode (PD) and VCSEL arrays that are directly flip-chip soldered to the CMOS IC. The substrate emitting/illuminated VCSEL/PD arrays operate at 985 nm and include integrated lenses.

As shown in Figure 13.3, flip-chip joining can be assembled by thermo-compression. In the thermo-compression bonding process, the bumps of the chip are bonded to the pads on the substrate by force and heat is applied from an end effector. The process requires gold bumps on the chip or the substrate and a correspondingly bondable surface (e.g., gold, aluminum). The bonding temperature is usually high (e.g., 300°C for gold bonding) to soften the material and increase the diffusion bonding process. The bonding force can be up to 1 N for an 80-mm diameter bump. Due to the required high bonding force and temperature, the process is limited to rigid substrates such as alumina or silicon. Additionally, the substrates must have high planarity. A bonder with high accuracy in the parallelism alignment is required. In an effort to avoid pre-damaging the semiconductor material, the bonding force must be applied with a gradient.

The complete transceivers are low-cost, low-profile, highly integrated assemblies that are compatible with conventional chip packaging technology such as direct flip-chip soldering to organic circuit boards. In addition, the packaging approach, dense hybrid integration, readily scales to higher channel counts, supporting future massively parallel optical data buses. All transmitter and receiver channels operate at speeds up to 15 Gbps for an aggregate bi-directional data rate of 240 Gbps. Inter-channel crosstalk was extensively characterized, and the dominant source was found to be between receiver channels, with a maximum sensitivity penalty of 1 dB measured at 10 Gbps for a victim channel completely surrounded by active aggressor channels. The transceiver measures 3.25 × 5.25 mm and consumes 2.15 W of power with all channels fully operational. The per-bit power consumption is as low as 9 mW/Gbps, and this is the first single-chip optical transceiver capable of channel rates in excess of 10 Gbps. The area efficiency of 14 Gbps/mm^2 per link is the highest ever reported for any parallel optical transmitter, receiver, or transceiver reported to-date.

13.4.2 Phase 2: Create an Integrated VLSI Photonic System

Phase 2 of the case study is to create an integrated VLSI photonic system, as shown in Figure 13.4. VCSELs as emitters and optical detectors are integrated onto a driver chip. This gives an optical communications transceiver on a single chip, with two of these chips able to communicate when connected with fiber-optic cable. The design of a single-chip optical transceiver to optimize the performance of a short-distance optical datalink is proposed.

As shown in Figure 13.4, the thermo-compression bonding process can be made more efficient using ultrasonic power to speed up the welding process. Ultrasonic energy is transferred to the bonding area from the pick-up tool through the back surface of the chip. The thermo-sonic bonding introduces ultrasonic energy that softens the bonding material and makes it vulnerable to plastic deformation. The main benefit of the method compared to thermo-compression is lower bonding temperature and shorter processing time. One potential problem associated with thermo-sonic bonding is silicon cratering. It is generally believed that such damage results from excessive ultrasonic vibration.

The transceiver includes an embedded hybrid automatic repeat request (ARQ) controller capable of operation at several gigahertz clock rates. The hybrid ARQ controller uses a combination of packet retransmission protocols and forward error correction (FEC) to minimize bit errors and achieve a transmitter power coding gain of several dB. Conventional FEC codes such as Reed–Solomon codes cannot be used due to their excessive hardware cost and delays. A practical multilevel coding scheme is explored. The inner codes consist of small linear block codes with reasonable FEC capability, such as small BCH codes, which can be encoded and decoded with reasonable hardware cost and delay. The outer code for a complete packet consists of a long linear block code with excellent error-detection ability, such as a cycle redundancy check code. Low-power pipelined on-chip FEC decoders with estimated throughputs of several hundred gigabits per second per square millimeter (Gbps/mm^2) are proposed. Mathematical analysis indicates that substantial coding gains are possible, which can be used to

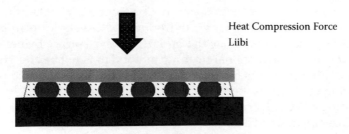

Heat Compression Force
Liibi

Thermo-sonic Bonding

FIGURE 13.4
Principles of flip-chip joining by a thermo-sonic process.

increase the data rate or the distance span of the link. The proposed designs can be used in short-distance optical transceivers for 10-Gb Ethernet, fiber-channel, and very-short-reach optical datalinks, and are scalable to future 2-D optical datalinks with terabits of capacity.

For Phase 1 of the case study, the goal is to develop an effective flip-chip bonding technique that can be used to bond optical devices to MESFET and hetero-junction bipolar transistor (HBT) logic circuits. With this goal in perspective, the following design objectives are identified:

- Most importantly, the flip-chip bonding technique developed must be viable at temperatures below 450°C, ensuring the protection of sensitive optical devices.

- Furthermore, it was important to consider an important trade-off, aimed at minimizing the resistance of bonds, while maintaining reasonable process area. The process designed would need to be reliable and low-cost, thus enhancing the value and usefulness of our techniques.

- Finally, it was desirable to use environmentally friendly materials, shunning the use of lead and cadmium in favor of less-toxic metals. This is a departure from many early techniques, which favored lead-based solders rather than lead-free solders.

These design objectives serve to create a practical and widely applicable technique for flip-chip bonding.

13.5 Solders for Flip-Chip Bonding

Alloys of the lead–tin (Pb–Sn) system are the most common solder alloys used today. However, there are environmental and health issues concerning the toxicity of Pb present in these Sn–Pb solder alloys. Also, flux residue removal is mandatory and leads to environmental threats. More importantly, the use of flux may contaminate the optically active surface by organic residue leftover, and a conventional cleaning method may not be effective for OE assemblies. Therefore, it is necessary to look for fluxless soldering processes for soldering OE systems. Researchers have conducted low-temperature flip-chip bonding of VCSEL arrays on a glass substrate that provides propagation paths of laser beams and also supports a polymeric waveguide. Considering both the die shear test and the spreading test, the appropriate bonding temperature and pressure using an indium (In) solder bump were found to be about 150°C/500 gf. The fracture occurred between the In solder bump and the VCSEL chip pad during the die shear test. It is inferred that both the low bonding temperature and the oxide layer formed on the surface of the In solder prevented the bump from interacting with the chip pad. Researchers expect the thin silver (Ag) layer coating on the In bump to protect the inner In solder from oxidation and to decrease the melting temperature of the In solder. Thus, researchers tried coating a thin Ag layer onto the In surface. A eutectic reaction occurs at 97 wt% In with a eutectic point of 144°C, and the outer Ag layer interacts with In to form an $AgIn_2$ compound layer due to the high inter-diffusion coefficient. As a result, the thin Ag layer coated on the solder bump is very effective in enhancing the adhesion strength between the In bump and the VCSEL chip pads by decreasing the melting temperature of the In solder bump locally.

Flip-chip bonding is a bonding technique in which a device and separate host substrate are electrically connected together through a physical bond of solder. The chip bonding technique using solder bumps, a type of flip-chip bonding, offers a number of advantages. The interconnection length is in the micrometer range and thus provides superior electrical characteristics at high frequencies. Another attractive feature of flip-chip bonding is the self-aligning effect. As the temperature of the solder is raised above its melting point, the molten solder starts wetting the metal pad and moving the chip. This chip movement is driven by surface tension in an effort to minimize the surface area to reach the lowest total energy of the assembly. Thus, the chip is bonded accurately without the need for positioning adjustments.

Conductive adhesives have also become a viable alternative to Sn–Pb solders in flip-chip joining. Adhesively bonded flip chip combines the advantages of thin structures and cost efficiency. The advantages of conductive adhesives include ease of processing, low curing temperatures, and elimination of the need to clean after the bonding process.

Figure 13.5 shows flip-chip bonding with isotropically conductive adhesives (ICAs). Isotropically conductive adhesives are pastes of polymer resin that are filled with conducting particles to create a content that ensures conductivity in all directions. Generally, the polymer resin is epoxy and the conducting particles are Ag.

There are many difficulties with flip-chip bonding of VCSELs because the commercial VCSELs in present use were not developed for flip-chip bonding, but rather for die bonding. Also, the commonly used solders such as Au–Sn alloy cannot be applied to OE hybrid systems because high-temperature

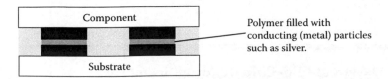

FIGURE 13.5
Flip-chip joint made with isotropically conductive adhesive (ICA).

processing degrades the VCSEL chip, microlens, or waveguide. The low temperature solders that can be processed below 160°C should be developed for this application. Therefore, it is necessary to investigate a new solder material suitable for optical device packaging and optimization of flip-chip bonding conditions such as the applied pressure, temperature, and environment. To achieve this goal, we considered five Sn-based, binary alloy solders:

- Sn–In
- Sn–Au
- Sn–Zn
- Sn–Cd
- Sn–Pb

As mentioned previously, the latter two, although initially promising, were eliminated from the Design Of Experiment (DOE) due to environmental concerns. The three remaining alloys trade off on desirable properties: melting temperature and electrical conductivity.

- The Sn–In solder has the lowest melting temperature, with a eutectic temperature of 117°C at 48Sn-52In, while Sn–Au has several melting points slightly below 400°C.
- Conversely, Sn–Au should give the highest electrical conductance because of the high conductance of Au, while Sn–In should be the poorest of the three.
- With regard to conductance, the quality of Sn–Zn should lie between those of its two counterparts, with a melting temperature near that of Sn–Au.

DOE studies the trade-offs and the relative advantages and disadvantages of each binary alloy solder. Vertical-cavity surface-emitting lasers (VCSELs) arranged in 2-D configurations of independently addressable elements hold great potential for short-reach fiber-optic as well as free-space networking applications. Given the continuous growth of required data rates in networks, parallel optical interconnects (POIs) based on VCSEL arrays are expected to be found more and more at increasingly deeper network levels in the future.

13.6 Design of Flip-Chip Bonding Structure

As Phase 1 of the case study focused on the metal choices for flip-chip bonding, the choice of substrate was not important in the design, as neither optical device nor conventional circuitry would be fabricated into a substrate. The three considerations made for a substrate were

1. Silicon (Si)
2. Gallium arsenide (Ga–As)
3. Glass

Of the three, Ga–As was the most expensive material and glass was the cheapest. In addition to the cost benefit, glass slides also have the side benefit of being clear, allowing for easier alignment during the final bonding process. As such, the fabrication of the flip-chip structure began on 1 cm × 1 cm glass slides.

Anisotropically conductive adhesives have the ability to connect fine-pitch devices. Figure 13.6 shows a flip-chip bonding with anisotropically conductive adhesives (ACAs). Anisotropically conductive adhesives are pastes or films of thermoplastics or b-stage epoxies. They are filled with metal particles or metal-coated polymer spheres to a content that ensures electrical insulation in all directions before bonding. After bonding, the adhesive becomes electrically conductive in the z-direction. The metal particles are typically Ni or Au, and these metals are also used to coat polymer spheres.

As with virtually all flip-chip processes, the aluminum bond pads must be re-metallized to eliminate nonconductive aluminum oxide. This ensures a low and stable contact resistance at the "bump/bond pad" interface. Several flip-chip interconnection methods have been compared by measuring the interconnect resistance before and after exposure to environments, including preconditioning, 85°C/85% RH exposure, 150°C storage, and 0°C to 100°C temperature cycling. The goal was to determine an acceptable low-cost, reliable method for bumping and assembling chips to flexible or rigid substrates using flip-chip assembly techniques. Alternative flip-chip bumping methods are compared to a traditional wafer solder bumping method. Flip-chip interconnection methods evaluated included high-Pb content solder, Ag-filled conductive adhesive, and Au stud bumps. Under-bump metallurgies (UBM) evaluated included bare Al, evaporated

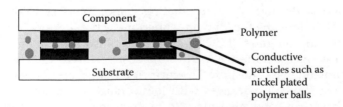

FIGURE 13.6
Flip-chip joint made with anisotropically conductive adhesive.

Cr/Cr–Cu/Cu, and electroless Ni plating. Resistance can be calculated using Equation 13.1:

$$R = \frac{(P * I)}{w * t}$$ (13.1)

where
  P is resistivity of the wire material
  L is the length of the wire
  w is width of the wire
  t is the thickness of the wire

The square pads are 250 μm on each side, while the round pads are 200 μm in diameter. While all the horizontal traces are 50-μm wide, the diagonal segments of the wires are approximately 35-μm wide. The length of the traces varies throughout the bottom sample.

- Alumina has long been recognized and accepted as a mature and robust substrate for high-density flip-chip applications. Alumina's limitations are its low CTE, which creates reliability issues for large outline packages (>35 mm) and its high conductor resistance and dielectric constant, which can challenge high-power or very-high-frequency designs.

- Organic laminate-based substrates are becoming very popular for FCIP (fiber channel over IP). These are based on high-density sequential built up and microvia substrate manufacturing technologies. Organic flip chip packages have low-resistance Cu interconnects and low dielectric constants. One limitation of organic package technology is its high coefficient of thermal expansion (CTE), which creates a large CTE mismatch between the die and substrate. This places heavy demands on the underfill to keep the structure together. Organic materials are also more challenged in environmental reliability tests. Typically, they cannot pass as demanding a test suite as ceramic packages are able to. Many of the limitations are due to moisture absorption.

An advantage of flip chip is that it provides access to the backside of the die to remove heat through a low-thermal-resistance interface to the lid/heat-spreader. Therefore, the flip chip has superior electrical characteristics.

In a flip-chip-on-glass (FCOG) assembly, anisotropic conductive film (ACF) is used as the adhesive to bind the desired interconnection between the flip chip and glass substrate. However, it remains a challenge to develop ACF bonded flip-chip packages with low contact resistance. Considerable research has been conducted recently to investigate the effect of different parameters on the contact resistance. Here we discuss the critical issues that can easily control the contact resistance of ACF joints in FCOG packages.

These mainly include surface cleanliness, bonding tracks, process parameters, and operating environmental-related issues. The findings can serve as a guide for minimizing the contact resistance of FCOG packages with ACF. By such minimization, ACF can be used as an environmentally friendly solder replacement in the very large scale integration (VLSI) industry.

13.7 Processing of Flip-Chip Bonding Structures

The processing of flip-chip bonding structures entails similar techniques to those used in the processing of semiconductor integrated circuits. Included in the process are

- Oxide growth
- Photolithography
- Metallization
- Etching

Because electrical devices are not fabricated, complex steps such as ion implantation and diffusion are not included. As such, flip-chip bonding is a relatively simple process that may be integrated into a larger manufacturing process. Most likely, bonding would be implemented in the final steps of a process. The design used for the flip-chip bonding structures involves two oxidation steps, two photolithography steps, and two metallization steps, in addition to the flip-chip bonding process. The concept of a flip-chip process where the semiconductor chip is assembled face-down onto circuit board is ideal for size considerations because there is no extra area needed for contacting on the sides of the component.

13.8 Solders for Flip-Chip Bonding

A variety of Pb-free solder alloys have been studied for use as flip-chip interconnects, including Sn-3.5Ag, Sn-0.7Cu, Sn-3.8Ag-0.7Cu, and eutectic Sn-37Pb as a baseline. The reaction behavior and reliability of these solders were determined in a flip-chip configuration using a variety of under-bump metallurgies (UBM; TiW/Cu, electrolytic nickel, and electroless Ni-P/Au). The solder microstructure and intermetallic reaction products and kinetics were determined. The Sn-0.7Cu solder has a large grain structure and the Sn-3.5Ag and Sn-3.8Ag-0.7Cu have a fine lamellar two-phase structure of

Sn and Ag-3Sn. The intermetallic compounds were similar for all the Pb-free alloys. On Ni, Ni3Sn4 formed and on copper, Cu6Sn5 and Cu3Sn formed. During reflow, the intermetallic growth rate was faster for the Pb-free alloys, compared to eutectic Sn–Pb. In solid-state aging, however, the interfacial intermetallic compounds (IMCs) grew faster with the Sn-Pb solder than with the Pb-free alloys. The reliability tests performed included shear strength and thermomechanical fatigue. The lower-strength Sn-0.7Cu alloy also had the best thermomechanical fatigue behavior. Failures occurred near the solder/intermetallic interface for all the alloys except Sn-0.7Cu, which deformed by grain sliding and failed in the center of the joint. Based on this assessment, the optimal solder alloy for flip-chip applications was identified as eutectic Sn-0.7Cu.

The effects of phase change (from solid to liquid) of flip-chip Pb-free solders during second-level interconnect reflow have been investigated. Most of the current Pb-free solder candidates are Sn-based solders, and their melting temperatures are similar in the range of 30°C. Thus, flip-chip Pb-free solders are melted again during subsequent second-level interconnect (BGA) reflow cycles. Like most other metals, a solder expands its volume during the phase change by as much as 4%. The volumetric expansion of solder in a confined space formed of chip, substrate, and underfill creates serious reliability issues. These issues include underfill fracture and delamination of the underfill from the chip or substrate. This leads to the shorting of neighboring flip-chip interconnects by the interjected solder through the underfill crack or delaminated interfaces. Accordingly, Pb-free flip-chip packages should have an additional reliability issue that is not a concern for Pb solder packages.

Thermomechanical analysis with conservative estimates of in-service power generation and forced cooling convective coefficients has determined the temperature profiles within a representative VCSEL assembly for four solder metallurgies, namely 48Sn-52In, 42Sn-58Bi, 63Sn-37Pb, 80Au-20Sn, and Sn(3–4)Ag(0.5–0.7)Cu (all wt%). The resulting temperature distributions were applied to structural models that examined the creep response of the solder alloys. In addition to the responses of the solder joints, the stresses imparted to the AlGaAs laser pad were also studied. Some important results are summarized in the following:

- The creep rate of Au–Sn is slower than that of Sn–Ag–Cu (SAC), Sn–Bi, Sn–Pb, and Sn–In. Also, the creep rate of SAC is slower than that of Sn–Bi, Sn–Pb, and Sn–In. In general, the creep rate of Sn–In is the largest.

- The modulus of Au–Sn is the largest and the moduli of Sn–Pb and Sn–In are the smallest among the solder alloys considered herein. Above room temperature, the modulus of Sn–In is larger than that of Sn–Pb. The modulus of Sn–Bi is smaller than that of SAC.

- The CTE of Sn–In is the largest and the CTE of Sn–Bi is the smallest of the alloys studied.

- The order (from small to large) of the steady-state temperature at the AlGaAs laser pad with different alloys is Au–Sn, SAC, Sn–Pb, Sn–In, and Sn–Bi.

- The order (from small to large) of the maximum creep strain at the solder joints with different alloys is Au–Sn, Sn–Bi, SAC, and Sn–Pb/Sn–In.

- The order (from small to large) of the maximum stress at the solder joints with different alloys is Sn–In, Sn–Pb, Sn–Bi, SAC, and Au–Sn. The stress relaxation of Au–Sn solder joints is very slow.

- The order (from small to large) of the maximum creep strain energy with different alloys is Au–Sn, Sn–In/Sn–Bi, Sn–Pb, and SAC.

- The order (from small to large) of the maximum stress at the AlGaAs laser pad with different alloys is Sn–In, Sn–Bi, Sn–Pb, SAC, and Au–Sn. The stress relaxation at the AlGaAs laser pad with Au–Sn is very slow.

- Based on the least creep damage of solder joints and the lower stress at the AlGaAs laser pads, Sn–In is the best choice among the solder alloys considered.

The VCSEL arrays were flip-chip bonded on a transparent glass substrate on which a polymeric waveguide is formed. In this structure for optical interconnection, requiring low-temperature processing, researchers conducted the flip-chip bonding of VCSEL arrays on the glass substrate using In solder bumps. It was found that the optimum condition for flip-chip bonding for VCSEL arrays using an In bump was 150°C, 500 gf. Also, the fracture between the In solder bump and chip pad occurred during the die shear test because the bonding temperature was low compared to the melting temperature of In and consequently indium oxide prevented good adhesion between the In bump and the VCSEL chip pad.

The most common method of removing the solder oxide is to apply acid rosin flux, which reduces the oxide and protects the solder against further oxidation. In this assessment, flux need not be used because a thin silver layer coating on the solder bump was very effective, both in preventing oxidation of the In bump and decreasing the melting temperature of pure In solder. Thus, the adhesion strength between the In bump and VCSEL chip pad could be enhanced.

Bibliography

Akay, H.U., Paydar, N.H., and Bilgic, A. (1997). Fatigue life predictions for thermally loaded solder joints using a volume-weighted averaging technique, *ASME Trans., J. Electronic Packaging*, 119 (December), 228–234.

Amagai, M., Watanabe, M., Omiya, M., Kishimoto, K., and Shibuya, T. (2002). Mechanical characterization of Sn–Ag based lead-free solders, *Micoelectronics Reliability*, 42, 951–966.

Antolovich, S.D., and Antolovich, B.F. (1996). *An Introduction to Fracture Mechanics in ASM Handbook 19 Fatigue and Fracture*, Materials Park, OH: ASM International®.

Arora, N.D., Raol, K.V., Schumann, R., and Richardson, L.M. (1996). Modeling and extraction of interconnect capacitances for multilayer VLSI Circuits, *IEEE Trans. Computer Aided Design of Integrated Circuits and Systems*, 15(1), 58–66.

Basaran, C., and Yan, C.Y. (1995). A thermodynamic framework of damage mechanics of solder joints, *J. Electronic Packaging*, 10(1), 365–376.

Bhattachaiya, B., and Ellingwood, B. (1998). Continuum damage mechanics-based model of stochastic damage growth, *J. Engineering Mechanics*, September, pp. 1000–1009.

Bilotti, A.A. (1974). Static temperature distribution in IC chips with isothermal heat sources, *IEEE Trans. Electron Devices*, ED-21 (March), 217–226.

Black, J.R. (1969). Electromigration failure models in aluminium metallization for semiconductor devices, *Proc. IEEE*, 57(9), 1587–1594.

Blech, I.A., and Herring, C. (1976). Stress generation by electromigration, *Appl. Phys. Lett.*, 29, 131–133.

Boresi, A.P., et al. (1993). *Advanced Mechanics of Materials, (5th edition)*. Canada: John Wiley & Sons.

Box, G.E.P., Hunter, W.G., and Hunter, J.S. (1978). *Statistics for Experimenters—An Introduction to Design, Data Analysis, and Model Building*. New York: John Wiley & Sons, Inc.

Brakke, K.A. (1994). *Surface Evolver Manual*. University of Minnesota, Geometry Center.

Chen, C., and Liang, S.W. (2007). Electromigration issues in lead-free solder joints, *J. Mater. Sci.*, 18, 259–268.

Coffin, L.F., Jr. (1954). A study of the effects of cyclic thermal stresses on a ductile metal, *ASME Transactions*, 76, 931–950.

Darveaux, R. (2000). Effect of simulation methodology on solder joint crack growth correlation, *Proc. 50th ECTC*, May 2000, pp. 1048–1058.

Darveaux, R. (1996). How to use finite element analysis to predict solder joint fatigue life, *Proc. VIII Int. Congress on Experimental Mechanics*, Nashville, TN, June 10–13, 1996, pp. 41–42.

Dayhoff, J. (1990). *Neural Network Architectures – An Introduction*. New York: Van Nostrand Reinhold, pp. 217–243.

Dreezen, G., Deckx, E., and Luyckx, G. (2003). Solder alternative: Electrically conductive adhesives with stable contact resistance in combination with non-noble metallization, *CARTS Europe 2003*, pp. 223–227.

Ferreira, F.K., Moraes, F., and Reis, R. (2000). LASCA-interconnect parasitic extraction tool for deep-submicron IC design, *Proc. 13th Symposium on Integrated Circuits and Systems Design*, 2000, pp. 327–332.

Frear, D.R., and Kinsman, K.R. (1991). *Solder Mechanics - A State of the Art Assessment*, Warrendale, PA: Minerals, Metals, and Material Society.

Gale, W.F., and Totemeier, T.C. (2004). *Smithells Metals Reference Book, (8th edition)*. Maryland Heights, MO: Elsevier.

Galyon, G.T. (2003). *Annotated Tin Whisker Bibliography*, Hearndon, VA: a NEMI Publication, July.

Gilat, A., and Krisha, K. (1997). The effects of strain rate and thickness on the response of thin layers of solder loaded in pure shear, *J. Electronic Packaging*, 119, 81.

Goel, A.K., and Au-Yeung, Y.T. (1990). Electro migration in the VLSI interconnect metallizations, *Proc. 32nd Midwest Symposium on Circuits and Systems*, 1989, Vol. 2, pp. 821–824.

Grunwald, J., and Schnack, E. (1995). Models for shape optimization of dynamically loaded machine parts, *Proc. WCSM01*, Oxford: Pergamon Press, pp. 307–310.

Guo, Q., Cuttiongco, E.C., Keer, L.M., and Fine, M.E. (1992). Thermomechanical fatigue life prediction of 63Sn/37Pb solder, *ASME Trans., J. Electronic Packaging*, 114, 145–150.

Haykin, S. (1997). *Neural Networks – A Comprehensive Foundation, (2nd edition)*. Upper Saddle River, NJ: Prentice-Hall, pp. 2–10.

Hong, B.Z. (1997). Finite element modeling of thermal fatigue and damage of solder joints in a ceramic ball grid array package, *J. Electronic Materials*, 26(7), 814–820.

Hunter, W.R. (1997), Self-consistent solutions for allowed interconnect current density. I. Implication for technology evolution, *IEEE Trans. Electron Devices*, 44(2), 304–309.

Hunter, W.R. (1997). Self-consistent solutions for allowed interconnect current density. II. Application to design guidelines, *IEEE Trans. Electron Devices*, 44(2), 310–316.

Hunter, W.R. (1995). The implications of self-consistent current density design guidelines comprehending electromigration and Joule heating for interconnect technology evolution, *Int. Electron Devices Meeting*, 1995, pp. 483–486.

Jerke, G., and Lienig, J. (2002). Hierarchical current density verification for electromigration analysis in arbitrarily shaped metallization patterns of analog circuits, *Proc. Design, Automation and Test in Europe Conference and Exhibition*, 2002, pp. 464–469.

Ju, T.H., Chan, Y.W., Hareb, S.A., and Lee, Y.C. (1995). An integrated model for ball grid array solder joint reliability, structural analysis in microelectronic and fiber optic systems, *ASME, EEP*-Vol. 12, 83–89.

Jung, W., Lau, J.H., and Pao, Y.-H. (1997). Nonlinear analysis of full-matrix and perimeter plastic ball grid array solder joints, *ASME Trans., J. Electronic Packaging*, 119 (September), 163–170.

Lall, P., Islam, N., Suhling, J., and Darveaux, R. (2003). Model for BGA and CSP Reliability in Automotive Underhood Applications, *Proc. 53rd Electronic Components and Technology Conference*, May 27–30, 2003, pp. 189–196.

Lall, P., Pecht, M., and Hakim, E. (1997). *Influence of Temperature on Microelectronic and System Reliability*. Boca Raton, FL: CRC Press.

Lau, J.H., (Editor) (1991). *Solder Joint Reliability—Theory and Application*. New York: Van Nostrand Reinhold, p. 279.

Lau, J.H., and Pao, Y.H. (1997). *Solder Joint Reliability of BGA, CSP, Flip Chip and Fine Pitch SMT Assemblies*. New York: McGraw-Hill.

Lee, S.-W.R. and Lau, J.H. (1998). Solder joint reliability of cavity-down plastic ball grid array assemblies, *Soldering & Surface Mount Technology*, 10(1), 26–31.

Lemaitre, J. (1996). *A Course on Damage Mechanics*. Berlin: Springer-Verlag, pp. 11–36.

Lienig, J., Jerke, G., and Adler, T. (2002). Electromigration avoidance in analog circuits: two methodologies for current-driven routing, *Proceedings of ASP-DAC Conference*, 2002; *Proc. 7th Asia and South Pacific and the 15th International Conference on VLSI Design*.

Manson, S.S. (1965). Fatigue: A complex subject—Some simple approximations, *Experimental Mechanics*, 5(7), 193–226.

Meekisho, L., and Nelson-Owusu, K. (1999). Stress analysis of solder joint with torsional eccentricity subjected to based excitation, Conference paper presented at the *12th Int. Conf. Mathematical and Computer Modeling and Scientific Computing,* Chicago, IL, August.

Muju, S., et al. (1999). Predicting durability, *Mechanical Engineering Magazine of ASME,* March, pp. 64–67.

Nagaraj, B., and Mahalingam, M. (1993). Package-to-board attachment reliability methodology and case study on OMPAC package, *ASME Advances in Electronic Packaging,* EEP-4-1, 537–543.

Ohring, M. (1998). *Reliability and Failure of Electronic Materials and Devices.* San Diego: Academic Press.

Pan, T.-Y. (1994). Critical accumulated strain energy (CASE) failure criterion for thermal cycling fatigue of solder joints, *ASME Trans., J. Electronic Packaging,* 116 (September), 163–170.

Pang, J.H.L., and Chong, D.Y.R. (2001). Flip chip on board solder joint reliability analysis using 2-D and 3-D FEA models, *IEEE Trans. Advanced Packaging,* 24(4), 499–506.

Pang, J.H.L., Xiong, B.S., and Low, H. (2004). Creep and fatigue characterization of lead free 95.5Sn-3.8Ag-0.7Cu solder, *Proc. 54th ECTC,* June, pp. 1333–1337.

Pao, Y.-H., Jih, E., Adams, R., and Song, X. (1998), BGAs in Automotive Applications, *SMT,* January, pp. 50–54.

Paydar, N., Tong, Y., and Akay, H.U. (1994). A finite element study of factors affecting fatigue life of solder joints, *ASME Trans., J. Electronic Packaging,* 116 (December), 265–273.

Racz, L.M., and Szekely, J. (1993). An analysis of the applicability of wetting balance measurements of components with dissimilar surfaces, *Advances in Electronic Packaging, ASME,* EEP-4-2, 1103–1111.

Sakimoto, M., Itoo, T., Fujii, T., Yamaguchi, H., and Eguchi, K. (1995). Temperature measurement of Al metallization and the study of Black's model in high current density, *IEEE Int. 33rd Annual Proc. Reliability Physics Symposium,* 1995, pp. 333–341.

Sampath, B.K. (2001). *Electromigration dependent MTTF Calculations,* Analog IC Research Group, University of Texas, Arlington.

Setlik, B., Heskett, D., Aubin, K., Briere, M.A. (1997). Electromigration investigations of aluminum alloy interconnects, *Proc. Twelfth Biennial University/Government/Industry Microelectronics Symposium,* 1997, pp. 159–160.

Shi, X.Q., Pang, H.L.J., Zhou, W., and Wang, Z.P. (1999). A modified energy-based low cycle fatigue model for eutectic solder alloy, *Scripts Material,* 41(3), 289–296.

Skrzypek, J.J., and Hetnarski, R.B. (1993). *Plasticity and Creep-Theory, Examples, and Problems,* Boca Raton, FL: CRC Press.

Steinberg, D.S. (2000). *Vibration Analysis for Electronic Equipment.* New York: John Wiley & Sons.

Strauss, R. (1998). *SMT Soldering Handbook, (Second edition).* Maryland Heights, MO: Elsevier/Newnes.

Suo, Z. (2004), A continuum theory that couples creep and self-diffusion, *J. Appl. Mechanics,* 71, 646–651.

Syed, A R. (2004). Accumulated creep strain and energy density based thermal fatigue life prediction models for SnAgCu solder joints, *Proc. 54th ECTC,* June 2004, pp. 737–746.

Syed, A.R. (1995). Creep crack growth prediction of solder joints during temperature cycling: An engineering approach, *Trans. ASME,* 117 (June), 116–122.

Teng, C.C., Cheng, Y.K., Rosenbaum, E., and Kang, S.M. (1997). iTEM: A temperature-dependent electromigration reliability diagnosis tool, *IEEE Trans. Computer-Aided Design of Integrated Circuits and Systems,* 16(8), 882–893.

Tu, K.N. (2003). *Solder Joint Technology: Materials, Properties, and Reliability.* Berlin, Germany: Springer.

Tu, K.N. (2003). Recent advances on electromigration in very-large-scale integration of interconnects, *J. Appl. Phys.,* 94, 5451–5473.

Tu, K.N. (1994). Irreversible processes of spontaneous whisker growth in bimetallic Cu-Sn thin-film reactions, *Phys. Rev. B,* 49(3), 2030–2034.

Tunga, K., Pyland, J., Pucha, R.V., and Sitaraman, S.K. (2003). Field-use conditions vs. thermal cycles: A physics-based mapping study, *Proc. 53rd Electronic Components and Technology Conference,* May 27–30, 2003, pp. 182–188.

Wiese, S., Schubert, A., Walter, H., Dudek, R., Feustel, F., Meusel E., and Michel, B. (2001). Constitutive behaviour of lead-free solders vs. lead-containing solders experiments on bulk specimens and flip-chip joints, *Proc. 51st Electronic Components and Technology Conference,* 2001, pp. 890–902.

Winter, P. R., and Wallach, E.R. (1997). Microstructural modeling and electronic interconnect reliability, *Int. J. Microcircuits and Electronic Packaging,* 20(2), 124–129.

Wu, W., Kang, S.H., Yuan, J.S., and Oates, A.S. (2000). Electromigration performance for Al/SiO2, Cu/SiO2 and Cu/low-k interconnect systems including Joule heating effect, *2000 IEEE Int. Integrated Reliability Workshop Final Report,* pp. 165–166.

Yeh, C.-P., Zhou, W.X., and Wyatt, K. (1996). Parametric finite element analysis of flip chip reliability, *Int. J. Microcircuits and Electronic Packaging,* 19(2), 120–127.

Zahn, B.A. (2002). Finite element based solder joint fatigue life predictions for a same die size-stacked-chip scale-ball grid array package, *SEMICON West, International Electronics Manufacturing Technology (IEMT) Symposium,* pp. 274–284.

Zahn, B.A. (2003). Solder joint fatigue life model methodology for 63Sn37Pb and 95.5Sn4Ag0.5Cu materials, *Proc. 53rd Electronic Components and Technology Conference,* May 27–30, 2003, pp. 83–94.

14

Let's Package a Lead-Free Electronic Design

In flip-chip packaging, the mismatch of thermal expansion coefficients among the silicon die, copper heat spreader, and packaging substrate induces a concentrated stress field around the edges and corners of the silicon die during assembly, testing, and service. The concentrated stresses result in delamination on various interfaces involving a range of length scales from hundreds of nanometers to millimeters. Among these failures, underfill delamination is a dominant failure mode.

Lead-free solder materials are emerging to replace tin-lead solders in electronic assemblies as the industry adapts to legislation and environmental lobbying to ban lead in electronic products. Implementation of lead-free solders in electronic packaging involves knowledge of lead-free solder materials characterization and mechanical properties, solder joint reliability tests and failure analysis, finite element analysis simulation, and fatigue life prediction. To successfully achieve lead-free electronics assembly, each participant in the manufacturing process, from purchasing to engineering to maintenance to quality/inspection, must have a solid understanding of the changes required of them. This pertains to considerations regarding design, components, PWBs, solder alloys, fluxes, printing, reflow, wave soldering, rework, cleaning, equipment wear and tear, and inspection.

14.1 Select the Package Type: Flip-Chip Packaging

Flip-chip packaging, in which the silicon die is directly attached to the substrate using solder bumps instead of wire bonds, provides the densest interconnect with the highest electrical and thermal performance. Flip-chip technology is used in a wide array of applications, ranging from consumer products to highly sophisticated ASICs (application specific integrated circuits), PC chipsets, graphics, and memory packages. Flip-chip interconnection provides the ultimate in miniaturization and reduced package parasitics, and enables new paradigms in the area of power and ground distribution to the chip, which are not feasible in other traditional packaging approaches.

Here, flip-chip packaging is chosen to house a 3.5-GHz die. Technologies such as quad flat pack and ball grid array are not recommended because wire bonding can introduce signal degradation at this high frequency. With careful planning, it is possible to use wire-bonding technology with the low I/O pins for this packaging; however, it may be best to eliminate all wire-bonding for higher-frequency applications in favor of the potentially cheaper and higher performance flip chip. The advantage of flip chip over quad flat pack or ball grid array becomes apparent when a large production volume is required or the number of I/O pins is high. In these situations, flip chip may provide a high number of I/O pins with low noise.

Flip-chip packaging has long been the highest-performance packaging solution available in the industry. It is the preferred solution for CPUs/MPUs (central processing units/micro processor units), high-end ASICs, and other high-I/O and high-speed devices. Flip-chip microelectronic assembly uses a direct electrical connection of a face-down die onto substrates, circuit boards, or carriers, by means of conductive bumps on the chip bond pads. As a result, flip chip requires a shorter length for electrical connection from the chips to the PCB (printed circuit board), which helps reduce the noise and decrease signal degradation. In contrast, face-up chips generally use wire-bonding, which may lead to longer trace and increase the chances of cross-talk and external electrical interferences.

The demand for flip-chip interconnect technology is being driven by a number of factors from all corners of the silicon industry. Flip-chip packaging is now in widespread use in computing, communications, and consumer and automotive electronics. The flip-chip packaging of today is drastically different from the technology of the 1990s. The demand for flip-chip technology will continue to grow to meet the need for products that offer better performance, are smaller, and are environmentally sustainable.

Flip chip (FC) is not a specific package (such as SOIC [small-outline integrated circuit]), or even a package type (such as ball grid array, BGA). Flip chip describes the method of electrically connecting the die to the package carrier. The package carrier, either substrate or lead-frame, then provides the connection from the die to the exterior of the package. In "standard" packaging, the interconnection between the die and the carrier is made using wire. The die is attached to the carrier face-up; then a wire is bonded first to the die, then looped and bonded to the carrier. Wires are typically 1 to 5 mm in length, and 25 to 35 μ in diameter. In contrast, the interconnection between the die and carrier in flip-chip packaging is made through a conductive "bump" that is placed directly on the die surface. The bumped die is then "flipped over" and placed face-down, with the bumps connecting to the carrier directly. A bump is typically 70 to 100 μ high and 100 to 125 μ in diameter. Using flip-chip interconnects offers a number of possible advantages, including

- *Reduced signal inductance.* Because the interconnect is much shorter in length (0.1 mm versus 1 to 5 mm), the inductance of the signal path is greatly reduced. This is a key factor in high-speed communication and switching devices.

- *Reduced power/ground inductance.* By using flip-chip interconnects, power can be brought directly into the core of the die, rather than having to be routed to the edges. This greatly decreases the noise of the core power, thus improving the performance of the silicon.

- *Higher signal density.* The entire surface of the die can be used for interconnect, rather than just the edges. This is similar to the comparison between QFP (quad flat package) and BGA packages. Because flip chip can connect over the surface of the die, it can support vastly larger numbers of interconnects on the same die size.

- *Die shrink.* For pad-limited die (die where size is determined by the edge space required for bond pads), the size of the die can be reduced, thereby saving silicon cost.

- *Reduced package footprint.* In some cases, the total package size can be reduced using flip chip. This can be achieved by either reducing the die to package edge requirements, as no extra space is required for wires, or by utilizing higher-density substrate technology, which allows for reduced package pitch.

14.2 Select Substrate or Die Attachment: FR4

Anisotropic conductive adhesive films (ACFs) were used to attach daisy-chained test chips on FR4 substrates. Because the operating frequency is located at the low end of the high-frequency signal, a single-core, double-sided FR4 substrate is selected for the die. The FR4 substrate has the advantages of being cheap and easily upgradeable for higher frequency demand in the future. For example, as the demand for higher frequency increases, a BT (bismaleimide triazine) substrate can be used in place of the FR4 to boost the performance of the chip. ACFs have been successfully used to join driver circuits on liquid crystal displays. Using ACFs, reliable high-density interconnections have also been made on flexible substrates. The use of rigid organic substrates in adhesive flip-chip attachment is an interesting alternative for making high-density interconnections. However, cost-effective manufacturing will require reduced bonding cycle times and/or use of multi-head joining equipment.

Long-term reliability of adhesive joints on organic substrates is a key factor to be ensured. Co-planarity of substrate and chip is important for achieving bonding quality. Warpage and bump height variation in the organic substrates may cause problems that do not occur when even glass substrates are used. The challenge, however, is to achieve a uniformly low contact resistance that is stable over time under environmental stresses.

Thermal characteristics of flip chip on FR4 boards have been investigated. The thermal resistances are determined for different die and board constructions, underfill material, and heat-sink applications. Thermal paths are analyzed to understand the flip-chip heat dissipation mechanism. It is realized that the junction to ambient thermal resistance is dominated by the system environment. The package resistance is only a trivial portion of the total resistance. Improvement in thermal performance should be concentrated on the system level. Thermal performance of flip chip is compared to that of PQFP (plastic quad flat pack) and PBGA (plastic ball grid array). Recommendations in flip-chip thermal management are given at the end. This work is based on both experimental and numerical studies. The use of anisotropic conductive adhesives (ACAs) in flip-chip interconnection technology has become very popular because of their numerous advantages. The ACA process can be used in high-density applications and with various substrates as the bonding temperature is lower than that in the soldering process.

14.3 Select Electrical Connections from Die to FR4

The electrical connections from the chip to the FR4 substrate is accomplished by a gold-to-gold thermosonic processes. Thermosonic flip-chip bonding is an emerging, solderless technology for area-array connections. The thermosonic approach is used to join ICs (integrated circuits) with gold bumps to gold plated pads on the substrate. It offers a range of features superior to soldering and ACF counterparts, that is, it simplifies the processing and assembly steps; reduces the levels of assembly temperature, loading pressure, and bonding time; and increases current-carrying capacity. Thermosonic bonding technology has the following advantages over the existing ACF mounting method:

- Metallurgical joining is more reliable than conductive particles and adhesive joining.
- Process cycle time can be reduced from several minutes to less than 10 seconds.
- Lower manufacturing cost per unit.

Thermosonic bonding is used because it reduces the processing time from several minutes to less than 10 seconds, thereby lowering the manufacturing cost. Furthermore, the metallurgical joining from the gold-to-gold thermosonic processes is also more reliable than conductive particles or adhesive joining. The gold bump produced in this process generally has a diameter of 75 μm and a height of 50 μm. Finally, gold-to-gold thermosonic bonding is proven to be useful for die with dimensions up to 5 × 5 mm and up to 68 I/Os.

Thermosonic bonding uses a micro-weld interconnection die attach method at lower bonding temperatures (150°C). The thermosonic metal-to-metal interconnection method is lead-free and the process does not use flux or solder alloys. The thermosonic flip-chip die attach process uses a robust individual die "scrubbing" process that reduces the number of assembly steps and eliminates the mass reflow oven commonly used in Controlled Collapse Chip Connection (C4) solder processes. Gold-to-gold thermosonic electrical connections from the die to FR4 substrate also improve signal quality by creating a short, constant electrical pathways directly through the via to the external circuits. Without a flip-chip configuration, the electrical pathway tends to be longer, thus increasing the chances of cross-talk or electromagnetic interference. When compared to the more traditional wirebond method, the electrical connection of the flip chip is much shorter because wires do not have to travel to the top surface of the chip. In addition, the flip chip does not have to account for the varying die size associated with the sawing processes.

Flip chip, also known as C4, is a method for interconnecting semiconductor devices, such as IC chips and Micro Electro-Mechanical Systems (MEMS), to external circuitry with solder bumps that have been deposited onto the chip pads. Gold-to-gold interconnection (GGI) flip-chip bonding technology has been developed to bond the drive IC chip on the integrated circuit suspension used in hard disk drives. GGI is a Pb-free process where the gold (Au) bumps and Au bond pads are joined together by heat and ultrasonic power under a pressure head. The use of a GGI flip-chip assembly process will help to eliminate equipment parts and processing steps of the traditional flip-chip C4 process and hence shorten the overall cycle time. With the IC suspension design, it becomes possible to assemble the drive IC chip nearby the magneto-resistive head slider on the suspension. Finally, gold-to-gold thermosonics is cost effective for low I/O die, and it is preferred over wafers by many companies. Other methods, such as overlay interconnect technology, can also be used to provide better performance, but at the present time they seem to be too expensive and the chips have a low I/O pin configuration. With time, it will probably become more economical to use the overlay interconnect method to connect dies directly together and ignore the first-level packaging processes; but until such time, the gold-to-gold thermosonic process may be the best process to connect this sixteen-pin die.

The thermal characteristics of flip chip on FR4 boards have been studied. Thermal resistances are determined for different die and board constructions, underfill material, and heat sink applications. Thermal paths are analyzed to understand the flip-chip heat dissipation mechanism. It is realized that the junction to ambient thermal resistance is dominated by the system environment. The package resistance is only a trivial portion of the total resistance. Improvement in the thermal performance should be emphsized at the system level. The thermal performance of flip chip is compared to that of PQFP and PBGA. Recommendations for flip-chip thermal management are given at the end. This work is based on both experimental and numerical studies.

14.4 Assess Impact of CTE Mismatch on Stress and Fatigue Life

As IC packages clock higher frequencies and pin counts, assembly houses are considering flip-chip packaging to meet high-volume demands. The underfill material used in this process in no small way determines the reliability of the package, irrespective of its nonmolded or overmolded design. Among the material properties, the coefficient of thermal expansion (CTE) of the underfill should match that of the solder bump connecting the die to the substrate. Typical solders used for flip chip are Pb5Sn (with a CTE of 29 ppm) and eutectic solder (with a CTE of 24.3 ppm). Thus, formulating an underfill calls for fine-tuning the material to have a CTE within this range. Because FR4 has a relatively high CTE, an underfill is used for this package. No Flow Underfill (NUF 2071E Cookson Semiconductor Packaging Materials) is chosen to distribute the stress between the die and the FR4 substrate and enhance the lifetime of the chips.

The underfill process is used in this situation because underfill tends to increase the fatigue life of solder bumps by at least an order of magnitude. Finally, underfill protects the underlying electrical connections by helping to eliminate short-circuits and prevent oven circuit failure due to corrosive chemical processes. The CTE of an underfill is adjusted by varying two parameters: the silica filler and the polymer chemistry of the material. The optimum filler loading is in the range of 60% to 65% for the two types of solders, and varies somewhat with other filter properties such as particle size and distribution. Filler loading also affects the "flow" of the material. Thus, we need to strike a fine balance when trying to optimize the material, taking into account both of these desired characteristics.

The CTE is optimized not only at room temperature, but also at all package reliability testing temperatures. This is because the CTE of polymer materials increases once the glass transition temperature (Tg) is exceeded. Other tests include humidity tests, which include 85°C/85% relative humidity temperature cycling, and thermal shock tests conducted between −55°C and +150°C. The best option is to manufacture the underfill with its Tg at about 150°C or higher to maintain a CTE match with the solder throughout the reliability tests.

14.5 Design Solder Balls for External Connection to PCB

One way to create the external electrical contact to connect the chip to a PCC (plastic chip carrier) is by some solder balls. Using flip-chip bonding techniques with micromachined conductive polymer bumps and passive alignment techniques with electroplated side alignment pedestal bumps, a prototype micro-opto-electromechanical system (MOEMS) structure for optical I/O couplers has been designed, fabricated, and characterized. A top MOEMS substrate has through-holes, contact metal pads, and side alignment pedestals with electroplated NiFe to align GaAs metal-semiconductor-metals (MSMs). Conductive polymer bumps have been formed on contact metal pads of GaAs MSMs using thick photoresist bump-holes as molding patterns. A diced GaAs (gallium arsenide) photodetectors die with micromachined conductive polymer bumps was aligned to the side alignment pedestals and flip-chip bonded onto the substrate. This conductive polymer flip-chip bonding technique allowed a very low contact resistance (~10 mΩ), a lower bonding temperature (~170°C), and simple processing steps. The GaAs MSM photodetectors flip-chip mounted on the top of OE-MCM (opto-electronic multi-chip module) substrate showed a low dark current of about 10 nA and a high responsivity of 0.33 A/W. Although any lead bumps will be sufficient for this chip, a cheaper solution and probably better process might be to use a polymer flip-chip (PFC) process to produce this solder ball.

Compared to the traditional bumping process, PFC bumping is performed in a one- or two-step process, compared to several depositions of metal required for traditional solder bumping. As a result, the equipment and processing cost can be lower without expensive evaporation equipment or mask aligners. Furthermore, replacing solder with polymers allows processing at temperatures lower than 160°C, which may decrease the chance of failure due to thermal cycling. A PFC technique also reduces the processing cost and enables packagers to bond chips to less costly substrates that cannot tolerate the temperatures needed for solder connections. Finally, the PFC process does not use Pb, flux, or CFC-laden solvents. Depending on the exact situation, the PFC process might also be used to attach the die directly to the FR4 substrate.

Compared with classic solder flip-chip technologies, the use of isotropic conductive adhesive for bumping and assembling results in a number of advantages, such as a simple yet versatile process, lower temperatures, and environmental friendliness. The PFC technology requires suitable bond pads. The commonly used aluminum alloys result in unstable contacts for the conductive adhesives, and hence there is an urgent need for an additional under-bump metallization (UBM) for PFC technology.

14.6 Thermal Analysis of Flip-Chip Packaging

The function of an electronic cooling package is to dissipate heat to ensure proper operation and reliability. The flip-chip ball grid array package is probably the most suitable package for high-level thermal performance applications. A high thermal performance flip-chip ceramic ball grid array (FC-CBGA) package with an aluminum silicon carbide (AlSiC) lid and one without the lid were evaluated using the computational fluid dynamics (CFD) technique. To better understand its heat dissipation and for device optimization, it is desirable to know the temperature distribution not only in the active device area, but also in the bulk substrate. This cannot be achieved using traditional thermal imaging techniques. Three-dimensional (3-D) thermal imaging was demonstrated by probing the temperature-dependent Raman shift of phonons at different depths within the bulk substrate using confocal micro-Raman spectroscopy. The thermal analysis is performed using the following assumptions:

1. The major source of heat is generated by the die; all other heat sources are insubstantial.
2. Dissipation of energy toward the side of the chip is insignificant.
3. Thermal conduction of the two-metals layer on the FR4 substrate is assumed insignificant because metal conducts heat very quickly compared to FR4.
4. The sputter gold used in the thermosonic process is assumed negligible or part of the bump's height.
5. Convection of the heat is insignificant.

Printed circuit board and package top-surface temperature patterns were measured using an infrared thermal camera. The usefulness of the thermal characterization parameter is demonstrated by system-level applications. Parametric studies were carried out to understand the effect of die size, radiation, grid size variations, and airflow rate on the die junction temperature and package thermal resistance. This thermal analysis also incorporates the effects of substrate, lid, die, and PCB temperatures for different die sizes in natural and forced convection environments. For this thermal analysis, the thermal resistance is given by

$$R = \frac{L}{\lambda A}$$

where
  λ is the thermal conductivity of the material
  A is the cross-sectional area
  L is the thickness

The trend in packaging microelectronic systems and subsystems has been to reduce size and increase performance, both of which contribute to increased heat generation. Evidence of this trend can be observed in the higher level of integration at the device/package level. Placing more functions in a smaller microelectronic device/package has resulted in nonuniform power distribution with extreme heat density, mandating that thermal management be given higher priority in the design cycle in order to maintain system performance and reliability. The thermal conductivity of the materials used in the packaging are as follows:

$$\lambda_{chip} = 0.150 \ \frac{W}{mmK}$$

$$\lambda_{FR4} = 0.00050 \ \frac{W}{mmK}$$

$$\lambda_{solder} = 0.050 \ \frac{W}{mmK}$$

$$\lambda_{underfill} = 0.00040 \ \frac{W}{mmK}$$

$$\lambda_{epoxy} = 0.00030 \ \frac{W}{mmK}$$

$$\lambda_{gold} = 0.317 \ \frac{W}{mmK}$$

$$\lambda_{air} = 0.026 \ \frac{W}{mmK}$$

The methodology for thermal analysis and characterization of a flip-chip microelectronic package with nonuniform power distribution has been developed. Analytical methods, employing multiple linear regression (MLR), temperature superposition, and Lagrangian interpolation techniques to predict the temperature distribution of nonuniform powered microelectronic devices for thermal analysis and genetic algorithms for optimization are introduced. These methods are useful in investigating the thermal interactions of heat sources within the silicon chip. Critical thermal parameters (i.e., the heat source placement distance, level of heat dissipation, and magnitude of convection heat transfer) are examined in more than 900 simulations. Optimal placement of heat sources within the silicon chip is being carried out using genetic algorithms. The locations of this placement have been verified with finite element analysis (FEA). The area and thickness for heat transfer are as follows:

$$A_{chip} = 25 \ mm^2 \qquad\qquad L_{chip} = 0.50 \ mm$$

$$A_{FR4} = 64 \ mm^2 \qquad\qquad L_{FR4} = 0.50 \ mm$$

$$A_{gold} = (16\pi r^2) = 16\pi(0.0375)^2 = 0.071 \ mm^2 \qquad L_{gold} = 0.050 \ mm$$

$$A_{\text{underfill}} = A_{\text{FR4}} - A_{\text{gold}} = 63.929 \text{ mm}^2 \qquad\qquad L_{\text{underfill}} = 0.050 \text{ mm}$$

$$A_{\text{epoxy}} = 64 \text{ mm}^2 \qquad\qquad L_{\text{epoxy}} = 0.50 \text{ mm}$$

$$A_{\text{solder}} = (16\pi r^2) = 16\pi(0.150)^2 = 1.131 \text{ mm}^2 \qquad\qquad L_{\text{solder}} = 0.30 \text{ mm}$$

$$A_{\text{air}} = A_{\text{FR4}} - A_{\text{solder}} = 62.868 \text{ mm}^2 \qquad\qquad L_{\text{air}} = 0.30 \text{ mm}$$

From the above equations, the thermal resistance of each of the components is calculated. An epoxy surrounding the chip with an FR4 PCB is used to simulate the worst situation possible. In both cases, the temperature inside the die did not go above the T_{jmax} of 125°C. However, temperature can vary from 77°C to 119°C, depending on the exact nature of the enclosure. The chip is small enough that a heat sink might not be necessary if the PCB or the packaging does not have extremely low thermal conduction. To be safe, the outside enclosure of the chip should have high thermal conduction to serve as a heat sink. However, adding a heat sink is probably not essential in most situations.

Advanced assessment employs the coupled field method of thermal analysis with the structural method to simulate the thermal conduction and mechanical behaviors of flip-chip packages. The full, one-quarter symmetrical, one-eighth symmetrical, and stripped finite element models are analyzed by thermal shock testing in the temperature loading range of 55°C to 125°C. Engelmaier's model is used to assess the issues of fatigue life of these models. The results reveal that the one-eighth symmetrical model can save CPU time and still retain the accuracy of analysis. Furthermore, advanced studies investigated the geometrical dimensions of bumps regarding the effects and the reliability of flip-chip bumps by means of a one-eighth symmetrical finite element model with CAE (computer-aided engineering) analysis under the loading of cyclic temperature. The Taguchi method is employed to find the optimum upper diameter, lower diameter, height, and width of bumps in this research. There is a novel concept that the climbing height of underfill is considered the noise factor because it has an impact on the fatigue life of flip-chip packages. Analysis of variance is used to examine the contribution of these geometric parameters with respect to the fatigue life of solder bumps. It is found that the influence of climbing height of underfill has a significant effect on the fatigue life of flip-chip bumps. Also the width of the solder bump is the most important factor.

14.7 RLC for Flip-Chip Packages

The continuous increase in demand for product miniaturization, higher package density, higher performance, and integration of different functional chips has led to the development of three-dimensional (3D) packaging technologies. Face-to-face silicon (Si) die stacking is one of the 3D packaging technologies used to form a high-density module.

Electrical simulations are carried out to obtain RLC (resistor, inductor, capacitor circuit) parameters of micro-bump interconnects and complete interconnection from daughter die to substrate. Mechanical simulations are also carried out to assess the stress analysis on micro-bumps and CSP (chip scale packaging) bumps in the package and parametric analysis of the stacked module package to assess the effect of substrate material, underfill material die thicknesses on package reliability, and warpage.

Test chips are designed and fabricated with daisy-chain test structures to access the reliability of the stack module. Pb-free Sn–Ag micro-bumps of 40 µm on daughter die wafers and eutectic Sn-Pb solder CSP bumps of 200 µm height on mother die wafers are fabricated. Mother die and daughter die bumped wafers were thinned to 300 µm and 60 µm, respectively, using a mechanical backgrinding method. These thin dies are stacked using chip-to-wafer flip-chip bonding and the underfill process is established for the micro-bump interconnects. The assembled Si die stacked modules are subjected to JEDEC (Joint Electron Devices Engineering Council) package-level reliability tests in terms of temperature cycle (TC) tests, high-temperature storage (HTS) tests, moisture sensitivity test level 1 (MST L1) and MST L3, and unbiased high accelerated stress test (uHAST). The best way to analyze the RLC characteristic is to use supposition and add the effect from all contributing factors. However, an easier and better way would be to download a program from the Internet and use it to simulate the transmission line for this circuit. Another alternative is to simplify the circuit with the following assumptions:

- The longest and shortest pair are next to each other.
- Contribution from other lines to the analyzed pair is negligible.
- Contribution from solder balls or gold bumps is insignificant.
- Capacitor edge effect is insignificant.
- Mutual inductance geometry is independent.
- Vias have negligible capacitance.
- All other inductances and capacitances are sufficiently far away and can be ignored for a rough analysis.

14.7.1 Capacitance

$$\frac{1}{C} = \frac{1}{C_{trace}} + \frac{1}{C_{wire}} + \frac{1}{C_{edge}}$$

$$C = C_{\text{trace}} = \varepsilon_r \varepsilon_o \frac{A}{h}$$

Assuming that the metal line is 40-μm wide, the length of the line on both sides of the chip is 3 mm total, and the FR4 substrate is 0.5-mm thick, the capacitance is calculated. Plugging these values into the equations above give a capacitance of 0.0085 pF for the short length and 0.0113 pF for the longest length.

14.7.2 Resistance

$R_{DC} = R_{trace} + R_{via}$, where $R = L/A\sigma$. Substituting $\sigma_{Cu} = 5.88 \times 10^5/\Omega cm$, $A_{Cu} = 400\mu m^2, \sigma_{via} = 5\times10^5/\Omega cm, A_{via} = 706\mu m^2, L_{via} = 500\mu m, L_{Cu\_shortia} = 3,000\mu m$, and $L_{Cu\_long} = 4,000$ μm into the resistor equation gives a short line resistance of 141 mΩ and a long path resistance of 184 mΩ. The value obtained seems reasonable because the pathway is half that of a wirebond and the total resistance reflects this change.

14.7.3 Inductance

The following equations were used to calculate the inductance of the two lines on the chip:

$$L_{11}, L_{22} = L_{trace\text{-}self} + L_{wire\text{-}self}$$

$$L_{12} = L_{Total\text{-}Mutual}$$

$$L = \frac{L_{11}L_{22} - L_{12}^2}{L_{11} + L_{22} - 2L_{21}}$$

In solving this problem, a mean length of 3,500 μm is used to simplify the calculation. The value of the inductance is calculated using a $d_{pitch} = 1,000$ μm, $\mu = 1$, $I_{trace} = 3,500$ μm, and $I_{via} = 500$ μm.

The following inductance is obtained for the first two inductance equations:

$$L_{11} = 1.89 \text{ nH}$$

$$L_{22} = 1.89 \text{ nH}$$

$$L_{21} = 0.41 \text{ nH}$$

When these values are plugged into the third equation, an inductance of 1.07 nH is obtained. Considering the short pathway used in this model, this was a reasonable value to obtain Z at 2.5 GHz for the data lines.

Using the equation

$$Z = \frac{87}{\sqrt{1.41 + \varepsilon_R}} \ln\left[\frac{5.98h}{0.8w + t}\right]$$

along with the values of $\varepsilon_r = 4.2$, $h = 50$ µm, and the same variable as above, the impedance is calculated as 75 Ω. Although a relatively sensible answer is obtained, this entire circuit should be simulated with more complex software. Furthermore, the value obtained may not be really precise because the answer obtained for the RLC calculation is small, which implied that more complex simulation is needed to predict this circuit.

The magnitude of the I/O requirements for modern ICs continues to increase due to the growing complexity and size of ICs. The large I/O count found on most ICs has forced most designers to use flip-chip packaging instead of wirebonded packaging. Unfortunately, the solder bumps in flip-chip packages are susceptible to failure, especially in the presence of high temperatures, which can cause large stresses and strains leading to mechanical failure of the bump. The flip chip is selected for this die because it has the potential of being cheaper when produced in large quantities. Furthermore, the flip-chip configuration can accommodate higher frequencies because the length of the data line is shorter. Newer technologies are also being developed that are making the flip chip faster, cheaper, and simply better than the old technology associated with QFP or BGA.

14.8 Drop Test of Flip-Chip Packaging

Flip chips are generally seen as a potential future "packaging" option providing an alternative to chip-scale packages. In this work, the reliability of flip-chip assemblies was analyzed using daisy-chain test components on a schematic test vehicle designed to emulate a cellular phone environment printed wiring board (PWB). The flip-chip components were assembled in a standard surface-mount technology process, where the flip-chip bumps were first dipped in a flux film. A test matrix consisting of a number of flip-chip test components with different input/output configurations, PWBs, fluxes, and underfills was built up. The assemblies were tested for potential damage to the flip chips and interconnect by thermal cycling and by mechanical shock in a drop. After testing, the root causes of the failures were analyzed. As a separate task, the stress/strain generation that occurs in the flip chips in the drop test was analyzed using simulation, in order to find the critical locations on the test PWB.

It was found that the molded package has a shorter lifespan when tested using drop test J as compared to drop test N; even though test N has a higher shock pulse than test J. From HSC (high-speed camera) data and strain measurements, higher board deflection and board strain were observed under test J conditions. This suggested that the strain metric is the key factor in determining solder joint reliability—and not the shock pulse. Further analysis with failure analysis (FA) revealed that the major failure mode at the solder joint-package interface is mechanical fatigue cracking, while at the board side failure is dominated by brittle fracture crack and broken board trace. Broken board trace failure was identified as the key factor attributed to lower drop performance. The dependency of drop performance on different board pad designs such as via-off-pad (VoP) and via-in-pad (ViP) were considered. The experiment was also complemented with detailed finite element analysis (FEA) to establish design sensitivity in the molded flip chip package, such as recommending a better board trace routing design. The assessment found that drop test J is more stringent than drop test N, and that optimizing the board trace design can improve solder joint reliability (SJR).

The reliability of electronic packages in mechanical drop tests is critical, especially for portable electronic devices, as such electronic packages are very vulnerable to solder joint failures caused by mechanical shock and PCB warping upon impact. Drop test studies are performed to investigate the solder joints' mechanical failure in electronic packages. Here, the mechanical impact on the solder joints of a flip chip in a simulated drop test is investigated. The drop test simulation consists of a typical flip chip on board (FCOB) that has forty-eight peripheral eutectic solder bumps modeled in CAD/CAM software. Flip-chip solder joint reliability under mechanical shock is studied using 3-D finite element simulation. Comprehensive design analyses are performed to assess three different models. The design models are varied in terms of substrate dimensions and the addition of encapsulation. The results of the stresses and strains in the solder joints are obtained using FEA in the drop test. The findings indicate that stress on the flip-chip corner solder joint decreases if the substrate is larger in dimension. In addition, the introduction of encapsulation helps reduce the stress experienced by the solder joint.

14.9 Weibull Analysis of Life Test Data

Lead-free 96.5Sn-3.5Ag flip-chip solder joints were subjected to both thermal shock (−55°C to 125°C) and thermal cycling (−40°C to 125°C) reliability tests. Through two-parameter Weibull distribution plots, the mean-time-to-failure (MTTF) life are compared for the thermal shock test and the thermal cycling test. Scanning electron microscopy (SEM) examination was performed on the cross-sectional surface of failed samples to observe the failure sites and modes. Finite element modeling and simulation of the thermal cycling and thermal shock tests were performed. An elastic-plastic-creep analysis model was implemented to simulate time-independent plasticity and time-dependent creep deformations in the solder joints. Solder joint fatigue models were used for life prediction analysis employing the inelastic strain parameters derived from the finite element results.

Die cracking in the assembly and reliability testing of flip-chip packages are often a major concern. A widely used die strength test is the so-called four-point bending (4PB) test. In the 4PB test, the die is under pure bending and the strength of the die is determined by its breaking tensile stress. Although the 4PB test has been widely used, a well-established relation between the 4PB result and die breaking in the FC package has not been reported. This section discusses the relation from a probabilistic mechanics point of view. The theory considers the following issues in material strength testing and the application loading conditions:

1. The die top in the 4PB test is under uniaxial tensile stress and the die in the FC package is under multi-axial stress.
2. The 4PB test only puts part of the die top under tension, and the die top in FC plastic package has almost 100% of the die top area under tension.
3. The die stress in the 4PB and in the package have different distributions, which contributes differently to die cracking.

Weibull distribution will be used to analyze the 4PB test data. A three-parameter Weibull distribution fitting procedure will be presented. The function form of the cumulative density function of Weibull distribution is specially modified to take the above three issues into consideration and reflect the stress distribution difference between the test and application. The three-parameter Weibull fitting is compared to a two-parameter fitting. It turns out that some systems need three-parameter fitting and some other systems only need the two-parameter fitting. For systems needing three-parameter fitting, a two-parameter fitting will be too conservative in design.

The effect of temperature and strain rate of the Pb-free solder alloy Sn99.3Cu0.7(Ni) was investigated. The tensile properties of this Pb-free solder at various temperatures and strain rates were determined and compared with those of the typical Pb-containing solder Sn63Pb37. During tensile tests at different temperatures and strain rates, the ultimate tensile strength (UTS) and 0.2% yield stress of the Pb-free solder alloy Sn99.3Cu0.7(Ni) decreased with increasing test temperatures and decreasing strain rate. It was noted that the mechanical properties of Sn99.3Cu0.3(Ni) showed strong temperature dependence and strain rate sensitivity in the tensile tests at various temperatures and strain rates. Temperature and strain rate impacted rigorously flowing behaviors of the Pb-free solder Sn99.3Cu0.7(Ni) in the course of tensile deformation. The microstructure and fracture morphology of the Pb-free solder Sn99.3Cu0.7(Ni) were analyzed by SEM. In the SEM micrograph of the fracture surface of the Pb-free solder specimen tested, it was noted that increasing temperature affected the microstructure and morphology of the fracture surface.

The evolution of area-array interconnects with high I/O counts and power dissipation has made thermal deformation an important reliability concern for flip-chip packages. Significant advances have been made in understanding the thermomechanical behavior of flip-chip packages based on recent studies using Moiré interferometry. Results from Moiré studies were reviewed by focusing on the role of the underfill to show how it reduces the shear strains of the solder balls but shifts the reliability concern to delamination of the underfill interfaces. The development of high-resolution Moiré interferometry based on the phase-shift technique provided a powerful method for quantitative analysis of thermal deformation and strain distribution for high-density flip-chip packages. This method has been applied to assess plastic flip-chip packages.

To know the probabilistic fatigue characteristics of materials, a lot of experimental data for statistical analysis are needed. The analysis of statistical properties and getting specific properties of distribution data by gathering a lot of experimental data is the best way to know the probabilistic fatigue characteristics of the materials. However, in the case of the fatigue problem, the experimental estimation of the life distribution is economically limited and, also, gathering lots of experimental data is not easy. Thus, methods to estimate the reliable probability distribution about fatigue fracture using a few fatigue tests are required. It is well known that a considerable amount of scatter is obtained in experimental results relating to the growth of a fatigue crack under constant-amplitude cyclic loading. The irregular nature of the crack growth has led workers to consider the crack length to be a random function of time, that is, a random process. A number of theoretical models of the fatigue process have been developed in an attempt to reduce the number of experiments required to determine the statistical properties of crack growth. Many experimental and theoretical studies on the randomness of fatigue crack growth (FCG) have been reported. Such a model assumes that fatigue crack growth is a random process and therefore there is a need to know the random nature of the material parameters before any practical application. For this purpose, not only the mean and variance of the crack growth rate, but also the spatial distribution of the material resistance to fatigue crack growth are necessary.

SAC (Sn–Ag–Cu) lead-free solder is currently the alloy of choice in the electronics industry for Pb-free applications. In this assessment, multiple WLPs (wafer-level packages) called the "Ultra CSPTM from Flip-Chip Technologies" were subjected to a TC (thermal cycling) test. The goal was to see if the current Al/NiV/Cu UBM (under-bump metallurgy) system that has been used for eutectic Sn–Pb solder Ultra CSP would be suitable for the SAC Pb-free solder version. Both SAC and eutectic Sn–Pb solders were tested together. In this TC test, two parts were taken out of the TC chamber after every 200 cycles for monitoring the characteristics of deformation and crack growth in the solder joints. The results showed that eutectic Sn–Pb solder joints might have a global and uniform deformation in the high-temperature regime. On the other hand, in the low-temperature regime, the deformation is localized only at the chip-side solder joint while maintaining the global deformed shape from the previous high-temperature regime. This localized deformation in the low-temperature regime created a large shear dislocation at the chip-side solder joint, with the crack initiating at the outside corner of the solder joint and growing toward the inside of the chip. On the other hand, the SAC solder joints did not show that kind of large sliding at the chip-side solder joint. Instead, two cracks initiated at both the outside and inside of the chip-side solder joint and grew at almost the same rate. There was very good agreement between Weibull life and the time that the cracked length (%) goes to 100% in eutectic Sn–Pb solder. Extending this correlation to SAC lead-free solder appears possible.

The reliability of the flip-chip package is strongly influenced by underfill, which has a much higher CTE compared with other packaging materials and leads to large thermomechanical stresses developed during the assembly processes. Thermal expansion mismatch between different materials causes interface delamination between the epoxy molding compound and silicon die as well as interface delamination between underfill and silicon die. The main objective of this assessment is to investigate the effects of underfill material properties, fillet height, and silicon die thickness on the interface delamination between the epoxy molding compound and silicon die during a Pb-free solder reflow process based on the modified virtual crack closure method. Based on FEA and experimental study, it can be concluded that the energy release rates at reflow temperatures are suitable criteria for the estimation of interface delamination. Furthermore, it is found that underfill material properties (elastic modulus, CTE, and chemical cure shrinkage), fillet height, and silicon die thickness can be optimized to reduce the risk of interface delamination between the epoxy molding compound and silicon die in the flip-chip ball grid array (FC-BGA) package.

Increasing die size and large CTE mismatching in FC-PBGA packages have made die fracture a major failure mode during reliability testing. Most die fractures observed previously were die backside vertical cracking caused by excessive package bending and backside defects. However, due to die edge defects induced by the singulation process and choice of underfill material, an increasing number of die cracks were found to initiate from the die edge and propagate horizontally across the die. To improve package reliability and performance, die edge cracking must be eliminated. An extensive FEA was completed to investigate die edge cracking and find its solutions. A fracture mechanics approach was used to evaluate the effect of various package parameters on die edge-initiated fracture. Strain energy release rate was found to be an effective technique for evaluating die edge-initiated fracture from singulation-induced flaws. The impacts of initial flaw size and a variety of package parameters were investigated. Unlike in die backside cracking, the dominant parameters causing die edge horizontal fracture are more closely related to local effects.

A strong interaction between material properties and flow voids was observed for flip-chip packages using low-viscosity underfills. Flow void counts increased significantly when substrates with high surface energy were used. A slight decrease in the surface energy of substrates or increase in the viscosity of the underfill can reduce flow voids. The fabrication process of a quartz wafer with dummy bumps and the attendant assembly process changes were developed. The initial flow video results showed the feasibility of using this method for flow void studies. Metrology can be used for capturing underfill capillary flow in real-time, studying voids formation mechanism, and providing flow pattern information. The adhesion between the bump and die in the quartz wafer requires improvements. This method seems beneficial to bump pattern design, underfill material development, and flip-chip process optimization.

In a flip-chip package, the mismatch of thermal expansion coefficients between the silicon die, copper heat spreader, and packaging substrate induces a concentrated stress field around the edges and corners of the silicon die during assembly, testing, and service. The concentrated stresses result in delamination on various interfaces involving a range of length scales from hundreds of nanometers to millimeters. Among these failures, underfill delamination is a dominant failure mode. Here, a full parametric 3-D model of the flip-chip package with heat spreader was developed with the capability of explicit modeling of 3-D cracks. The crack driving force was computed as a function of underfill properties, including CTE and Young's modulus, as well as underfill fillet dimensions. The impact of different shapes of crack front was also investigated. The results showed that the underfill properties must be optimized in order to minimize the occurrence of underfill delamination at the die corner. The results also showed that there exists an optimal range of underfill fillet height to balance the manufacturability and reliability.

Die cracking and interfacial delamination are major failure modes in IC packages. For these crack-related problems, design approaches based on traditional strength theories are inadequate and insufficient. To achieve a more reliable and robust design, a fracture mechanics approach is needed. Here, fracture mechanics was applied to the FC BGA design, and it was assumed that the major potential failure mode was cracking of the die from its backside. Fracture mechanics was integrated with FEA and design of virtual experiments to analyze the effects of the location and length of a die crack, and the effects of some key material properties and package dimensions on FC BGA reliability in terms of die cracking. The stress intensity factor and the strain energy release rate are used as the design indices. The FEA is used as a numerical tool to calculate the fracture parameters. And the virtual DOE (design of experiment) was employed to determine the contributions of each design parameter to die cracking and their acceptable design windows. The investigation consisted of the following parts:

1. The first part was the relationships among the length and location of a die crack and the fracture parameters. The relationships were established through sweeping along crack length for a crack located in the center of die backside, and along the die backside surface. The critical crack length was determined.

2. The second part was the virtual DOE based on fracture mechanics. Several key material properties and cracking were calculated. From it, some generic design guidelines were made.

3. The third part compared the virtual DOE design results between using the maximum normal stress (MNS) theory and using the fracture mechanics approach. The comparison gives a clear picture of the applicable range of MNS theory.

It was concluded that design optimization is a must in order to achieve a robust package design. Substrate and die thicknesses are the two most significant factors in die cracking of FC BGA. Increasing substrate thickness and reducing the die thickness are the most effective measures to design packages with a high resistance to die cracking. The fracture mechanics approach will produce more accurate design results than the MNS theory.

The effects of phase change of Pb-free flip-chip solders during board-level interconnect reflow were investigated using a numerical technique. Most current Pb-free solder candidates are based on Sn, and their melting temperatures are in the range of 220°C to 240°C. Thus, Pb-free flip-chip solders melt again during the subsequent board-level interconnect (BGA) reflow cycle. Because solder volume expands as much as 4% during the phase change from solid to liquid, the volumetric expansion of solder in a predefined volume by chip, substrate, and underfill creates serious reliability issues. One issue is the shorting between neighboring flip-chip interconnects by the interjected solder through underfill crack or delaminated interfaces. The interjection of molten solder and the interfacial failure of underfill during solder reflow process have been observed. Here, a flip-chip package is modeled to quantify the effect of the volumetric expansion of Pb-free solder. Three possible cases were investigated. One is without the existence of micro-cracks, and the other two are with the interfacial crack between the chip and underfill and the crack through the underfill. The strain energy release rate around the crack tip calculated by the modified crack closure integral method was compared with interfacial fracture toughness. Parametric studies were carried out by changing the material properties of the underfill and interconnect pitch. Also, the effects of solder interconnect geometry and crack length were explored. For the case with interfacial cracks, the configuration of a large bulge with small pitch is preferred for the board-level interconnect, whereas a large pitch is preferred for cracks in the mid-plane of the underfill.

Flip-chip technology offers a number of advantages over conventional packaging techniques, such as smaller size and efficient high-speed signal transmission. However, when assembled on organic substrates, the flip chip should be underfilled with a suitable adhesive to enhance the thermomechanical reliability of its solder bumps. When such flip-chip assemblies are subjected to thermal excursions, the underfill material may delaminate, resulting in premature solder bump fatigue failure. The available literature has extensively focused on underfill delamination propagation due to monotonic loading conditions. The fracture toughness of the passivation-underfill interface has been characterized using the single leg bending test. In addition, a fatigue delamination propagation experiment has been performed, and a Paris law type model for delamination propagation has been developed. In parallel. numerical models have been developed to determine the available energy release rate under monotonic loading conditions, as well as the range of energy release rate range under thermal cycling conditions. Mode mixity calculations have been carried out using the crack surface displacement (CSD) method. Using models and experimental data, guidelines against the delamination of the underfill material were developed. Solder joint reliability of the flip chip on various thicknesses of PCB with imperfect underfill is presented here. Emphasis is placed on the determination of the temperature-dependent stress and plastic strain at the corner solder joint with different crack (delamination) lengths. Also, the strain energy release rate and phase angle at the crack tip of the interface between the underfill and solder mask are obtained by fracture mechanics.

Solder joint fatigue failure is a serious reliability concern in area array technologies, such as FC and BGA packages of IC chips. The selection of different substrate materials could affect solder joint thermal fatigue life significantly. The reliability of solder joints in real flip-chip assembly with both rigid and compliant substrates was evaluated by the accelerated temperature cycling test and thermal mechanical analysis. The mechanism of substrate flexibility on improving solder joint thermal fatigue lifetime was investigated by fracture mechanics methods. Two different methods—crack tip opening displacement (CTOD) and the virtual crack closure technique (VCCT)—are used to determine crack tip parameters, which are considered as indices of the reliability of solder joints, including the strain energy release rate and phase angle for the different crack lengths and temperatures. It was found that the thermal fatigue lifetime of solder joints in flip-chip-on-flex assembly (FCOF) was much longer than that of the flip-chip-on-rigid-board assembly (FCOB). The flex substrates could dissipate energy that would otherwise be absorbed by solder joints; that is, substrate flexibility has a great effect on solder joint reliability and the reliability improvement was attributed to flex buckling or bending during thermal cycling. The problem of an overmolded flip chip containing a die/underfill interface corner crack subjected to thermomechanical loading was investigated using finite element simulation.

The fracture mechanics parameters, including the stress intensity factors and strain energy release rate along the corner crack front, were determined using a 3-D virtual crack closure technique. The effect of the interface corner crack shape on the propensity for crack propagation has been studied.

Fracture mechanics approaches have been used to assess reliability problems in electronic packages, in particular, adhesion-related failure in flip-chip assembly. It was verified in this work that the J-integral with a special flat rectangular contour near the crack tip can be used as the energy release rate at the interface between chip and underfill. Meanwhile, the delamination propagation rates at the interface were measured using C-mode scanning acoustic microscope (C-SAM) inspection of two types of flip-chip packages under thermal cycle loading. Finally, the half-empirical Paris equation, which can be used as a design base of delamination reliability in flip-chip packages, has been determined from the crack propagation rates measured and the energy release rates simulated.

Bibliography

Akay, H.U., Paydar, N.H., and Bilgic, A. (1997). Fatigue life predictions for thermally loaded solder joints using a volume-weighted averaging technique, *ASME Trans., J. Electronic Packaging*, 119 (December), 228–234.

Amagai, M., Watanabe, M., Omiya, M., Kishimoto, K., and Shibuya, T. (2002). Mechanical characterization of Sn–Ag based lead-free solders, *Microelectronics Reliability*, 42, 951–966.

Antolovich, S.D., and Antolovich, B.F. (1996). *An Introduction to Fracture Mechanics in ASM Handbook. 19. Fatigue and Fracture*, Materials Park, OH: ASM International®, 1996.

Arora, N.D., Raol, K.V., Schumann, R., and Richardson, L.M. (1996). Modeling and extraction of interconnect capacitances for multilayer VLSI circuits, *IEEE Trans. Computer Aided Design of Integrated Circuits and Systems*, 15(1), 58–66.

Basaran, C., and Yan, C.Y. (1995). A thermodynamic framework of damage mechanics of solder joints. *J. Electronic Packaging*, 10(1), 365–376.

Bhattachaiya, B., and Ellingwood, B. (1998). Continuum damage mechanics-based model of stochastic damage growth, *J. Eng. Mechanics*, September, pp. 1000–1009.

Bilotti, A.A., (1974). Static temperature distribution in IC chips with isothermal heat sources, *IEEE Trans. Electron Devices*, ED-21 (March), 217–226.

Black, J.R. (1969). Electromigration failure models in aluminium metallization for semiconductor devices, *Proc. IEEE*, 57(9), 1587–1594.

Blech, I.A., and Herring, C. (1976), Stress generation by electromigration, *Appl. Phys. Lett.*, 29, 131–133.

Boresi, A.P., et al. (1993). Advanced Mechanics of Materials, *(5th edition)*. Canada: John Wiley & Sons.

Box, G.E.P., Hunter, W.G., and Hunter, J.S. (1978). *Statistics for Experimenters: An Introduction to Design, Data Analysis, and Model Building*. New York: John Wiley & Sons, Inc.

Brakke, K.A. (1994). *Surface Evolver Manual*. University of Minnesota, Geometry Center.

Chen, C., and Liang, S.W. (2007). Electromigration issues in lead-free solder joints, *J. Mater. Sci.*, 18, 259–268.

Coffin, L.F., Jr. (1954). A study of the effects of cyclic thermal stresses on a ductile metal, *ASME Trans.*, 76, 931–950.

Darveaux, R. (2000). Effect of simulation methodology on solder joint crack growth correlation, *Proc. of 50th ECTC*, May 2000, pp. 1048–1058.

Darveaux, R. (1996). How to use finite element analysis to predict solder joint fatigue life, *Proc. VIII Int. Congress on Experimental Mechanics*, June 10–13, 1996, Nashville, TN, pp. 41–42.

Dayhoff, J. (1990). *Neural Network Architectures – An Introduction*. New York: Van Nostrand Reinhold, pp. 217–243.

Dreezen, G., Deckx, E., and Luyckx, G. (2003). Solder alternative: Electrically conductive adhesives with stable contact resistance in combination with non-noble metallization, *CARTS Europe 2003*, pp. 223–227.

Ferreira, F.K., Moraes, F., and Reis, R. (2000). LASCA-interconnect parasitic extraction tool for deep-submicron IC design, *Proc. 13th Symp. Integrated Circuits and Systems Design*, 2000, pp. 327–332.

Frear, D.R., and Kinsman, K.R. (1991). *Solder Mechanics — A State of the Art Assessment*, Warrendale, PA: Minerals, Metals, and Material Society.

Gale, W.F., and Totemeier, T.C. (2004). *Smithells Metals Reference Book, (8th edition)*. Maryland Heights: MO: Elsevier.

Galyon, G.T. (2003). *Annotated Tin Whisker Bibliography*, Hearndon, VA: a NEMI Publication, July.

Gilat, A., and Krisha, K. (1997). The effects of strain rate and thickness on the response of thin layers of solder loaded in pure shear, *J. Electronic Packaging*, 119, 81.

Goel, A.K., and Au-Yeung, Y.T. (1990). Electro migration in the VLSI interconnect metallizations, *Proc. 32nd Midwest Symp. Circuits and Systems*, 1989, Vol. 2, 821–824.

Grunwald, J., and Schnack, E. (1995). Models for shape optimization of dynamically loaded machine parts, *Proc. WCSM01*, Oxford: Pergamon Press, pp. 307–310.

Guo, Q., Cuttiongco, E.C., Keer, L.M., and Fine, M.E. (1992). Thermomechanical fatigue life prediction of 63Sn/37Pb solder, *ASME Trans., J. Electronic Packaging*, 114, 145–150.

Haykin, S. (1997). *Neural Networks – A Comprehensive Foundation, (2nd edition)*. Upper Saddle River, NJ: Prentice-Hall, pp. 2–10.

Hong, B.Z. (1997). Finite element modeling of thermal fatigue and damage of solder joints in a ceramic ball grid array package, *J. Electronic Materials*, 26(7), 814–820.

Hunter, W.R. (1997). Self-consistent solutions for allowed interconnect current density. I. Implication for technology evolution, *IEEE Trans. Electron Devices*, 44(2), 304–309.

Hunter, W.R. (1997). Self-consistent solutions for allowed interconnect current density. II. Application to design guidelines, *IEEE Trans. Electron Devices*, 44(2), 310–316.

Hunter, W.R. (1995). The implications of self-consistent current density design guidelines comprehending electromigration and Joule heating for interconnect technology evolution, *Int. Electron Devices Meeting*, 1995, pp. 483–486.

Jerke, G., and Lienig, J. (2002). Hierarchical current density verification for electromigration analysis in arbitrarily shaped metallization patterns of analog
circuits, *Proc. Design, Automation and Test in Europe Conference and Exhibition,*
pp. 464–469.

Ju, T.H., Chan, Y.W., Hareb, S.A., and Lee, Y.C. (1995). An integrated model for ball
grid array solder joint reliability, structural analysis in microelectronic and fiber
optic systems, *ASME,* EEP-12, 83–89.

Jung, W., Lau, J.H., and Pao, Y.-H. (1997). Nonlinear analysis of full-matrix and perimeter plastic ball grid array solder joints, ASME Trans., *J. Electronic Packaging,* 119
(September), 163–170.

Lall, P., Islam, N., Suhling, J., and Darveaux, R. (2003). Model for BGA and CSP reliability in automotive underhood applications, *Proc. 53rd Electronic Components
and Technology Conference,* May 27–30, pp. 189–196.

Lall, P., Pecht, M., and Hakim, E. (1997). *Influence of Temperature on Microelectronic and
System Reliability.* Boca Raton, FL: CRC Press.

Lau, J. H., (Editor) (1991). *Solder Joint Reliability—Theory and Application,* New York:
Van Nostrand Reinhold, p. 279.

Lau, J.H., and Pao, Y.H. (1997). *Solder Joint Reliability of BGA, CSP, Flip Chip and Fine
Pitch SMT Assemblies.* New York: McGraw-Hill.

Lee, S.-W.R., and Lau, J.H., 1998. Solder joint reliability of cavity-down plastic ball
grid array assemblies, *Soldering & Surface Mount Technology,* 10(1), 26–31.

Lemaitre, J. (1996). *A Course on Damage Mechanics.* Berlin: Springer-Verlag, pp. 11–36.

Lienig, J., Jerke, G., and Adler, T. (2002). Electromigration avoidance in analog
circuits: Two methodologies for current-driven routing, *Proc. Design Automation
Conference (ASP-DAC 2002),* and *Proc. 15th International Conference on VLSI
Design.*

Manson, S.S. (1965). Fatigue: A complex subject—Some simple approximations,
Experimental Mechanics, 5(7), 193–226.

Meekisho, L., and Nelson-Owusu, K. (1999). Stress analysis of solder joint with
torsional eccentricity subjected to based excitation, Paper presented at the
12th Int. Conf. on Mathematical and Computer Modeling and Scientific Computing,
August, Chicago, IL.

Muju, S., et al. (1999). Predicting durability, *Mechanical Engineering Magazine of ASME,*
March, pp. 64–67.

Nagaraj, B., and Mahalingam, M. (1993). Package-to-board attachment reliability
methodology and case study on OMPAC package, *ASME Advances in Electronic
Packaging,* EEP-4-1, 537–543.

Ohring, M. (1998). *Reliability and Failure of Electronic Materials and Devices.* Academic
Press, San Diego, CA.

Pan, T.-Y. (1994). Critical accumulated strain energy (CASE) failure criterion for thermal cycling fatigue of solder joints, ASME Trans., *J. Electronic Packaging,* 116
(September), 163–170.

Pang, J.H.L., and Chong, D.Y.R. (2001). Flip chip on board solder joint reliability
analysis using 2-D and 3-D FEA models, *IEEE Trans. Advanced Packaging,* 24(4),
499–506.

Pang, J.H.L., Xiong, B.S., and Low, H. (2004). Creep and fatigue characterization of
lead free 95.5Sn-3.8Ag-0.7Cu solder, *Proc. 54th ECTC,* June 2004, pp. 1333–1337.

Pao, Y.-H., Jih, E., Adams, R., and Song, X. (1998). BGAs in automotive applications,
SMT, January, pp. 50–54.

Paydar, N., Tong, Y., and Akay, H.U. (1994). A finite element study of factors affecting fatigue life of solder joints, *ASME Trans., J. Electronic Packaging,* 116 (December), 265–273.

Racz, L.M., and Szekely, J. (1993). An analysis of the applicability of wetting balance measurements of components with dissimilar surfaces, *Adv. Electronic Packaging, ASME,* EEP-4-2, 1103–1111.

Sakimoto, M., Itoo, T., Fujii, T., Yamaguchi, H., and Eguchi, K. (1995). Temperature measurement of Al metallization and the study of Black's model in high current density, *33rd Annual Proc., IEEE Int. Reliability Physics Symposium,* pp. 333–341.

Sampath, B.K., *Electromigration Dependent MTTF Calculations,* Analog IC Research Group, University of Texas, Arlington.

Setlik, B., Eskett, D., Aubin, K., and Briere, M.A. (1997). Electromigration investigations of aluminum alloy interconnects, *Proceedings of the Twelfth Biennial University/Government/Industry Microelectronics Symposium,* 1997, pp. 159–160.

Shi, X.Q., Pang, H.L.J., Zhou, W., and Wang, Z.P. (1999). A modified energy-based low cycle fatigue model for eutectic solder alloy, *Scripts Material,* 41(3), 289–296.

Skrzypek, J.J., and Hetnarski, R.B. (1993). *Plasticity and Creep-Theory, Examples,* and Problems. Boca Raton, FL: CRC Press.

Steinberg, D.S. (2000). *Vibration Analysis for Electronic Equipment.* New York: John Wiley & Sons.

Strauss, R. (1998). *SMT Soldering Handbook, (Second edition).* Maryland Heights, MO: Elsevier/Newnes.

Suo, Z. (2004). A continuum theory that couples creep and self-diffusion, *J. Appl. Mechanics,* 71, 646–651.

Syed, A.R. (2004). Accumulated creep strain and energy density based thermal fatigue life prediction models for SnAgCu solder joints, *Proc. 54th ECTC,* June 2004, pp. 737–746.

Syed, A.R. (1995). Creep crack growth prediction of solder joints during temperature cycling: An engineering approach, *Trans. ASME,* 117 (June), 116–122.

Teng, C.C., Cheng, Y.K., Rosenbaum, E., and Kang, S.M. (1997). iTEM: A temperature-dependent electromigration reliability diagnosis tool, *IEEE Trans. Computer-Aided Design of Integrated Circuits and Systems,* 16(8), 882–893.

Tu, K.N. (2003). *Solder Joint Technology: Materials, Properties, and Reliability.* Berlin, Germany: Springer.

Tu, K.N. (2003). Recent advances on electromigration in very-large-scale integration of interconnects, *J. Appl. Phys.,* 94, 5451–5473.

Tu, K.N. (1994). Irreversible processes of spontaneous whisker growth in bimetallic Cu–Sn thin-film reactions, *Phys. Rev. B,* 49(3).

Tunga, K., Pyland, J., Pucha, R.V., and Sitaraman, S.K. (2003). Field-use conditions vs. thermal cycles: A physics-based mapping study, *Proc. 53rd Electronic Components and Technology Conference,* May 27–30, 2003, pp. 182–188.

Wiese, S., Schubert, A., Walter, H., Dudek, R., Feustel, F., Meusel E., and Michel, B. (2001). Constitutive behaviour of lead-free solders vs. lead-containing solders experiments on bulk specimens and flip-chip joints, *Proc. 51st Electronic Components and Technology Conference,* 2001, pp. 890–902.

Winter, P. R., and Wallach E.R. (1997). Microstructural modeling and electronic interconnect reliability, *Int. J. Microcircuits and Electronic Packaging,* 20(2), 124–129.

Wu, W., Kang, S.H., Yuan, J.S., and Oates, A.S. (2000). Electromigration performance for Al/SiO2, Cu/SiO2 and Cu/low-K interconnect systems including

Joule heating effect, *2000 IEEE Int. Integrated Reliability Workshop Final Report,* pp. 165–166.

Yeh, C.-P., Zhou, W.X., and Wyatt, K. (1996). Parametric finite element analysis of flip chip reliability, *Int. J. Microcircuits and Electronic Packaging,* 19(2), 120–127.

Zahn, B.A. (2002). Finite element based solder joint fatigue life predictions for a same die size-stacked-chip scale-ball grid array package, *SEMICON West, International Electronics Manufacturing Technology (IEMT) Symposium,* pp. 274–284.

Zahn, B.A. (2003). Solder joint fatigue life model methodology for 63Sn37Pb and 95.5Sn4Ag0.5Cu materials, *Proc. 53rd Electronic Components and Technology Conference,* May 27–30, 2003, pp. 83–94.

Index